Handbook of Computer and Computerized System Validation for the Pharmaceutical Industry

D1352360

Stephen Robert Goldman

ISBN: 1-4107-3744-6 (Paperback)

Library of Congress Control Number: 2003092224

This book is printed on acid free paper.

Printed in the United States of America
Bloomington, IN

1[st]Books – rev. 05/02/03

Preface

As the US Food and Drug Administration (FDA) steadfastly increases its requirement for compliance with 21 CFR Part 11, "Electronic Records; Electronic Signatures," the pharmaceutical industry must move regimented computer and computerized system validation to the forefront of laboratory, clinical, manufacturing and distribution operations. The regulation not only requires that electronic records and electronic signatures be guaranteed secure, it requires that systems in general be validated.

Much has been written about "why" to validate. This book will explain the step-by-step process of "how" to validate.

Through it's investigation of problems in areas such as pharmaceutics, biologics and medical devices, the FDA has discovered many instances where the root cause of adverse experiences and other problems with pharmaceutics and medical devices can be traced back to early stages of discovery as well as clinical trial data and manufacturing issues. In an effort to promote Total Quality Management, the practices put forth by the FDA were initially applied to specific areas. This is no longer the case. For example, 21 CFR Part 11, originally intended for applicability in clinical trials, is now applied across the board to all areas of discipline. Also, the guidance document, "General Principles of Software Validation; Final Guidance for Industry and FDA Staff" (issued on January 11, 2002, originally intended for the medical device community), is now being applied in all disciplines.

Compliance with 21 CFR Part 11, Electronic Records; Electronic Signatures, means that all computer and computerized systems used in environments that are required to comply with current Good Laboratory Practices (cGLP), current Good Clinical Practices (cGCP), current Good Manufacturing Practices (cGMP) and current Good Distribution Practices (cGDP), collectively known as cGxP, must also be validated and sustained in a validated state. The first requirement of 21 CFR Part 11 is "Validation of systems to ensure accuracy, reliability, consistent intended performance, and the ability to discern invalid or altered records." This book explains how to comply with this and other requirements of 21 CFR Part 11.

This book provides valuable information for both the executive level reader as well as the individual charged with defining, creating, and executing the validation program. It provides high-level information for the decision-making process and details for the front line staff to perform the actual work. Starting with an overview of the process, it includes a brief history, regulatory and guidance information and helps to define all of the roles and responsibilities involved. Specifics of Regulations, Policies, Guidelines and SOPs are covered in minute detail. Sustainable compliance with the regulations means that validation, processes and procedures meet or exceed best industry practices and can be demonstrated by both documentation and activities under observation. All staff must follow the procedures, and the procedures must accurately reflect the actions that occur. The ability to point to a set of books while undergoing an FDA inspection will not satisfy the requirements for sustainable compliance. The documentation and guidelines produced within the architecture must take on a life of their own in the control of the operation and they must stay in close agreement. The very existence of the operation must be governed by the documentation that enables it. Remember the words of every quality assurance and quality control person you have ever encountered. "If it wasn't documented, it didn't happen."

Table of Contents

Introduction

The manner in which the pharmaceutical and medical device industries use computers has lagged behind the world of computers and information systems for some time now. Methods commonplace in the computing environment have been generally ignored in the pharmaceutical and medical device arenas in the hopes that they will go away – like the Year 2000, they will not and cannot. They are engrained and intertwined in virtually every aspect of life on this planet. The enactment of 21 CFR Part 11 in 1997 forces the issue to the forefront of the industry as a whole. There is no escaping the necessity, and now the requirement, for applying the fundamentals of computer system and computerized system validation in these industries.

Investigations by the U.S. Food and Drug Administration (FDA) have revealed deficiencies that have caused serious adverse experiences and events resulting in illness, disease and death. These deficiencies trace their causes back to inadequate or non-existent computing practices in the industry. Information acquired during research, manufacturing, clinical trials and distribution has been tainted. This was sometimes due to a conflict of interest, malicious intent, or fraud, and at other times just due to a lack of understanding the necessity for good practices in the area involving computers, their application, and use.

I have collected current and proposed guidance and practice into this handbook to present a logical and methodical means for complying with the validation requirements of 21 CFR Part 11. This handbook deals primarily with the validation of software applications, computer systems and computerized systems. In Chapters 23 and 24, I have included all of the elements of 21 CFR Part 11 from a requirements and an assessment perspective.

In the many years that I have been performing computer validation work in this industry, I have been hard pressed to locate sources of information that explain the "how to" of computer system and computerized system validation for this industry. The vast majority of coursework and printed material deal with the "whys and wherefores" of compliance. This handbook provides the methods required to accomplish sustainable compliance with GxPs and 21 CFR Part 11 validation requirements regarding computer and computerized systems within the pharmaceutical and medical device industry.

The System Development Life Cycle methodology that I present is not new. It is a tried and proven method. I have stated it in the context of this industry and in a form that is best industry practice, that of the Standard Operating Procedure.

To the many people who have kindly helped, listened, and generally provided a path for me in the writing of this handbook, thank you! To my wife, Terry, for her support and encouragement. To my friends Stan Shore, Workflow, Inc., and Steven Field for entry into the industry. To Bill Wood, Stephen Daluisio, and Kimberly DiBartolomeo of GlaxoSmithKline Pharmaceuticals for the opportunity and assistance in my early days of validation. To Jeff Kosterich for his eminent wisdom and advice. To Samuel Yosen for his encouragement and advice. To Peter Lancy, Tunnell Consulting Co., for the opportunities he created. To Maria Scaglione for her unwavering and invaluable help.

Stephen Robert Goldman
West Chester, PA
March 2003

Part 1
Overview

Chapter 1 – A Brief History

Origins of the Modern FDA in Brief

One of the earliest recorded controls of food substances is *The Assize of Bread and Beer* (including *the Lucrum Pistoris* or *Gain of the Baker*) dating from the time of King John in England in 1202. The authorities strictly controlled the price, weight, content and profit with specific penalties for non-compliance.

In the United States in the 18th and 19th Centuries, the individual states controlled domestically produced and distributed foods and drugs. There was great disparity of requirements and controls from state to state. Federal authority was limited to non-domestic products. During this period adulteration and misbranding was common practice.

The first federal law dealing with consumer protection and therapeutic substances, known as the Vaccine Act, was passed in 1813. It was, however, short lived.

In 1862, during the Civil War, President Abraham Lincoln established the Division of Chemistry with the appointment of Charles M. Wetherill as its head. Operating within the Department of Agriculture, its initial charge was related to the treatment of the wounded. As the government had become a major consumer of drugs, it found out all too quickly that what was on the label was not necessarily what was in the product. It began investigating the adulteration of agricultural products around 1867. In 1870, the United States Pharmacopoeia (USP), the first drug regulation, was begun as a result of this activity. This was followed shortly thereafter by the National Formulary (NF).

Harvey Washington Wiley became chief chemist in 1883 and published a ten-part series titled *Foods and Food Adulterants* during the period 1887 to 1902. Wiley's "poison squad" experiments, volunteers ingesting questionable food additives, helped determine the health impact of those products.

In July of 1901, the Division of Chemistry became the Bureau of Chemistry with substantially the same function and responsibility.

Born out of the tragic death of 13 children from contaminated antitoxin in 1901, the Biologics Control Act came into being in 1902. The Act provided for the regulation of biological products to ensure the safety of the public. The Department of Health and Human Services (HHS), Food and Drug Administration (FDA), and Center for Biologics Evaluation and Research (CBER) continue this work today.

In 1906, the Pure Food and Drugs Act was passed in response to Upton Sinclair's novel, The Jungle, which was published around the turn of the century. This factual account of the practices of the meat packing industry in Chicago stirred a great emotional outcry from the public at the time.

In 1912, Congress passed the Sherley Amendment, which prohibits medicines from being labeled with false therapeutic claims. This was as a result of a lawsuit in 1911, U.S. vs. Johnson, in which it was claimed that the 1906 Food and Drugs Act does not prohibit such false labeling.

In 1927, the Food, Drug, and Insecticide Administration was formed from the regulatory wing of the Bureau of Chemistry. This was the predecessor of the FDA, which reorganized under United States Department of Agriculture from Food, Drug and Insecticide Administration in 1930. In 1940 the FDA left the Department of Agriculture for the Federal Security Agency. Following this in 1953, it was part of the Department of Health, Education, and Welfare, subsequently the Public Health Service in 1968 and finally where it is today under HHS in 1980.

Congress passed the Food, Drug and Cosmetics Act in 1938 after 107 people died from pediatric Elixir of Sulfanilamide that had been made using ethylene glycol, a deadly poison. This was the birth of the FDA, as we know it today, and the establishment of the current New Drug Application (NDA) process – pre-market approval. Cosmetics and medical devices were now controlled and drugs were required to be labeled with directions for safe use. Inspection and enforcement were now authorized. HHS, FDA, and the Center for Drug Evaluation and Research (CDER) carry on this work currently.

In 1951, the passing of the Durham-Humphrey Amendment established what constituted a prescription and what constituted an over-the-counter drug.

The Proviso – Food Additives Amendment, passed in 1958, requires the establishment of safety by producers of new food additives and prohibits carcinogens in foods.

In 1961/62, Thalidomide, a sleep-inducing agent that caused birth defects is banned in the U.S. by the FDA and causes pressure to provide stronger controls over drugs.

Congress enacted the Kefauver-Harris Amendment in 1962, requiring proof of efficacy and safety by drug manufacturers. This was the beginning of Good Manufacturing Practices (GMPs).

The Medical Device Amendments went into effect in 1976. These amendments ensue efficacy and safety of medical devices and diagnostics. This included registration, pre-market approval, quality control and other standards.

Resulting from cases of poor laboratory practices, lack of management and fraud, 21 CFR 58, Good Laboratory Practice (GLP) For Nonclinical Laboratory Studies came into being in 1978. The regulation forced the organization, accountability and responsibility of pharmaceutical laboratories and contract facilities. GLP regulations are intended to control the practices used in proving the safety of potential drugs.

The Orphan Drug Act of 1983 encourages research, development and marketing of drugs for the treatment of rare diseases.

The Drug Price Competition and Patent Term Restoration Act of 1984, also known as the Waxman-Hatch Act, was passed in 1984. It permitted approval of generic drugs by the FDA without the need for costly duplicate clinical trials. It also allows for the extension of patent protection for the original producer for up to 5 years. The Waxman-Hatch Act resulted in the proliferation of generic products. However, in 1989, scandal was uncovered. Payoffs, fraud and other illicit activities were discovered within the FDA to acquire quick approval and favoritism. As a result, testing was increased and in 1992, the Generic Drug Enforcement Act, Abbreviated New Drug Application (ANDA) Process for Generic Drugs went into effect. The Office of Generic Drugs, a part of FDA's Center for Drug Evaluation and Research, has responsibility for this work today.

The Food Quality Protection Act, passed in 1996, changed the applicability of the Delaney Proviso regarding pesticides in favor of risk-based testing.

The Food and Drug Administration Modernization Act (FDAMA) of 1997 reauthorized the Prescription Drug User Fee Act of 1992. Additionally, it mandated reforms in agency practices including accelerated review of devices, regulation of advertising of unapproved uses of approved drugs and devices, and regulation of health claims for foods.

21 CFR Part 820, Quality System Regulation (QSR) was passed October 7, 1996, went into effect on June 1st, 1997, with Design Control requirements coming into effect on June 1st, 1998. This major revision imposing significant additional requirements on device manufacturers is used to ensure that medical devices are safer

and more effective. Much of the QSR is slowly being applied to areas other than medical devices for Total Quality Management.

21 CFR Part 11, Electronic Records; Electronic Signatures was passed on March 20, 1997 and went into effect on August 20, 1997. To quote the FDA, "This regulation provides the criteria for acceptance by FDA, under certain circumstances, of electronic records, electronic signatures, and handwritten signatures executed to electronic records as equivalent to paper records and handwritten signatures executed on paper. These regulations, which apply to all FDA program areas, are intended to permit the widest possible use of electronic technology, compatible with FDA's responsibility to promote and protect public health. The use of electronic records as well as their submission to FDA is voluntary."

However, when it comes to the use of software applications, computers and computerized systems that produce data required by the FDA for proof of efficacy and safety, there is no longer any option as to the applicability of 21 CFR Part 11. It has become mandatory.

As we have seen in this brief historical outline, the FDA has been shaped as a result of fraud, deceptive advertising, collusion, malfeasance, honest mistakes, and many other activities and occurrences affecting the health and safety of the public. The outcome of all of this is excellent protection of the public's health and safety with respect to food, drugs, cosmetics and medical devices.

The Increasing Role of Computers

Since the invention of the digital computer, its use has been increasing exponentially in the pharmaceutical, medical device, biotechnology, and health care arenas. Computers and computerized systems are used for just about every imaginable type of work. From research and discovery, initial manufacturing, through all phases of clinical trial work, to post-approval manufacture and tracking of drugs, biologics, medical devices and diagnostics, computers and computerized systems are essential. Health care in general uses computers to track patients in: doctor's offices, hospitals, laboratories, drug distribution, and insurance companies.

With the advent of such modern day miracles as combinatorial chemistry, pacemakers, insulin pumps, defibrillators, Magnetic Resonance Imaging (MRI) equipment, and advanced monitoring and diagnostics, computers pervade our very lives. Therefore, it is in the best interest of the public to ensure that these creations of man do what they are supposed to do and work the way in which they were intended.

Up until the past few years, the area of discovery did not mandate compliance with FDA regulations. It wasn't considered necessary because the work wouldn't always result in a marketable discovery. Recent activities indicate that in the investigation of problems surrounding adverse experiences and events, many of the root cause issues started during preclinical work and were the result of improperly or non-compliant systems. The dotted line that separates research and discovery phases from the rest of the approval process has gone away. It is now considered a requirement for research and discovery to comply with all FDA regulations if the outcome of the work will be used in any filing with the FDA. The only exception would be if all of the work performed in a non-compliant environment were repeated in a compliant environment. This is clearly a business decision, which should be based on costs, return on investment and other factors beyond the scope of this handbook.

Computer systems and computerized systems with their associated software applications are critical tools in the processes that support business operations today. In this day and age, without these tools businesses would cease to operate.

Advent of 21 CFR Part 11

On March 20, 1997, 21 CFR Part 11, Electronic Records; Electronic Signatures was passed and went into effect on August 20, 1997. This sweeping regulation covers every aspect of computer and computerized system operation from a company's organization and how it operates, the software, computers and equipment, security, and integrity of the information, to the people doing the actual work

The pharmaceutical industry as a whole is just realizing the need *and* the requirement for sustainable, compliant operation of the computer side of the industry. Contrary to some beliefs that the pharmaceutical and medical device industry is leading the way when it comes to computers and systems, this industry as a whole is about 5-10 years behind the computer and software industry in validation and other activities. Many companies have historically made the business decision to deal with it when it happens. Just like the issues surrounding Y2K, the time has come. 21 CFR Part 11 affects every aspect of discovery, research and development, manufacturing, clinical trials, clinical and non-clinical laboratories, and distribution of drugs, biologics, diagnostics, and medical devices.

The FDA is enforcing the requirements of 21 CFR Part 11. Witness the contents of warning letters, 483's, and consent decrees issued by the agency. Look at the guidance documents surrounding Part 11 that are being developed. Although industry and the FDA are still uncertain of many aspects of the regulation because the state of the art changes almost daily, concrete determinations for compliance have been established. Not the least of which is the application of the computer industry's best practices to the pharmaceutical and medical device industry. The goal of the process is total quality management to protect the health and safety of the public.

Current Practice and Regulation

Current best practices and FDA regulations surrounding the use of computers and computerized systems encompass, at a minimum, the following areas:

- Formal Organizational Structure – Policies, Guidelines, SOPs and Training for the IT/IS Unit and Contractors
- IS Steering Committee
- GxP Training for IT/IS Staff and Contractors
- Computer Validation Group
- Guidelines, Policies, SOPs and Training in Support of System Development Life Cycle
- System Owner/End User Takes Ownership of Their Systems
- Active Control of Contractors
- Job Aides and Training
- Document Management Procedures
- Configuration Management Procedures
- Formal Disaster Recovery Program
- Physical and Logical Security Practices and Procedures
- Infrastructure Qualification
- Validation Master Plan
- Maintenance
- Environmental Controls
- Electronic Records
- Electronic Signatures

As can been seen in the Appendix 2 – "GLP Attachment A," as well as in the various inspection guidelines issued by the FDA, a company's structure, organization and infrastructure are just as important as validation and compliance with GxPs and 21 CFR Part 11. As discussed earlier, total quality management is the goal. This is expanded to operation of the software, hardware, people, processes and procedures in general where Part 11 is involved. It means that the system as a whole must work as defined and purported. Including the people following the written procedures.

Electronic record keeping and electronic signatures are at the forefront of current regulations. The scope of Part 11 indicates that the agency's intent is to consider, "...electronic records, electronic signatures, and handwritten signatures executed to electronic records to be trustworthy, reliable, and generally equivalent to paper records and handwritten signatures executed on paper." For example, in order to prevent an individual from claiming that an electronic signature or a manual signature tied to an electronic record is not of their making, 21 CFR Part 11 mandates specific requirements.

Great emphasis has been placed on eliminating conflict of interest and potential fraudulent situations. It is no longer acceptable for the System Owner or a person having a vested interest in the outcome of activities to be the System Administrator of their own system. 21 CFR Part 11 requires that measures be in place that necessitates two or more people working together to commit fraud.

The implementation of system and procedural controls are mandated to reduce operator error. In many areas it is currently too easy for a mistake to occur because of human error. The intent is to remove as many conditions as possible that afford this possibility.

Validation of *all* systems is now required. The agency has issued various guidelines regarding computers and computerized systems. Although many of these are in draft form at the time of this writing, they are an excellent resource and will, after approval, become practice.

Chapter 2 - Planning, Assessment, Implementation and Remediation

This chapter discusses sustainable compliance with 21 CFR Part 11, and effective planning, assessment, implementation, and remediation of software applications, computer and computerized systems. The areas included are:

- Roles and Responsibilities
- Jargon
- System Development Life Cycle Stages
- Setting Up Planning
- Preparing For Basic Computer Validation
- Identifying Procedural Controls
- Identifying System Controls
- Implementing Measures
- Preparing Reports
- Effective Closure
- Next Steps

A computer system, as defined by the U.S Food and Drug Administration (FDA), is a functional unit, consisting of one or more computers and associated peripheral input and output devices, and associated software. These systems use common storage for all or part of a program and also for all or part of the data necessary for the execution of the program; executes user-written or user-designated programs; performs user-designated data manipulation, including arithmetic operations and logic operations. Included in this classification are computer systems that can execute programs that modify themselves during their execution. A computer system may be a stand-alone unit or may consist of several interconnected units.

A computerized system as opposed to just the computer system includes hardware, software, peripheral devices, personnel, technical manuals, operating procedures, training materials and compliance/validation documentation.

No one part of this process is difficult. However, there is much to learn and do. Take it one step at a time. Remember, slow and steady wins the race!

Roles and Responsibilities

First, let's talk about Roles.

- You are probably associated with one or more of the following groups or departments in one or more environments:
- IS or IT department
- Quality Assurance Group or Unit
- System Owners
- End-Users
- System Administrators
- Developers
- Subject Matter Experts (SME)
- Vendors, Contractors or Consultants
- Customers or Consumers

In reality everyone is a customer or consumer. You purchase goods or have services performed for you or on your behalf. Customers and consumers demand quality goods and services. This puts you in a good position to understand the needs of your clients. And everyone serves an internal or external client of some kind.

Consider for a moment that you are a retail customer. You are out shopping. Would you buy a bruised head of lettuce? A radio with a broken knob? A torn dress? Then why would you want a computer or computerized system that isn't proven to be working properly and securely to create or process your information?

These exercises imposed by the FDA and other regulatory agencies are all about quality control.

An esteemed colleague and friend once said,

> "Quality Control begins with the person doing the work."
>
> - Brint Bodycott

Since all of us are doing the work, Quality Control begins with us!

Validation in accordance with Part 11 requirements involves using the full System Development Life Cycle methodology, which we will discuss shortly. It also involves Planning, Assessment, Gap Analysis, Prospective Validation, Ongoing Operations, Remediation and Retrospective Validation.

Jargon

First, let's define some technical terms or jargon:

21 CFR Part 11 – the actual regulation. The formal title for Part 11 is "Title 21 of the Code of Federal Regulations, Part 11, Electronic Records; Electronic Signatures." All of Title 21 is for the FDA. Part 11 is specifically Electronic Records and Electronic Signatures.

cGxP – Current Good Laboratory, Clinical, Manufacturing and Distribution Practice.

Remediation – the process of correcting deficiencies and effecting sustainable compliance of software, computer and computerized systems used in GxP environments with 21 CFR Part 11, all cGxP and other regulatory requirements.

Sustainable Compliance – the process of maintaining compliance with the regulations on an on-going basis.

SDLC – System Development Life Cycle Methodology

Project Plan - (NIST) A management document describing the approach taken for a project. The plan typically describes work to be done, resources required, methods to be used, the configuration management and quality assurance procedures to be followed, the schedules to be met, the project organization

Project in this context is a generic term. Some projects may also need integration plans, security plans, test plans and quality assurance plans.

Validation Master Plan – A plan used to catalog and track hardware, software and systems that need to be maintained in a validated state. It specifies what is to be validated, where it is located, and who is responsible for its validation. It also specifies when to schedule for validation, re-validation or periodic review independent of periodic review triggers. The Validation Master Plan may be separated into multiple plans based on areas of responsibility. For example, the IT/IS Unit may be responsible for software, patches and drivers, whereas Computer Operations may be responsible for hardware and environmental items in a data center. A support group would cover desktops and laptops. However, one person or group must be responsible for the overall coordination of the master plans so that nothing is overlooked. This includes both infrastructure and systems.

Validation Plan – A plan that details how the computerized system is to be validated. It defines the scope of the validation effort, the method by which it will be validated, the validation documentation deliverables, and roles and responsibilities and sequences of events. A formal validation plan may not always be necessary, but may be described in a change control document or as part of the overall project plan for minor system changes. The goal of the validation plan is to detail the activities in order to demonstrate that the computer system is developed, validated and maintained in a cGxP-sustainable compliant manner.

IQ – Installation Qualification – (FDA) Establishing confidence that process equipment and ancillary systems are compliant with appropriate codes and approved design intentions, and that manufacturer's recommendations are suitably considered.

("...compliant with appropriate codes..." One way the FDA ensures overall compliance is to invoke these requirements in the regulations. All regulations require compliance with all other codes and regulations. Don't be caught off guard by these requirements! This includes fire codes, OSHA, BOCA)

OQ – Operational Qualification – (FDA) Establishing confidence that process equipment and sub-systems are capable of consistently operating within established limits and tolerances.

This applies to operation of the software, hardware, people, processes and procedures in general where Part 11 is involved. It means that the system as a whole must work as defined and purported, including the people that follow the written procedures.

PQ – Performance Qualification – (FDA) Establishing confidence that the process is effective and reproducible.

Summary Reports – A summary of the outcome of a specified qualification procedure.

Prospective Validation – (FDA) Validation conducted prior to the distribution of either a new product, or product made under a revised manufacturing process, where the revisions may affect the product's characteristics.

Retrospective Validation – (FDA) (1) Validation of a process for a product already in distribution based upon accumulated production, testing and control data. (2) Retrospective validation can also be useful to augment initial premarket prospective validation for new products or changed processes. Test data is useful only if the methods and results are adequately specific. Whenever test data are used to demonstrate conformance to specifications, it is important that the test methodology be qualified to assure that the test results are objective and accurate.

As applied to Part 11, Retrospective Validation involves all legacy systems. Nothing is grandfathered!

Retrospective Validation is not a way of life. Prospective Validation is!

System Development Life Cycle (SDLC) Stages

Planning, Specification, Design, Construction, Testing, Installation, Acceptance Testing, Operation and Retirement all constitute the approach applied to the System Development Life Cycle of a computer or computerized system. SDLC stages are covered in detail in Chapter 3.

How To Set Up Planning

In order to perform remediation planning, it is best to have a plan that details what is to be done, who is to do it and the time frames necessary to affect a speedy resolution.

Create a project plan in both narrative and chart format that contains work breakdown structures for collecting information about each application, computer or computerized system. This information will become part of the Validation Master Plan describing the details of the information collected.

Include the following types of information:

Project Overview

1. **Facilities**

Describe the physical facilities to be surveyed.

2. **Work Breakdown Structures**

Specify what is to be done and by whom. For example:

Identify / Inventory at the site:

- System / Application
- Client Computers
- Backup Frequency
- Description
- IS Manger Contact
- Backup Type
- Validation Status
- IS Responsibility
- Run Frequency
- GxP
- Platform
- Recovery Procedures Documented
- System Owner
- Operating System
- Input Dependencies
- System Administrator
- Validation Date
- Output Dependencies
- Corporate System
- Criticality
- Comments

- Server(s)
- Recovery Priority

3. **Deliverable**

A Validation Master Plan containing the above information.

4. **Scope Management**
5. **Time Management**
6. **Cost Management**
7. **Human Resource Management**
8. **Communication Management**
9. **Risk Management**

How To Prepare For Basic Computer Validation

The outcome of the work from the Project Plan is a Validation Master Plan. From this, the various projects and their priorities are identified and created. Gap analysis or retrospective validation is performed. Procedural and System Controls are identified and put into place.

See Chapter 38 – "Retrospective Validation and Gap Analysis."

Identifying Procedural Controls

For each of the requirements of the regulation, formulate a series of questions that can be answered with yes, no or not applicable. For each system being assessed, apply these questions and document all observations and comments. The areas to be covered are from the regulation itself and are covered in Chapters 23 and 24.

For *each* software application, computer or computerized system:

- Perform a 21 CFR Part 11 assessment (see Chapter 24 – "21 CFR Part 11 Assessment")
- Perform a Gap Analysis (see Chapter 38 – "Gap Analysis and Retrospective Validation")

Use the responses to the 21 CFR Part 11 assessments and the results of the Gap Analyses to create:

- An upgrade plan with specified actions and timelines to bring each system into compliance.
- A validation plan.

Using the validation plan, gap analyses and other analysis, create the User and Functional Requirements, IQ, OQ and PQ protocols.

Create and imp lement Standard Operating Procedures to control the work that will be performed and ensure the security and integrity of the data. This process applies to everyone working in the GxP and 21 CFR Part 11 environment – Employees, Vendors, Contractors and Consultants. Everyone must be qualified prior to performing work. This is covered in detail in Chapters 4 and 6.

In order to accomplish the task expeditiously, it may not be feasible initially to create a suite of SOPs. Under these circumstances, it is acceptable to specify in each document the best industry practice instead of referencing an SOP. However, the SOPs must be planned, scheduled and created.

Do not reference non-existent or future documents in any document. Documents are not permitted to precede or pre-exist the documents that govern them. Chronology is determined by approval dates.

Remember, you are not expected to get everything done at once. Show intent, document that intent, and maintain the written evidence that demonstrates continued reasonable progress toward your goal.

Identifying System Controls

In performing the 21 CFR Part 11 assessment and the Gap Analysis, the system controls required by 21 CFR Part 11 are determined and addressed. The first one, Validation, answers the questions:

- Was the system built right?
- Was the right system built?

21 CFR Part 11 also requires the following:

- Accurate records
- Protection of the records and system
- Limited Access to the records and system
- Audit Trails of all activities where records are involved
- Operational, Authority and Device Checks
- Education, Training and Experience to qualify the people using the system
- Accountability
- Documentation Controls
- Revision and Change Control
- Authenticity, Integrity, Confidentiality in Open Systems

See Chapter 3 – "Overview of Computer and Computerized System Validation."

Implementing Measures

Implement the development, remediation or upgrade. If it is a retrospective validation or a remediation and it will take too long to implement, proceed with what can be validated retrospectively. Be sure to document reasonable progress.

Using the validation plan and qualification protocols for each software application, computer or computerized system, execute the qualification protocols and tasks.

Prepare and approve the summary report for each executed protocol before going on to the next one.

Also, demonstrate that the work follows the written procedures and vice versa, through documentation. This applies to the validation work as well as ongoing operation and use. Remember, if it wasn't documented, it didn't happen!

Preparing Reports

Best industry practices and FDA Guidance documents indicate that in order to demonstrate that a computer or computerized system performs as it purports and is intended to perform, interim and final Summary Reports are required to demonstrate, in a summarized format, the outcomes of qualification testing.

An SOP should contain instructions and procedures that enable the creation of a Summary Report for Unit Testing, Installation Qualification (IQ), Operational Qualification (OQ) and Performance Qualification (PQ) protocols.

Summary Reports should be prepared as stand-alone documents for all projects. It is not good practice to include them as part of other documents in an attempt to reduce paperwork. This would only create more work over the length of the project, from misunderstandings and confusion. Clear, concise documents that address specific parts of any given project are best.

Items to include are:

- An overview of the process
- The participants and their roles
- The accomplishments made
- The project's impact on the company or department
- Knowledge gained from the process
- A discussion of any re-work or delays that occurred
- A summary of the project
- The conclusions drawn at completion

Effective Closure

Effective Closure is accomplished by having completed each of the following:

- Validation Master Plan – Living Document
- Implementation of System Controls
- Completed Upgrades
- Completed Validation Packages
- Archival of Validation Packages
- Completed Training
- Reports to Management, Quality Assurance (Unit)

What's Next

The remainder of this book is primarily SOP-centric. Chapter 3, "Overview of Computer and Computerized System Validation" and Chapter 4, "Roles and Responsibilities in Computer and Computerized System Validation," explain in detail the requirements and steps involved in validation. The remaining chapters, Part II - Standard Operating Procedures for Validation, define the Standard Operating Procedures used to accomplish each set of required tasks or functions in the validation process and for sustainable compliance with 21 CFR Part 11.

Upon completion of this book, the reader will have a thorough understanding of the "how to" of computer and computerized system validation and be able to facilitate sustainable compliance with 21 CFR Part 11 and GxP regulations.

Chapter 3 - Overview of Computer and Computerized System Validation

This Overview of Computer and Computerized System Validation is an introductory explanation of the System Development Life Cycle (SDLC) methodology, which serves as the central framework to validate computer systems. SDLC, a generally accepted quality approach, is used throughout the information technology area in general, including the pharmaceutical industry. It is a structured method to plan, design, implement, test, roll out, and operate a system from its conception to its termination. It includes revision, change control and configuration management. This chapter describes the procedures for validating a computer or computerized system using SDLC methodology. It provides the framework to ensure that all computer systems governed by current Good Laboratory Practices (cGLP), current Good Clinical Practices (cGCP), current Good Manufacturing Practices (cGMP) and current Good Distribution Practices (cGDP) are examined for overall quality and reliability in a consistent fashion. It also ensures that computer systems are maintained in a controlled state throughout their useful life.

Note: Combinations of cGLP, cGMP, cGCP and cGDP are abbreviated as "cGxP" throughout this book.

SDLC provides a standard methodology that meets current industry practice. It enables the people responsible for computer and computerized system validation and operation to understand and employ a quality framework in a sustainable, compliant manner.

Validation throughout the life cycle of a system, in part, establishes documented evidence that a system does what it purports to do.

The SDLC methodologies and other good-quality practices should be followed in the course of everyday operations, and should become part of the culture of the organization. This is achieved through the SOPs, project management, business objectives, and other processes related to the SDLC methodology.

It is highly recommended that this chapter become a Standard Operating Procedure, be incorporated into the documentation system. The documentation provides a means of presenting a process overview in an official manner.

This overview is intended for senior and middle managers, and also those directly involved in the operation of computers or computerized systems. This includes systems that utilize a software application, computer or computerized system that controls equipment and/or produces electronic data or records in support of all regulated laboratory, manufacturing, clinical or distribution activities. This chapter explains the overall requirements and necessary conditions to maintain a sustainable, compliant computing environment.

Although this chapter focuses on the validation process itself, the IT/IS Infrastructure requirements are outlined here to clarify the requirements of both the validation process and 21 CFR Part 11. Details of these requirements are not covered in this handbook. Theoretically, the infrastructure should be in place before a validation process begins, but this may not always be possible. It is important to understand that the although the FDA recognizes it is not possible to do everything at once, they do require demonstration of commitment as well as demonstration of progress towards the goal of that commitment.

IT/IS Infrastructure

This section discusses the Policies, Guidelines, SOPs, and structure of Training and Organization that must be in place for operation of an IS/IT Unit and Computer Operations (Data Center).

Formal Organizational Structure - Guidelines, Policies, SOPs and Training for IT/IS Unit and Contractors / Consultants

The IT/IS Unit is formally organized with a clear reporting structure. The organization, along with specific roles and responsibilities within the unit, is defined and documented. There are formal Policies, Guidelines, SOPs, documentation, training, training aids, assessments and record keeping that all ensure the operation's sustainable compliance with cGxP.

The IT/IS Unit is an internal consultant and support arm to all the other departments within the company, for computer and computerized system validation in a cGxP environment. Formal processes, procedures, training, record keeping, and other compliance initiatives are implemented by each department with the aid of the IT/IS Unit.

All of the requirements incumbent upon the IT/IS Unit are also incumbent upon outside contractors and consultants.

Computer Validation Group

The Computer Validation Group is within the IT/IS unit. It coordinates validation activities and prepares Policies, Guidelines and SOPs, validation protocols and reports. It can also accept them from consultants and End-Users to assist System Owners with the SDLC process.

IS Steering Committee

An independent IS Steering Committee that meets regularly provides guidance to improve the processes and communications within and among the various departments. It creates and maintains methodologies for computer services and practices. It issues communications that state in a clear manner the roles and responsibilities of the IS/IT Unit, as well as the responsibilities of the System Owners and End-User community.

GxP Training for IT/IS Staff and Contractors / Consultants

The IT/IS Unit and its contractors must have general GxP training. Quality Assurance, in association with the IT/IS Unit, develops training curricula for each job function that includes general GxP training, corporate guidelines procedures, and appropriate site/facility Policies, Guidelines and SOPs.

Validation Master Plan

The IT/IS unit creates, implements and maintains a Validation Master Plan, which is made available electronically to the various company departments. The Validation Master Plan details all systems and their validation status.

See the section "Validation Master Plan" in "Ongoing Operations" later in this chapter.

Guidelines, Policies, SOPs and Training in Support of SDLC

Policies, Guidelines and SOPs in support of the full System Development Life Cycle (SDLC) Methodology are written and implemented. These include, but are not limited to, a Validation Plan, User Requirements, Functional Requirements, Design Specification, 21 CFR Part 11, Vendor Audits, Installation Qualification

(IQ), Operation Qualification (OQ), Performance Qualification (PQ), Summary Reports, Authorization for Use and Decommissioning.

System Ownership

System Ownership must be clear. In many situations, System Owners / End-Users are provided with limited or no information about the systems they are using. There is no formal process that assures applications are maintained in a validated state. Policies, Guidelines and SOPs are created and implemented to fill the gaps of these deficiencies. Overall responsibility for computer validation is assigned to the System Owner / End-User. The System Owner / End-User is responsible for, and participates in, the design, execution, and review of the documentation for their system(s). They must also ensure that all computer validation documentation and processes are followed, regardless of who performs the work. The Overview and Roles & Responsibilities SOPs are created and implemented. These provide a general understanding of SDLC for the company's End-User community as well as a vehicle for implementation of that methodology.

Active Control of Contractors / Consultants

There must be a system in place that maintains active control over the contractors and consultants performing work within the GxP environment. Policies, SOPs, training and other documentation are specifically created and implemented to govern this aspect of operations. . The IT/IS Unit and all other departments within the organization are responsible for the supervision and control of resources external to the company.

Job Aides and Training

All policies, procedures and guidelines have associated job aids, training curricula, training manuals and assessment documents generated and implemented.

Configuration Management Procedures

A configuration management and change control system is put in place to support the validation process. This system provides additional methods to support sustainable compliance with cGxPs. These methods control the configuration and management of a system.

Formal Disaster Recovery Program

Computer Operations (Data Center), as well as the company, should have a formal disaster recovery program. The specifics of this program are documented and distributed to all parties concerned.

Physical and Logical Security Practices and Procedures

Policies, Guidelines and SOPs are created and implemented in the areas of back-up, restoration and disaster recovery. Both physical and logical security practices and procedures are written and implemented.

Logs

A log is created and maintained for each of the servers, applications, and system owners, that details areas of responsibility, service level agreements, points of contact, and other pertinent information. The IT/IS Unit prepares a database with associated SOPs for Computer Operations, which contains this log and other information. Use logs are maintained for equipment and instruments to record usage (by whom, when and why), and other pertinent data.

Infrastructure Qualification

An overall plan is developed and implemented that migrates the obsolete operating systems of the benchtop and desktop computers from their current operating system environment, to a compliant operating system. A major effort is required to create a project plan, maintain SDLC documentation, develop the various images that meet the requirements for each hardware configuration, and perform a rollout. This process, based on current regulatory requirements necessitates the application and implementation of full SDLC methodologies. This work constitutes a major portion of the validation of the IT/IS infrastructure.

The remaining infrastructure work includes qualification of the servers, desktop computers, benchtop computers, network cabling, servers, hubs, routers, switches, uninterruptible power sources, generators, and other hardware. This includes an analysis of, the physical environment, electromagnetic interference (EMI), radio frequency interference (RFI), and qualification of all non-GxP software used throughout the facility.

Intellectual Property Database and Procedures

The IS/IT Unit physically retains and stores licenses for application software and other items of intellectual property. An SOP is written and implemented to properly handle licenses for intellectual property. This includes the creation and implementation of a database to track these types of items, as many licenses require periodic renewal or upgrade. A designated staff member is assigned to maintain this database.

Validation

This section discusses the different aspects of validation: its meaning as a classification of systems, the roles and responsibilities associated with validation, the procedures and documentation required for its performance, and sustainable compliance within a full System Development Life Cycle methodology.

Introduction

> Verification: "Was the product built right?"
>
> Validation: "Was the right product built?"
>
> – Janis V. Halvorsen, Director, SIRMO Office, FDA

It is the responsibility of each company to ensure that all computer and computerized systems required to be cGxP–compliant, are qualified and validated in accordance with standardized procedures. Such systems should be maintained in a validated and compliant state throughout their lifetime.

Validation, as defined by the FDA - Establishing documented evidence that provides a high degree of assurance that a specific process will consistently produce a product meeting its predetermined specifications and quality attributes.

21 CFR Part 11, "Electronic Records; Electronic Signatures", brings to the forefront the need to properly validate and qualify computer software, hardware, computers and computerized systems. This need can be met with the use of a clear and concise methodology, such as SDLC. Utilization of the full System Development Life Cycle methodology will ensure continued compliance with Part 11.

As its title indicates, the focus of 21 CFR Part 11 is on electronic records and electronic signatures, but its scope is much broader. Chapters 23 and 24 will demonstrate that fact. The System Development Life Cycle

(SDLC) methodology is a process that prescribes the phases through which a computer system progresses. These include:

- Planning
- Specification
- Design
- Construction
- Testing
- Installation
- Acceptance Testing
- Operation
- Retirement

Table 1 lists the activities surrounding a general computer system, and its associated validation documentation.

Because of the complexity of different computer systems, the International Society of Pharmaceutical Engineers (ISPE), Good Automation Manufacturing Practices (GAMP), developed a five-level classification scheme. It is used as a guide in matching validation requirements to the complexity of a computer system. Each level or category of computer system has its own suggested validation requirements, ranging from a minimal base for Category 1 systems, up to the most stringent requirements for Category 5 systems. Appendix 3, "Roles and Responsibilities – Validation", details the roles, responsibilities and requirements for each document. The following are the GAMP4 categories and related validation requirements:

Operating Systems (GAMP4 Category 1)

"These are established, commercially-available operating systems. Applications are developed to run under the control of these operating systems. Established operating systems are not subject to specific validation although their features are functionally tested and challenged indirectly during testing of the application. The name and version number of the operating system should be documented, and verified during Installation Qualification (IQ).

Change Control should be applied to manage upgrades to the operating system. The impact of new, modified, or removed features should be determined. Application verification and re-testing should reflect the degree of impact. Examples include Unix®, Windows NT® and VMS®."

Firmware (GAMP4 Category 2)

"Instrumentation and controllers often incorporate firmware. Configuration of this firmware may be required in order to set up runtime environment and process parameters. The name, version number, and any configuration or calibration for the firmware should be documented and verified during IQ. Functionality should be tested during Operational Qualification (OQ).

Change control should be applied to manage any change to firmware or configuration parameters. Standard Operating Procedures (SOPs) should be established and training plans implemented. Supplier Audits may be needed for highly critical or complex applications. Custom firmware should be considered as Category 5."

Firmware is the combination of a hardware device; e.g., an IC; and computer instructions and data that reside as read only software on that device. Such software cannot be modified by the computer during processing.[1]

---

[1] IEEE Definition.

Stage	Development Activities	Validation Activities / Documents
Planning	Project Planning	Project Plan – Narrative, Gantt, PERT, CPM[2]
Specification	User Requirements Functional Requirements	Project/Validation Plan Vendor Audits & Selection Specification Reviews
Design	Design Specification	Design Reviews
Construction	Hardware Manufacture & Assembly Software Module Development Equipment Manufacture & Assembly Network Manufacture & Assembly	Code Reviews
Testing	Hardware Testing Software Module Testing Software Integration Testing Equipment Testing	Unit Testing Protocols Module Testing Protocols System Testing Protocols
Installation	Hardware Installation Software Installation Equipment Installation	Installation Qualification (IQ)
Acceptance Testing	System Acceptance Testing	Operational Qualification (OQ) Performance Qualification (PQ) Final Validation Report Authorization for Use
Operation	Use Maintenance Change Control	SOPs Change Control Periodic Review
Retirement	Data Retrieval	Decommissioning Plan

Table 1 - Validation Approach Applied to the SDLC of a Computer System

Not all activities and documents listed in Table 1 may be necessary for each system. The extent of documentation and testing depends upon the complexity and nature of the system and its environment.

---

[2] See Project Management, *A Systems Approach to Planning, Scheduling, and Controlling*, Sixth Edition by Harold Kerzner, Ph.D. published by John Wiley & Sons, Inc. for an explanation of Gantt, PERT (Program Evaluation and Review Technique) and CPM (Critical Path Method) chart types.

Standard Software Packages (GAMP4 Category 3)

"These are commonly available, standard software packages, providing an 'off-the-shelf' solution to a business or manufacturing process. The package is not configured to define the business or manufacturing process itself. Configuration should be limited to establishing the runtime environment of the package (e.g., network and printer connections). Process parameters may be input into the application. The name of the package and the version number should be documented and verified during IQ.

To satisfy validation requirements, user requirements (e.g., security, alarm and event handling, calculations and algorithms) should be documented, reviewed, and tested during OQ. Supplierdocumentation, such as user and technical manuals, should be assessed and used wherever possible. Supplier Audits may be needed for highly critical or complex applications or where experience with the application in a GxP regulated environment is limited. SOPs should be established and training plans implemented.

Examples of standard software packages may include statistical analysis packages, and software within laboratory instruments such as HPLC."

The following list contains documents recommended for GAMP4 Category 3 validation. Some of these documents may be able to be combined into one (e.g. IQ and OQ), but this approach should be pre-approved by QA(U).

Documents for Category 3

- Letter of Notification/System Proposal
- Validation Plan
- User Requirements
- Functional Requirements
- Change Control Document
- IQ Protocol
- IQ Summary Report
- OQ Protocol
- OQ Summary Report
- PQ Protocol
- PQ Summary Report
- Traceability Matrix
- Project Validation Final Report & Authorization for Use

Configurable Software Packages (GAMP4 Category 4)

"Configurable software packages provide standard interfaces and functions that enable configuration of user specific business or manufacturing process. This involves configuring predefined software modules and possibly developing further bespoke or customized modules. Complex systems often have layers of software, with one system including several software categories. Software packages and the platform should be well known and mature before being considered Category 4 software, otherwise Category 5 may be more appropriate.

A Supplier Audit is usually required to confirm that the software package has been developed using appropriate quality systems and that application development and support organizations are robust and competent. In the absence of a documented quality system, suppliers should use this guide to provide the foundation for establishing a suitable quality system to control package development and support. Under such circumstances the software should be handled as Category 5. Users are responsible for ensuring the quality of the software and hardware, and the fitness for purpose of the complete system.

Validation should ensure that the software package meets the user requirements with particular focus on the configured business or manufacturing process. Bespoke or customized modules should be handled as Category 5.

The Validation Plan should define a structured approach to the validation of the application covering the full life cycle, including assessment f the supplier and the configurable package. The approach should address the layers of software involved and their respective categories. The Validation Plan should reflect the assessment of the supplier and any audit observations, application criticality, size and complexity. It should define strategies for the mitigation of any weakness identified in the supplier's development process.

Since each application of the standard product is specific to the user process, support of such systems needs to be carefully managed. For example, when new versions of software products are introduced, serious issues can arise from the dependency of bespoke or customized code on features of the standard product that may have changed.

Examples of Category 4 packages include Distributed Control Systems (DCS), Supervisory Control and Data Acquisition packages (SCADA), Manufacturing Execution Systems (MES), and some LIMS, ERP, and MRPII packages."

The following table lists recommended documents for GAMP4 Category 4 validation:

Documents for Category 4

- Letter of Notification/System Proposal
- Change Control Document
- Project Plan
- Validation Plan
- User Requirements
- Functional Requirements
- Vendor Audit
- Design Specification
- Factory Acceptance Test Documents
- IQ Protocol
- IQ Summary Report
- OQ Protocol
- OQ Summary Report
- PQ Protocol
- PQ Summary Report
- Traceability Matrix
- Project Validation Final Report & Authorization for Use

Custom (Bespoke) Software (GAMP4 Category 5)

"These systems are developed to meet the specific needs of the user company. Custom developments may be a complete system or extension to an existing system. Complex systems of ten have layers of software, with one system including components of several software categories.

A Supplier Audit is usually required to confirm that appropriate quality systems are established to control development and ongoing support of the application. In the absence of a documented quality system, suppliers should use this Guide to provide the foundation for establishing a suitable quality system to manage application development and support.

The Validation Plan should define a full life cycle approach to the validation of the application based on guidance in this document. The approach should address the layers of the software involved and their respective categories. The Validation Plan should reflect the assessment of the supplier and any other audit observations, application criticality, size and complexity. It should define strategies for the migration of any weaknesses identified in the supplier's development process."

The following table lists recommended documents for GAMP4 Category 5 validation:

Documents for Category 5

- Letter of Notification/System Proposal
- Change Control Document
- Project Plan
- Validation Plan
- User Requirements
- Functional Requirements
- Vendor Audit
- Design Specification
- Unit Testing Documentation
- Unit Testing Summary Report
- Factory Acceptance Test Documents
- IQ Protocol
- IQ Summary Report
- OQ Protocol
- OQ Summary Report
- PQ Protocol
- PQ Summary Report
- Traceability Matrix
- Project Validation Final Report & Authorization for Use

After category determination of a computer system, validation of the computer system can proceed. The project team, consisting of the System Owner, End-Users, CVG and QA(U), makes this determination. Creation of the Validation Plan is the initial step in computer system validation.. The plan details the scope of the validation effort, the method by which it will be validated, validation documentation deliverables, and responsibilities and sequence of events. Project success depends on the accuracy and completeness of this plan.

When a computer system is validated at the time of development or when it is initially installed, the process is known as *prospective validation*. However, in cases where a computer system has been installed and utilized for some time without having been systematically tested and documented, then validation of the computer system is known as *retrospective validation*. Retrospective validation is not acceptable for general use. That is, the practice of implementing a system and then validating it retrospectively is not acceptable. However, it may be applicable for legacy systems that did not previously require validation, to set a baseline for future system changes. After the computer system is validated prospectively or retrospectively, all changes to the computer system must be performed using standardized configuration management and change control procedures. Minor changes to the computer system that are part of its maintenance must be documented following the maintenance procedure written specifically for the system. By documenting each phase in the life cycle of a computer system, evidence is provided that the system is being maintained under control from development through to retirement.

Roles and Responsibilities

Each computer or computerized system validation is a team effort. At a minimum, the team consists of, the End-User and/or System Owner, Information Systems and Technology (IST), Computer Validation Group (CVG) and the Quality Assurance (Unit) (QA(U).

The System Owner, the Department Manager, and End-User(s) are responsible for overseeing the qualification and validation of GxP computer and computerized systems. The process is started by notifying QA(U) and the CVG of new computer projects and/or changes to existing computer systems. The End-User(s) generally performs and/or witnesses the actual validation work and formally accept the system when the validation has been successfully completed.

CVG coordinates validation activities and serves as an internal validation consultant to assist the End-Users during the project. The CVG serves as the central coordinator for all validation work.

The QA(U) in all organizations has the ultimate responsibility for assuring compliance with departmental, site, and company policies, practices and procedures along with government regulations. All computer system validation documentation, including protocols, reports, deviation reports and SOPs require the QA(U) to act as signatory for final authority over the process.

After the qualification and validation of a computer or computerized system, the System Owner/End-User is responsible for conducting periodic reviews of computer systems to ensure compliance with all appropriate policies, practices, and procedures.

Procedure

The following procedure is used to prospectively validate a computer system. The procedure required to retrospectively validate a computer system is covered at the end of this chapter. The individuals or groups responsible for the tasks listed below are detailed in Chapter 3, "Roles and Responsibilities" and Appendix 3, "Roles and Responsibilities – Validation."

Initiate "Letter of Notification"

The System Owner / End-User, with assistance from CVG, initiates the validation process by sending a "Letter of Notification" to QA(U) and CVG. It is sent as notification of an impending new computer project and/or changes to an existing system.

Scope Out Validation Effort

The System Owner / End-User then arranges a meeting consisting of QA(U), appropriate users and the CVG to discuss the scope of the project, the category of the computer system and the participants of the validation team.

Define Roles and Responsibilities

Once the validation team has been determined, the role of each department and individual in the team is assigned according to his or her roles and responsibilities. The roles and responsibilities of each department/individual in the validation project must be defined in the validation plan.

See Chapter 4 – "Roles and Responsibilities" and Appendix 3 – "Roles and Responsibilities – Validation."

Project Plan

Although a project plan is not a validation requirement, it is highly recommended that a project plan be prepared, progress through the formal approval process, and be used as a living document to control the validation process of each project.

See the section "Project Plan" under "Validation Documents" and Chapter 7 – "Project Plan."

Prepare and Execute the Validation Plan

The validation plan is developed as a collaborative effort between the System Owner / End-User and the CVG. The plan details the scope of the validation effort, the method by which it will be validated, the documentation deliverables, responsibilities, and sequence of events. The typical validation documents produced on a validation project are listed in the section "Validation Documents."

See the section "Validation Plan" under "Validation Documents" and Chapter 8 – "Validation Plan."

Archive the Final Validation Package

Completed and approved computer system validation documents are stored in a validation library or other designated location. The location of the original copy of the validation documents and other documents related to the computer system (e.g., vendor manuals and SOPs) must be recorded in the validation plan. In addition, electronic files used in support of the validation effort are also included and stored using suitable electronic storage media, such as CD-ROMs, tapes SOPs control the security and orderly storage of validation documents.

See Chapter 28 – "Document Management and Storage."

Maintain Computer System in Validated State

Once the computer system is validated and approved for use, the computer system is maintained in a sustainable and compliant validated state. This is in accordance with appropriate cGxPs, which include, but are not limited to: Change Control, Document Management and Storage, Configuration Management and Periodic Review.

See the following chapters: Chapter 27 – "Change Control," Chapter 28 – "Document Management and Storage," Chapter 29 – "Configuration Management" and Chapter 30 – "Periodic Review."

Perform Periodic Review

A periodic review of the validation status of each validated system is mandatory. This review assesses whether the system has changed and if it continues to operate as it was previously validated. Review of the system and its accompanying validation must be documented. Review frequency is discretionary, but should be less than a three-year review cycle, and should be stated in the validation plan or other official document. If not formally stated, all systems should default to a two-year review cycle. Periodic reviews are conducted by the System Owner / End-User in accordance with a Periodic Review Procedure that is specified in an SOP.

See Chapter 30 – "Periodic Review."

Decommissioning Plan

When it is determined that a computer system is no longer functionally useful, the system is retired. Decommissioning follows an approved plan prior to initiating the shutdown process. When a computer or computerized system or any software version is decommissioned, validation documentation and a backup copy of the software, computer or computerized system must be retained. These items are kept for an indefinite period of time, unless otherwise determined on advice from counsel. Under no circumstances should this time period be less than 5 years. Any data files present on the system must be archived and retained in accordance with appropriate procedures, based on the data and records that were generated.

See the section "Decommissioning and Retirement" in "Ongoing Operations" later in this chapter.

Validation Documents

The validation documents are the procedures followed to systematically progress through the System Development Life Cycle and to create records of the events and actions of the process. They become the defensible evidence that the computer or computerized system has been produced and maintained in a sustainable and compliant manner, and does what it is intended and purports to do.

These documents are discussed briefly here, and in the corresponding chapters that detail their requirements as SOPs.

Project Plan

Although a Project Plan is not required for validation purposes, this document serves an important role in defining a project and specifying the validation efforts.

A clearly defined plan guides the implementation and is the means for measuring its success. Depending on the complexity of the project, either separate project and validation plans, or a combined project/validation plan is prepared. In cases where an existing computer system is undergoing a minor modification, Change Control documentation could substitute for a Project Plan.

The Project Plan defines the activities in the form of work breakdown structures that will be performed throughout the SDLC for the computer system. Each of these plans provides a blueprint for ensuring that the installed computer system performs as designed and intended. The project plan includes milestones, activities, timelines, responsibilities, project procedures and other information that helps define the project. In essence, it summarizes the project and is one of the first documents that a manager or auditor examines to understand the system and the validation package.

See Chapter 7 – "Project Plan."

Also see *Project Management, A Systems Approach to Planning, Scheduling, and Controlling,* Sixth Edition by Harold Kerzner, Ph.D. It is published by John Wiley & Sons, Inc. and is an invaluable aid and guide to project planning.

Validation Plan

The Validation Plan details how the computerized system is to be validated. It defines the scope of the validation effort, the method by which it will be validated, the validation documentation deliverables, roles and responsibilities and sequences of events. A formal validation plan may not always be necessary, but the process may be described in a change control document or as part of the overall project plan for minor system changes. The goal of the validation plan is to detail the activities in order to demonstrate that the computer system is developed, validated, and maintained in a cGxP sustainable and compliant manner.

See Chapter 8 – "Validation Plan."

Back-Out Plan

In certain circumstances, such as upgrades, it is necessary to have a back-out or contingency plan. This is used in the event that the work does not function as anticipated and authorization for use is not given. The Back-Out Plan defines the method, with validation, to restore the system to its previous validated working state. The Back-Out Plan is contained in the Validation Plan.

User Requirements

Written in non-technical general terms by the System Owner and End-User, the User Requirements document describes and documents the functions, operations, process workflow and expectations of a desired computer system in order to meet the needs of the End-User.

See Chapter 10 – "User Requirements."

Functional Requirements

The Functional Requirements document describes and documents, at a high-level, the operations, process workflow and expectations, hardware, software, and peripherals of the required system and indicates whether the system will meet each of the User Requirements. It references the data model, if applicable, and defines the performance requirements. It provides a vehicle to assure the development and delivery of a desired system that is designed to meet those needs.

To clearly define the Functional Requirements and tie them to the User Requirements, it is necessary to represent the current and proposed procedures, workflow and processes in such a way that the developer can create the design specifications of the software, hardware or system. The Functional Requirements must enable an understanding of the business role or function that the computer system is intended to perform without limiting the selection of hardware or software solutions.

Functional Requirements must be documented, complete and accurate prior to initiation of the development and/or procurement process. Functional Requirements include or address the following:

- Key objectives and benefits of the system
- Description of the operation or process for which the computer system will be involved
- Process flow diagram
- Business/process improvement
- Inputs/outputs (types and quantity)
- Specific user activities/functions to be addressed by the system
- Operator interfaces (displays, input) to the system
- System interfaces
- Query and report functions
- Security
- Archiving
- Communications/network considerations
- High-level description of the system
- Main system

- Sub-systems
- Individual functions
- Facilities provided
- Modes of operation
- Performance requirements
- Critical calculations and algorithms
- Design assumptions
- Implementation assumptions
- Non-conformance with User Requirements
- Data definition, handling, access and management

Typically the System Owner or End-User is responsible for preparation of the Functional Requirements document. Items are indicated as either "Must Have" or "Desirable." Often not all requirements can be met or may need to be phased in over time.

See Chapter 11 – "Functional Requirements."

Vendor Audit (Evaluation and Selection)

When a computer or computerized system is custom built by a third party or selected from a group of products, a vendor evaluation is performed. Usually a corporate group or function performs the vendor audit and provides the findings to the System Owner / End-Users and CVG. In the absence of a current vendor audit for a particular company, the System Owner or End-User submits a formal request to audit a vendor. Vendor evaluations are typically not performed for custom-built (bespoke) systems developed in-house.

The act of formally contacting the corporate group usually fulfills the requirement of notifying the corporate group of a validation project. In cases where an audit is not required, the CVG notifies the appropriate corporate groups about the validation activities. This may be separate from QA(U) notification.

The vendor audit assists in the selection of the appropriate group to develop or supply a computer or computerized system, or one of its components. Vendor selection should consider some of the following points:

- The ability of the vendor to achieve requirements
- The technical competence and qualifications to supply and support hardware and/or software for the proposed system
- Vendor acceptance of the validation requirements
- Vendor ability to provide system/program testing and documentation
- Timeframe required

When a vendor is selected, preferably from a group of candidates, the selection process is documented and the documentation is stored as part of the validation package.

See Chapter 12 – "Vendor Audit."

Software Design Specification Document

The Software Design Specification document or protocol, which is normally prepared by the developer or system integrator, identifies all measurable or determinable parameters affecting system performance together with the acceptable values and each critical parameter characteristic. Specifications address essential areas to be verified by installation checks and functional tests. These include engineering drawings, component specifications, design schema, logic flow diagrams , system variables, mathematical algorithms Design documents tend to be dynamic to reflect changes to the system as more about the system is learned through development and testing activities.

It must provide sufficient technical information on what a system must do and how it is to do it. It must be based on the User Requirements and Functional Requirements against which the system will be tested. The Software Design Specification becomes the standard that the system developer uses to create the system.

See Chapter 13 – "Software Design Specification."

Hardware Design Specification Document

The Hardware Design Specification document or protocol, which is normally prepared by the developer or system integrator, identifies the method to be used for specifying the design that a computer operations and a procurement group will use in acquiring, configuring and implementing computer and computer related hardware.

It provides technical information on what a computer, printer or other peripheral must do and how it is to do it. It is based on the User and Functional Requirements against which the system will be tested. The Hardware Design Specification becomes the standard that is used to acquire, configure and deploy a specific hardware item.

See Chapter 14 – "Hardware Design Specification."

Design Qualification

In order to objectively determine the applicability and completeness of a design specification prior to its implementation, it is highly recommended that the design be formally reviewed in accordance with a Standard Operating Procedure. In performing this process, assurance may be had that the design adequately addresses the User and Functional Requirements of the project and is qualified for use for its intended purpose.

See Chapter 10 – "Design Qualification."

Unit Testing Documentation

Unit testing is the basic detailed examination and testing of a new or changed program at the programming level. This most basic level of testing is usually the responsibility of the developer/programmer. Its purpose is to verify that an application program is structurally sound and conforms to design specifications and appropriate programming standards.

Unit Testing is generally performed for all code produced under GAMP4 Categories 4 and 5.

See Chapter 15 – "Testing" and Chapter 16 – "Unit Testing."

Factory Acceptance Testing

Factory Acceptance Testing (FAT) combines units that were previously tested. Although the responsibility of the system developer, the client (i.e., System Owner or End-User) may be present to assure that the computer system being developed meets the requirements. FAT is often called system testing. FAT normally takes place at the vendor facility. On larger systems, often simulators are used for testing rather than actual hardware. The purpose of this testing is to demonstrate that the combined units and modules operate correctly together. This testing is performed in accordance with an approved protocol (normally approved by vendor management) and executed in accordance with good documentation practices.

See Chapter 15 – "Testing" and Chapter 17 – "Factory Acceptance Testing."

End-User Testing

User qualification testing begins once development of the computer system is complete and the system is installed and operating in a qualified environment. The System Owner/End-User is responsible for the validation of the computer system. Test protocols are written specific to the computer system and the appropriate groups prior to use and are reviewed and approved. The protocols are then executed, after which, QA(U) reviews the results. This review consists of two parts, verification that the tests were completed in accordance with the procedure and that the entire protocol has been completed; and that all deviations are properly documented and remediated.

End-User validation testing consists of Installation, Operational and Performance Qualification testing. Installation Qualification (IQ) provides documented verification of the installation of the software and hardware components and that these are in accordance with the approved design. Operational Qualification (OQ) provides documented verification that each unit or subsystem operates within the defined operating range. Performance Qualification (PQ) provides documented verification that the integrated system, as tested by users, performs in its normal operating environment.

See Chapter 15 – "Testing" and Chapters 18 through 22.

Installation Qualification

Installation Qualification consists of testing and documenting the installation of the hardware and software components of the integrated system. The IQ, at a minimum, consists of the following:

- Inventorying the hardware and software components
- Inventorying all documentation
- Verifying that the system design is accurate
- Verifying that the physical environment adheres to the manufacturer's and design specifications
- Verifying that the system has been installed in accordance with the installation procedures
- Documenting configuration settings in the software or hardware

In cases where the computer system relies on other existing components, such as a server or network managed by the IT/IS Unit or Computer Operations that is outside the control of the user, these separate elements are:

- Individually qualified by the appropriate group in accordance with approved procedures
- Documented and available for review
- Maintained in accordance with the appropriate procedures

Although the System Owner / End-User is not expected to duplicate these efforts, it is their responsibility to reference the individual qualifications and other appropriate documents in the validation package.

Where the computer system is interfaced to instrumentation or manufacturing equipment, these devices are normally qualified prior to or as part of the computer validation IQ. Qualification of these devices consists at minimum of:

- Inventorying each component

- Inventorying all documentation
- Verifying that the installation drawings and specifications are accurate
- Verifying that the installation and physical environment adheres to the manufacturer's specifications
- Calibrating the devices in accordance with the manufactures specifications and appropriate departmental/site SOPs
- Verifying that the required IQ, OQ and PQ tests have been satisfactorily performed

See Chapter 15 – "Testing," Chapter 18 – "Hardware Installation Qualification" and Chapter 20 – "Software Installation Qualification."

Operational Qualification (OQ)

Operational Qualification consists of testing the integrated system against the Functional Requirements and Design Specifications. The purpose of this testing is to verify that the modules within the system operate as expected. At a minimum, the testing consists of:

- Boundary Testing - testing all system functions on, outside and inside normal operating limits
- Verifying calculations and algorithms
- Testing peripherals and other external devices interfaced to the system
- Verifying compliance in accordance with the Functional Requirements and Design Specifications

See Chapter 15 – "Testing," Chapter 19 – "Hardware Operational Qualification" and Chapter 21 – "Software Operational Qualification."

Performance Qualification (PQ)

Performance Qualification consists of verifying the correct operation of the integrated system in the production environment including human use and interaction with the system. The Performance Qualification procedure includes records and procedures surrounding the use of the system and verification that users and administrators have been trained on the use and administration of the system prior to the system being placed in production. Training is documented in accordance with company policies and procedures. PQ testing also includes, at a minimum, verification that:

- The system can handle the anticipated data volume
- The system responds to error conditions (e.g. data outside of boundary values, incorrect data types) or problem situations (e.g., power failures)
- The system security is present and properly configured
- The system records changes to configuration or data (i.e., audit trails)
- The system can handle the potential load of multiple concurrent users (stress testing)
- The response times of the system are reasonable
- The backup and recovery procedures are functional
- The data integrity of information loaded into the system is demonstrated
- Data can be transferred from another like system, converted from a different data source or input by a user

See Chapter 15 – "Testing" and Chapter 22 – "Performance Qualification."

Summary Reports

After completing each phase of the validation testing, a summary report is written that documents the main points of the testing. This report summarizes all of the failures, deviations, and unexpected results along with remediation accomplishments. The summary report is written and approved prior to executing any additional test protocols.

See Chapter 25 – "Summary Reports."

Traceability Matrix

The Traceability Matrix ties the User Requirements, Functional Requirements and System Design Specifications to each other and to the Validation Protocols (IQ, OQ, PQ). The Traceability Matrix lists the User Requirements in the left most column with the remaining columns representing activities in subsequent requirement/specification documents and validation protocols. The matrix is completed by indicating in the appropriate columns where the requirements of each preceding document have been met. This is a freestanding document.

See Chapter 26 – "Traceability Matrix."

Manuals and SOPs

All manuals for the system, such as, but not limited to, hardware, software, installation and operating instructions must be retained as part of the validation package.

SOPs are created to govern the use and management of system. These, too, must be retained as part of the validation package.

A checklist of the SOPs necessary to manage the operation and maintenance of the validated system is put in place. The checklist includes signatures for approval indicating that the SOPs have been written and are available for use. The following is a list of SOPs that should be available before the system is authorized for production use: Training, Security, Change Control, Periodic Review, Backup/Restore/Archive, Disaster Recovery, System Use and Preventative Maintenance.

See Chapter 31 – "Manuals and SOPs."

Miscellaneous Documents

Documents that must be retained as part of the validation package include, but are not limited to, procurement documents, service level contracts, vendor contracts and safety notices.

Training

Participants in the validation process must be trained on the validation procedures and system SOPs. Users and maintainers of the system must be trained as appropriate before the system is placed into production. Training is documented in accordance with company policies and procedures.

See Chapter 34 – "Project Validation Final Report."

Project Validation Final Report

The Project Validation Final Report summarizes all testing results. It is in the format of an executive report that analyzes the entire validation process and forms the basis for the claim of system validation. This report serves as the basis for authorizing or commissioning the computer system for use in production. A statement in this report usually reads to the effect, "Approval of this report indicates acceptance of the system and authorization for its use in the production environment."

The appropriate management and QA(U) personnel approve or disapprove the Project Validation Final Report containing the authorization for use. This approved document must be in place before the computer system can be used in a GxP capacity.

See Chapter 34 – "Project Validation Final Report."

Ongoing Operations

After a computer system is placed into operation, it needs to be maintained in a validated state. This requires that appropriate SOPs be observed, that the system is periodically reviewed to verify that the system has not been changed or has not degraded with time and that any changes to the system have been performed in accordance with appropriate change control procedures. The computer system is reviewed on a periodic basis as specified in its validation documentation. Generally the review cycle should not exceed two years.

Document Management and Storage

The management of documentation is critical to the success of any GxP operation. It is important that the documentation in use to control a given function or process be the most current and appropriate for the activity it governs. Document changes and storage are tightly controlled. Document revisioning, issue and storage are governed by standard operating procedures during their entire life cycle. This applies to items such as SOPs, protocols and manuals for all computer and computerized systems, software, hardware and associated items.

When and how a document is revised is critical because it ensures that whatever process is taking place is being performed in accordance with the most current methods and practices.

Documents, both paper and electronic, need to be stored in an organized, secure, controlled and systematized manner. A database that controls and tracks who has what document is necessary so that new, changed, expired or obsolete documents may be distributed and recalled in an organized and timely manner.

See Chapter 28 – "Document Management and Storage."

Change Control

Change Control applies to computer or computerized systems, hardware, software, firmware and operating systems. It is used in conjunction with a configuration management method to ensure that an item is managed in accordance with a standard operating procedure. Changes may not be made without going through a properly controlled process that documents what is undergoing a change, what is being changed, how the change is to be effected, what the change affects, the roles and responsibilities of the people involved, when it is to happen and why it is necessary. This occurs through a specified review and approval process that takes into account the configuration management aspect of the change. A standard form is used that, at a minimum, tracks the following information:

- Title of Change

- System Name
- Description of Change
- Reasons for the change
- Benefits of the change
- Previous functionality that may be lost
- Requestor and Date
- Approved for Evaluation By and Date
- Base Line Affected
- Raw Data Affected with Explanation
- Effect on Logistics, Support, Interfaces
- Is Revalidation Required
- Validation Team Approval and Date
- Programmer(s) / Team Members Assigned
- Files, Designs, Documents Affected
- Files, Designs, Documents Checked Out
- Change Plan Approved By and Date
- SOPs Updated and Date
- Documentation Updated and Date
- Files, Designs, Documents Affected Checked In and Date
- Testing Completion Date
- Training Completion Date
- Installed to Production Date
- Final Approval Date

See Chapter 27 – "Change Control."

Configuration Management

An internal configuration management system is used for the control of all configuration documentation, physical media, and physical parts representing or comprising the system. For computer or computerized systems, hardware, software, firmware, operating systems, and documentation, it addresses the various and changing configurations and support environments (engineering and design, implementation and test) used to generate and test the system, subsystem or component. The configuration management system consists of the following configuration elements:

- Identification
- Control
- Status Accounting
- Audits

Plan a configuration management program tailored appropriately for the particular configuration item (CI), its scope and complexity and the phase(s) of its life cycle. Planning should be consistent with the objectives of a continuous improvement program that is sustainedly compliant and that includes the analysis of identified problem areas and correction of procedures as necessary to prevent reoccurrence. Configuration management planning includes:

- Program Objectives
- Each Configuration Item
- Organization and Relationships
- Roles and Responsibilities
- Resources (tools, techniques, and methodologies)

See Chapter 29 – "Configuration Management."

Decommissioning and Retirement

When a computer system is to be decommissioned, a project plan is created to document the requirements for decommissioning the system. The plan includes how the data is to be archived and/or transferred to another system. A Decommission Report is issued upon the decommissioning of the software or system.

See the section "Decommissioning Plan" in this chapter.

It is important to carefully plan the decommissioning of a system so that the data are preserved in a useable fashion for future use. If product is involved, the data and the system may need to be reconstructed to its original final state for FDA audit purposes.

See Chapter 36 – "Decommissioning and Retirement."

Validation Master Plan

The Validation Master Plan is used to inventory and track hardware, software and systems that are required to be maintained in a validated state. It specifies what is to be validated, where it is located, who is responsible for its validation and when it is scheduled for validation, re-validation or periodic review independent of periodic review triggers. The Validation Master Plan may be separated into multiple plans based on areas of responsibility. For example, the IT/IS Unit may be responsible for software, patches and drivers whereas Computer Operations may be responsible for hardware and environmental items in a data center. A support group would cover desktops and laptops. However, one person or group should be responsible for the overall coordination of the master plans so that nothing is overlooked.

See Chapter 37 – "Validation Master Plan."

Review of Equipment Qualification Status

As part of the Validation Master Plan, associated equipment should be inventoried and tracked. A periodic review of equipment qualification status should be performed. It is not possible for a computerized system to be validated if the equipment with which it is associated is out of calibration or not qualified for use.

Retrospective Validation

Retrospective validation is used for legacy systems that were not previously or sufficiently validated. It is the process of establishing documented evidence that a system does what it purports to do based on an analysis of historical information.

See Chapter 38 – "Retrospective Validation."

Chapter 4 - Roles and Responsibilities in Computer and Computerized System Validation

This chapter covers Roles and Responsibilities for validation. It is a comprehensive description of the major roles and responsibilities of individuals involved in computer validation using the full System Development Life Cycle (SDLC) methodology and on-going daily operations in a regulated computing environment. This chapter describes the roles and responsibilities necessary to ensure that all computer and computerized systems governed by current Good Clinical, Laboratory, Manufacturing and Distribution Practices (cGxP) are validated in a consistent fashion and that computer systems are maintained in a validated state throughout their life cycle.

See Appendix 3 – "Roles and Responsibilities – Validation"

A framework for identifying major roles and responsibilities is needed so that a standard methodology that meets current industry practice can be understood and employed by the people responsible for computer and computerized system validation and operation in a sustainable compliant manner.

The information in this chapter is used to establish roles and responsibilities for people who perform daily activities in a sustainable compliant cGxP environment using computer and computerized systems and when they are involved in the validation process. It is highly recommended that it be made into a Standard Operating Procedure and incorporated into the documentation system in order to have a means of presenting an overview of the process in an official manner.

This overview is used by senior and middle managers and those who are directly involved in the operation of a software application, computer or computerized system that controls equipment and/or produces electronic data or records in support of all regulated laboratory, manufacturing, clinical or distribution activities. It is intended to provide an understanding of what is involved and required to maintain a sustainable compliant computing environment.

Roles and Responsibilities

System Owner

The System Owner is the designated manager who is the principle stakeholder involved in the design and implementation of a computer or computerized system and normally is the principle decision maker throughout the system's lifetime. This person serves as a bridge between the business users – End-Users – and the technical implementation and support team.

The System Owner is the most important person in the process, having the most responsibility and authority. Everyone involved is answerable to the System Owner because the System Owner is the client or customer. In addition, the System Owner is answerable to everyone as the person who needs to provide all of the answers. Not personally, but by means of orchestration. Success or failure rests on the shoulders of this individual.

Overall Management

All projects require or involve acquisition and management of resources, funding, progress reports to upper management, documentation, vendor or developer selection and control, and sustainable compliance with the policies, procedures and regulations.

The System Owner obtains and manages funding and resources. He or she works with other departments and external resources (i.e., system integrator, validation consultants, subject matter experts) to have staff assigned to the project.

For corporate system deployments – multi-site, the System Owner works with local management to fulfill these responsibilities.

In conjunction with the guidance and approval of QA(U), the System Owner is responsible for the level and type of training and technical support that is provided to the users of the system. The System Owner also decides when and how to implement the system and system changes, and when and how to upgrade or decommission the system.

Overall Maintenance

The computer system must be maintained in a validated state – sustainable compliance. This includes, but is not limited to: security, training, equipment qualification, document management, configuration management, change control, periodic review and use of the system.

Security of the system must be accomplished and be compliant with 21 CFR Part 11 regulations.

Training is an ongoing process. Each person involved must be trained appropriate to his or her job function.

Equipment qualification must be performed on a scheduled basis to show ongoing continuous credibility of data and/or process control.

Document management must be performed in accordance with established procedures.

When changes are made to the computer system or associated devices, *configuration management and/or change control* must be performed. The changed computer system must be returned to a fully validated state prior to subsequent use.

Periodic review of the computer system must be completed to demonstrate sustainable compliance.

Only properly trained people may have access to and use the computer or computerized system in accordance with approved SOPs.

Liaison

Providing the connectivity between the technical and the business needs of the project is paramount and a prerequisite to the success of any project and ongoing operations. It is a difficult task, at best, for all of the people involved in any project to reach a level of understanding of each other's wants and needs. The System Owner must provide the means to accomplish the goal of achieving understanding by all team members of what is required to complete the project and support its ongoing operation. He or she is responsible for acquiring senior management buy-in and support.

Delegation to Project Manager

Much of the work involved in the management of any project is an add-on to the normal work and responsibilities of the System Owner. It is easy for the System Owner to become overwhelmed with the tasks required to maintain order in day-to-day operations without the responsibility of one or more projects at the same time. The appointment of a Project Manager (VPM) to oversee the daily work of the project is, therefore, a viable option. The VPM serves as the designee of the System Owner and is responsible for representing and communicating the System Owner's goals. In the absence of a VPM, the System Owner assumes the Project Manager responsibilities.

Training

Training is required to ensure that individuals (e.g., users, vendors and contractors) who prepare and/or execute validation protocols and actually use the system know and understand the cGxP requirements and relevant SOPs. This is essential in order to fulfill the regulatory requirements surrounding the validation process and maintenance of sustainable compliance. In coordination with a company trainer and QA(U), training is accomplished to provide each team member with the proper level and type of instructions/knowledge/skills required for fulfillment of their designated/officia l responsibilities.

Approval Authority

Documentation review and approval ensures that the project meets its business goals. It must also address the usability of the software, hardware, computer or computerized system.

End-User

End-Users routinely use the computer system for its intended purpose. They operate the system using standard procedures created specifically for that system. The End-User or their managers usually report to the system owner.

As participants in the system design, creation and/or selection and validation process, their responsibilities are important.

Documentation Preparation

End-Users participate in almost all of the documentation preparation and execution. In many cases, through the auspices of the System Owner, they will have the final say in what policies are adopted and what decisions are made because the ultimate purpose of the system is to fulfill their business needs.

End-User Testing

End-User Testing involves Installation Qualification (IQ), Operational Qualification (OQ) and Performance Qualification (PQ). End-Users are primarily concerned with OQ and PQ, as these tests will qualify the system for use. They maintain a communication through the System Owner with QA(U) and upper management as to the status of the validation process. The actual work is performed with a developer, vendor or contractor. However, the End-User is responsible to ensure that the computer system is tested relative to the user requirements, which include, but are not limited to, their needs and those of their department.

System Improvement

As the project progresses through it phases to completion, subsequent implementation and use, suggestions for improvement and problem remediation are collected and reported to the System Owner. This may occur directly or through a Help Desk function operated by Information Systems / Information Technology Unit.

Approval Authority

The End-User has no approval authority. The End-User only reviews documents and provides feedback in the decision-making process.

Quality Assurance (Unit) (QA(U))

The Quality Assurance (Unit) is the final authority. Only after all other parties involved on the project are satisfied with the outcomes does QA(U) become involved. Although they should be part of any review process from the beginning, they are the last to approve all policies, procedures, actions, and decisions providing they are acceptable to QA(U). After all, why would QA(U) believe anything is acceptable if the System Owner, End-Users or CVG doesn't?

Guidance

Communication, instruction, and training of End-Users in the appropriate regulations that affect their area must be accomplished throughout each phase of the project.

Documentation

QA(U) participates in the review and approval cycle for all documentation (e.g., validation plans, protocols, summary reports) to ensure that the documentation suitably demonstrates computer system compliance with regulations, corporate guidelines and SOPs. QA(U) is last in the document approval process. By approving a document, QA(U)'s approval indicates that the document is complete and in compliance.

All executed protocols are reviewed to ensure accuracy, completeness and correctness. All validation protocol deviations are approved or disapproved based on the impact assessment and the corrective action activity, both before and after protocol execution.

Audit

The QA(U) conducts periodic audits to ensure that the system is being maintained in a validated state and operated in sustainable compliance with all polices, procedures and regulatory requirements.

Vendor audits, both internal (e.g., Information Services software development groups) and external, assess compliance with all policies, procedures and regulatory requirements, as well as best industry practices and accepted quality standards.

Approval Authority

Approves or disapproves all validation documents (e.g., validation plans, protocols, summary reports) to ensure that the contents comply with all polices, procedures and regulations.

Approves or disapproves the system for use (i.e. Final Report).

Approves or disapproves all change control documentation.

Computer Validation Group (CVG)

The Computer Validation Group (CVG) is a support arm of the IS/IT Unit and a major participant in the validation process. Its primary function is to assure that the validation process is performed in accordance with best industry practices, regulatory requirements, company policies and guidelines, and SOPs.

Guidance

The Computer Validation Group maintains current information about company, government and cGxP policies and requirements. In this manner the CVG can provide guidance based on their validation expertis e to keep the project on track and in sustainable compliance.

In providing technical guidance to the System Owners, End-Users, consultants, software and system developers, the CVG ensures that the development efforts comply with the current validation practices.

Documentation

The CVG participates in the review and approval cycle for all documentation (e.g., validation plans, protocols, summary reports) to ensure that the documentation suitably demonstrates that the computer system complies with regulatory requirements, corporate policies, guidelines and SOPs. By being routinely involved in the preparation and review of SOPs, guidance documents and training materials for the computer validation program, the CVG will expedite the SDLC.

Validation Master Plan

The CVG is responsible for the Is/IT Unit Validation Master Plan, which, as part of a site Validation Master Plan, ensures that the site is maintained in sustainable compliance with all policies, procedures and regulations.

Liaison

By communicating the status of computer validation projects to the System Owner, End-Users and senior management, the CVG establishes and maintains itself as liaison for each project and system operation.

Training

The CVG assists the System Owner, End-Users and training department in the development and maintenance of computer validation training modules and curricula.

Review

Works closely with project management teams of strategic corporate systems to ensure QA(U) acceptance of validation documentation.

The tracking, management and review of periodic reviews of validated systems demonstrates responsible management of sustainable compliance initiatives. By issuing and maintaining a document that identifies and summarizes software applications and systems, the System Owner, System Administrator and validation status, the CVG is able to plan its activities and keep everyone apprised of the status of each system.

Approval Authority

The CVG is one of the three primary approvers of most documents. The System Owner and QA(U) are the other two, thus maintaining consistency across the site and ensuring proper and timely completion of validation projects.

System Administrator (SA)

The System Administrator is the person who maintains the system without necessarily using or having responsibility for it. Activities include administering user security, documenting and remediating system issues, performing database maintenance, data backup and archiving In certain cases, two types of system administrators are employed: one to administer the server and network hardware (sometimes referred to as computer infrastructure) and the other administrator to manage business applications and systems.

Maintenance

Computer and computerized systems as well as applications and hardware must be maintained in accordance with documented procedures. This role may be divided into two areas, one for maintaining the computer infrastructure and the other for maintaining the business application. Part III of this book addresses infrastructure. Maintenance procedures include:

- Back-up operations
- Restoration Operations
- Report generation
- Security administration
- Preventive maintenance
- Logging system problems and failures
- Monitoring technical response to system problems and failures
- Initiation of system change control procedures
- Periodic reviews

Training

Providing gatekeeper access to systems ensures that users are trained prior to acquiring access.

Configuration Management and Change Control

Standardized configuration management and change control procedures are used during the validation process and ongoing operations. This assures that only authorized, documented, and known configurations of the system and the supporting and controlling documentation are in use.

Notification of system availability to the appropriate user/management community allows users to be efficient in the use of their time and the system.

Approval Authority

The SA has no approval authority.

Subject Matter Expert (SME)

The Subject Matter Expert is a person who has acquired sufficient training, knowledge and experience in an area to be considered an expert in a particular subject.

Guidance

The technical aspects of the validation project must meet the project goals while staying in compliance with polices, procedures and regulatory requirements. This responsibility is often shared among different individuals having varied technical expertise. The Subject Matter Expert (SME) may provide guidance with such areas as:

- Engineering
- Manufacturing
- Analytical Chemistry
- Biochemistry
- Bioinformatics
- Clinical Trials
- Information Technology
- User interactions with the system
- Graphical user interface
- Process procedures
- Response times for various operations
- Data entry operations
- Report formats
- Interfaces between the computer and equipment/instruments
- Validation
- Quality Control

Assessment

The affect of changes to a system can have a multitude of outcomes or consequences. The careful accurate assessment of proposed changes is critical to these outcomes. Assessments are used, in part, to determine whether or not to make a system change and how this change will affect the overall computer system. The SME assists the System Owner, CVG and QA(U) in determining what and how the proposed changes should be and what regression testing is necessary.

Training

The SME assists the System Owner, End-Users and training department in the development and maintenance of computer validation training modules and curricula to keep them up to date with information about current technology.

Documentation

The SME assists in the review and approval cycle for all documentation (e.g., validation plans, protocols, summary reports) to ensure that the documentation suitably demonstrates that the computer system complies with corporate guidelines and SOPs. By being routinely involved in the preparation and review of SOPs, guidance documents and training materials for the computer validation program, the SME can expedite the SDLC. Depending to the nature of the task at hand, one or more SMEs may be utilized.

Approval Authority

The SME has no approval authority.

Vendor / Supplier

The Vendor or Supplier is a person or an organization that provides software and/or hardware and/or firmware and/or documentation to the user for a fee or in exchange for services. Such a firm could be a medical device manufacturer. (FDA)

The Vendor or Supplier must be qualified, trained and follow all policies, procedures and regulations, as related to the compliant processes used by the End-User or other company employees.

Goods and Services

The Vendor or Supplier provides software and/or hardware and/or firmware and/or documentation and associated services and documents for installation, maintenance and troubleshooting.

Guidance

See "Subject Matter Expert – Guidance" in this chapter.

Training

The Vendor or Supplier assists the System Owner, End-Users and training department in the development and maintenance of computer validation training and operation manuals, modules and curricula to keep them up to date with current technology. They provide training to users on equipment use and operation.

Approval Authority

Vendors and Suppliers have no approval authority.

System Developer (SD)

A person, or group, that designs and/or builds and/or documents and/or configures the hardware and/or software of computerized systems.[3]

The System Developer (SD) works with the System Owner, Project Manager and/or Project Team throughout the project life cycle to design, develop, implement, modify and maintain the computer system at the highest quality level achievable within the timeline and financial resources of the project and in accordance with policies, procedures and regulatory requirements.

Development

Using best industry practices, develops, configures or modifies the computer software and/or hardware to meet User and Functional Requirements and Design Specifications in a way that assures a problem-free system that can be readily validated. This includes, but is not limited to, maintaining regular code reviews, unit or modular testing and acceptance testing.

---

[3] FDA, Glossary of Computerized System and Software Development Terminology

Uses written procedures for programming conventions that specifies rules governing the use of individual constructs provided by the programming language, and naming, formatting, and documentation requirements which prevent programming errors, control complexity and promote understandability of the source code. These documents are maintained in accordance with document management and change control procedures during the development process and delivered to the System Owner or designee on a regular basis.

Documentation

Works with the System Ownerand Project Team to devise a clear User Requirements document that can become the basis for the Functional Specifications document.

Works with the System Ownerand Project Team to devise clear Functional Requirements and Design Specifications.

Maintains the documentation for all source code.

Participates in the review and approval cycle for their associated documentation (e.g., protocols, test procedures) ensures that the documentation suitably demonstrates that the computer system complies with corporate guidelines and SOPs. By being routinely involved in the preparation and review of SOPs, guidance documents and training materials for the computer validation program, they will expedite the SDLC.

Liaison

Regular communication with the System Owner and Project Team on the development status helps keep projects on time and within budget.

Approval Authority

No approval authority.

Project Champion

The Project Champion assists the System Owner and Project Team to obtain the necessary resources for the computer system project.[4]

Liaison

The Project Champion represents the needs of the computer system project to managers that are not normally available to the System Owner or Project Team.

Approval Authority

The SD has no approval authority.

---

[4] For the downside risks and an excellent comparison of Project Champions vs. Project Managers, see the dissertation in Project Management, *A Systems Approach to Planning, Scheduling, and Controlling*, Sixth Edition by Harold Kerzner, Ph.D. published by John Wiley & Sons, Inc.

Project Manager

The Project Manager is the designee of the System Owner.

Project Management

Responsible for the portion of the project assigned to them by the System Owner, the Project Manager has the same responsibilities as the System Owner for the delegated portion of the project.

Approval Authority

As delegated by the System Owner, the Project Manager has approval authority for the assigned portion of the project documents to ensure that the project meets the business goals of the project and complies with policies, procedures and regulatory requirements.

Procedure

Introduction

The demanding GxP environment necessitates that computer and computerized systems be quite complex because they impact multiple departments and disciplines as well as outside organizations. Validation assures that all aspects of a computer or computerized system have been qualified and meet their business and technical needs. Various individuals from within and outside the organization are involved in the validation, operation and maintenance of each system. Clearly defined roles and responsibilities are important in ensuring that the correct qualified individuals fulfill all requirements of the validation process.

This chapter is used in concert with Chapter 3 – Overview of the Computer Systems Validation Process to form the basis for computer and computerized system validation using the full System Development Life Cycle methodology.

Procedures

The System Owner, CVG and QA(U) determine the roles and responsibilities required for the validation and maintenance of a computer system based on the complexity of the computer system. The minimum number of participants required for validation are the System Owner / End-User, CVG, and QA(U).

Individuals involved in the computer validation process must fulfill their specific responsibilities as delineated in the Roles and Responsibilities Section of this procedure. Table I is a chart of the recommended roles and responsibilities of each individual for each stage of the process.

(Refer to Appendix 3 for a comprehensive table of Roles and Responsibilities in Computer Validation).

Table 2 - Roles and Responsibilities in Computer Validation

Document / Validation Activities	Prepared By	Reviewed By	Pre-Execution Approval	Executed By	Post-Execution Approval
Letter of Notification / System Proposal	**U**	**O, Q, G**	**O, Q, G**	**N**	**N**
Change Control Documents	**U**	**O, Q, G**	**O, Q, G**	**N**	**N**
Project Plan	**U**	**O, Q, G**	**O, Q, G**	**N**	**N**
Validation Plan	**U**	**O, Q, G**	**O, Q, G**	**N**	**N**
User Requirements	**U**	**O, Q, G**	**O, Q, G**	**N**	**N**
Functional Requirements	**U**	**O, Q, G**	**O, Q, G**	**N**	**N**
Vendor Audit	**Q**	**Q**	**Q**	**Q**	**Q**
Design Specification	**D**	**U, Q, O**	**O, Q***	**N**	**N**
Unit Testing Documentation	**D**	**U, Q, O**	**O, Q***	**D**	**O, Q**
Factory Acceptance Test Documents	**D**	**U, Q, O**	**O, Q***	**U, D**	**O, Q**
IQ Protocol	**U, D or C**	**O, Q, G**	**O, Q, G**	**U, D or C**	**O, Q**
IQ Summary Report	**U, D or C**	**O, Q**	**O, Q**	**N**	**N**
OQ Protocol	**U, D or C**	**O, Q, G**	**O, Q, G**	**U, D or C**	**O, Q**
OQ Summary Report	**U, D or C**	**O, Q**	**O, Q**	**N**	**N**
PQ Protocol	**U, D or C**	**O, Q, G**	**O, Q, G**	**U**	**O, Q**
PQ Summary Report	**U, D or C**	**O, Q**	**O, Q**	**N**	**N**
Traceability Matrix Document	**U, D or C**	**O, Q, G**	**O, Q, G**	**N**	**N**
Manuals	**V, D or C**	**O, Q, G**	**O, Q, G**	**N**	**N**
Miscellaneous Documents	**U, V, D or C**	**O, Q, G**	**N**	**N**	**N**
Training Documents	**U, D or C**	**O, Q, G**	**O, Q, G**	**N**	**N**
SOPs	**U, D or C**	**O, Q, G**	**O, Q, G**	**N**	**N**
Project Validation Final Report	**U, D or C**	**O, Q, G**	**O, Q, G**	**N**	**N**
Decommissioning Plan & Report	**U**	**O, Q, G**	**O, Q**	**N**	**N**

Key: C = Consultant; D = System Developer; G = Computer Validation Group; N = Not Applicable; O = System Owner; Q = Quality Assurance (Unit); U = End User; V = Vendor/Supplier

Note: The asterisk (*) signifies that the QA(U) responsibility is limited to auditing the work of the Developer's Compliance Unit.

Part 2
Standard Operating Procedures for Validation

Chapter 5 – Regulatory Requirements, Policies, Guidelines, and Standard Operating Procedures

In this chapter we are going to review the mission of the FDA in their own words and explain regulations. Then, within the corporate environment, we will explain the 3 levels of control in general practice: Corporate Policies, Corporate Guidelines, and local Standard Operating Procedures.

FDA's Mission

"The FDA Modernization Act of 1997 (PL 105-115) affirmed FDA's public health protection role and defined the Agency's mission:

> To promote the public health by promptly and efficiently reviewing clinical research and taking appropriate action on the marketing of regulated products in a timely manner;
>
> With respect to such products, protect the public health by ensuring that foods are safe, wholesome, sanitary, and properly labeled; human and veterinary drugs are safe and effective; there is reasonable assurance of the safety and effectiveness of devices intended for human use; cosmetics are safe and properly labeled, and; public health and safety are protected from electronic product radiation;
>
> Participate through appropriate processes with representatives of other countries to reduce the burden of regulation, harmonize regulatory requirements, and achieve appropriate reciprocal arrangements; and,
>
> As determined to be appropriate by the Secretary, carry out paragraphs (1) through (3) in consultation with experts in science, medicine, and public health, and in cooperation with consumers, users, manufacturers, importers, packers, distributors and retailers of regulated products."[5]

Regulations

A *regulation* is a rule or requirement that holds the same weight as a law issued by an executive agency to help administer and enforce the laws that have been passed by a legislative body. U.S. Federal regulations are published in the *Federal Register* and the Code of Federal Regulations.

A *regulatory agency* is an agency that enforces, defines and carries out regulations. The U.S. Food and Drug Administration (FDA), is such an agency.

In addition to its primary charge, the Federal Food, Drug, and Cosmetic Act, the laws enforced by the FDA are too numerous to list here. They may be found on their website at http://www.fda.gov or by contacting the FDA. The regulations enforced by FDA are the Code of Federal Regulations Title 21- Food and Drugs.

Because much of what this handbook covers is as a result of 21 CFR Part 11, *Electronic Records; Electronic Signatures,* the scope of that regulation is set forth in full below.

21 CFR Part 11, Electronic Records ; Electronic Signatures

> § 11.1 Scope.

---

[5] From the FDA website.

(a) The regulations in this part set forth the criteria under which the agency considers electronic records, electronic signatures, and handwritten signatures executed to electronic records to be trustworthy, reliable, and generally equivalent to paper records and handwritten signatures executed on paper.

(b) This part applies to records in electronic form that are created, modified, maintained, archived, retrieved, or transmitted, under any records requirements set forth in agency regulations. This part also applies to electronic records submitted to the agency under requirements of the Federal Food, Drug, and Cosmetic Act and the Public Health Service Act, even if such records are not specifically identified in agency regulations. However, t his part does not apply to paper records that are, or have been, transmitted by electronic means.

(c) Where electronic signatures and their associated electronic records meet the requirements of this part, the agency will consider the electronic signatures to be equivalent to full handwritten signatures, initials, and other general signings as required by agency regulations, unless specifically excepted by regulation(s) effective on or after

(d) Electronic records that meet the requirements of this part may be used in lieu of paper records, in accordance with Sec. 11.2, unless paper records are specifically required.

(e) Computer systems (including hardware and software), controls, and attendant documentation maintained under this part shall be readily available for, and subject to, FDA inspection.

The term "Predicate Rule" appears throughout this handbook and in literature pertinent to FDA activities. The FDA defines Predicate Rule as "Requirements set forth in the Act, the PHS Act, or any FDA regulation, with the exception of 21 CFR Part 11."

Policies

A *policy* is a high-level overall plan embracing the general goals and acceptable procedures or course of action intended to influence and determine decisions, actions, and other matters. Policies define the scope, execution, and intent of various topics and areas within a company along with associated expectations in terms of conduct and practice. This provides a vehicle to ensure compliance with regulations. Polices are official directives that clearly state, in a high level manner, the minimum requirements and standards to be met by the company as a whole. Some example subjects for policy are:

- Audits
- Quality Assurance
- Change Control
- Sick Leave
- Documentation
- SOPs
- Electronic Records
- Testing
- Electronic Signatures
- Training
- Guidelines
- Vacation
- Investigations
- Validation

Guidelines

A *guideline* or *guidance document* is an official recommendation on how to do something, the action to take in a particular circumstance, or conduct.

The FDA has issued, at the time of the writing of this handbook, several draft guidance documents relating to 21 CFR Part 11. Additionally, many other guidelines have been published within the industry addressing 21 CFR Part 11 and validation in general.

Most industry associations and groups have issued various guidelines and guidance documents for use within the pharmaceutical and medical device industry. For example, ISPE's sub-group GAMP has issued GAMP4, "GAMP Guide for the Validation of Automated Systems," which is referenced in several places throughout this handbook.

As a responsible company operating in a regulated environment, it is important to issue guidelines with associated training. In this way, a company can recommend clear strategies and methods to meet the expectations required by regulation and policy.

Standard Operating Procedures

Standard Operating Procedures (SOP) are established or prescribed methods that are followed in carrying out a specific operation or in a given situation. They are used in the creation, implementation, distribution, control, maintenance and decommissioning of all processes and procedures used for software applications, computers and computerized systems. They are needed to establish and maintain standards within specific areas to achieve uniform operation, use, maintenance and sustainable compliance with company policies, guidelines and regulations.

SOPs are used to define the procedures that meet the business needs and requirements of the company. They are used by all levels of personnel involved in the operation of computers or computerized systems that produce electronic data or records in support of all regulated laboratory, manufacturing, clinical or distribution activities.

This handbook is primarily SOP-centric in its presentation.

Chapter 6 - The SOP on SOPs

In this chapter we will discuss the Standard Operating Procedure (SOP) for the creation, implementation, distribution, control, maintenance and decommissioning of all Standard Operating Procedures.

The SOP is needed to establish and maintain norms within the affected group(s) to achieve consistency in content, format, and depth in a systematic manner that will best achieve compliance to the combined set of SOPs.

This SOP is used to define the procedures surrounding the format & structure, forms & attachments, initiation, change, deletion, maintenance, implementation, distribution, tracking and security, and all other business procedures. It is the SOP that governs all other SOPs.

Everyone in the organization uses this SOP to produce Standard Operating Procedures. An SOP on SOPs may be generic for the entire organization or may be specific to a department or group based on business requirements. If an overlying SOP on SOPs does not exist, then this SOP is used by those involved in the operation of a software application, computer or computerized system that controls equipment and/or produces electronic data or records in support of all regulated laboratory, manufacturing, clinical or distribution activities

Title Block

This SOP should have a title block, title or cover page along with a header and/or footer on each page of the document that clearly identifies the document. The minimum information that should be displayed is as follows:

The SOP on SOPs should have a title block, title or cover page along with a header and/or footer on each page of the document that clearly identifies the document. The minimum information that should be displayed for the SOP on SOPs is as follows:

- Document Title
- Document Number
- Document Revision Number
- Project Name and Number
- Site/Location/Department Identification
- Date Document was issued, last revised or draft status
- Effective Date
- Pagination: page of total pages (e.g., Page x of y)

The Document Title should be in large type font that enables clear identification. It should clearly and concisely describe the contents and intent of the document. Since this document is the governing document for all other SOPs, it should be a model SOP. It should specify that it is the SOP on SOPs. The title should appear on every page of the document.

The Document Number should appear in large type font that enables clear identification. Since this document will be part of a series of documents that controls the operation of a company, division or department, it should be consistent with other documents in this category. It should be on the every page of the document.

The Document Revision Number, which identifies the revision of the document, should be on each page of the document. Many documents will go through one or more changes over time, particularly during the draft stage and the review and approval cycle. It is important to maintain accurate revision control because even a subtle change in a document can have an immense impact on outcomes of activities. This SOP on SOPs should describe and govern the numbering methodology.

The Project Name and Number, if applicable, should be indicated on the cover or title page. This information identifies the overall project to which this document belongs.

The Site/Location/Department Identification should be indicated on the cover or title page. It identifies the Site/Location/Department that has responsibility for this document and its associated project.

The Date that the Document was issued, last revised or issued with draft status should be on each page of the document (e.g., 12-Dec-01). This will clearly identify the date on which the document was last modified.

The Effective Date of the Document should be on the title or cover page of the document (e.g., 12-Dec-01). This will clearly identify when to begin using the current version of the document.

Pagination, page of total pages (e.g., Page x of y), should be on each page of the document. This will help identify whether or not all of the pages of the document are present. Pages intentionally left blank should clearly be designated as such.

Approval Section

The Approval Section should contain a table with the Author and Approvers printed names and titles and a place to sign and date the SOP. This SOP on SOPs should govern who the approvers of the document should be. The functional department (e.g., Production / Quality Assurance / Engineering) of the signatory should be placed below each printed name and title along with a statement associated with each approver indicating the person's role or qualification in the approval process. Indicate the significance of the associated signature (e.g. This document meets the requirements of Corporate Policy 9055, Electronic Records; Electronic Signatures, dated 22-Mar-2000).

The completion (signing) of this section by all parties indicates that the contents have been reviewed and approved by the listed individuals.

Table of Contents

The Table of Contents provides a way to easily locate various contents of the document and to provide the reader with a quick check of the contents of the document. The SOP should contain a Table of Contents, which is located towards the beginning of the document.

Purpose

This introduction to the SOP concisely describes the document's purpose in sufficient detail so that a non-technical reader can understand it.

The Purpose section clearly states that the intention of the SOP on SOPs is to provide a Standard Operating Procedure to be used in governing all other SOPs. This introduction to the SOP concisely describes the document's purpose in sufficient detail so that a non-technical reader can understand the reason for the document's existence.

Scope

The Scope section identifies the extent of applicability and limitations, if any, of an SOP with respect to the performed procedure. The section should limit the scope of the SOP to all GxP systems and should apply to all Sites, Locations and Departments within the company. It should include electronic records, electronic signatures and handwritten signatures executed to electronic records and apply to all records that are created, modified, maintained, archived, retrieved or transmitted in electronic form under any requirements set forth in FDA regulations. The scope or range of control and authority of the SOP are provided. State who, what, where, when, why and how the SOP is to control the processes or procedures specified in the Purpose section. For example:

"This procedure applies to all departments and divisions at all locations within the Example Pharmaceuticals Corporation that perform GxP-related activities."

Equipment / Materials / Tools

This section lists any required or recommended equipment, materials, tools, templates or guides. Provide the name and a brief description of the equipment, instruments, materials and/or tools to be used in the procedures or testing covered by the SOP. This is for reference purposes and requires the user to utilize the specific item or a designated equivalent. The purpose and use is explained elsewhere in the document. Substitution of comparable equipment may be made in accordance with procedures for Change Control.

If the information is included in a job aid or operating manual, a description of where the information may be found with a specific reference to the document number, title, date and revision number should be specified.

Warnings / Notes / General Information / Safety

Warnings, notes, general or safety information applicable to the SOP are listed in this section.

Policies, Guidelines, References

This section lists references to appropriate regulatory requirements, SOPs, policies, guidelines, or other relevant documents referred to in the SOP. The name and a brief description of applicable related SOPs, corporate policies, procedures and guidelines, Quality Control monographs, books, technical papers, publications, specific regulations, batch records, vendor or system manuals, and all other relevant documents that are required by the SOP should be included. This is for reference purposes and requires the user to use the specific item or a designated equivalent. The purpose and use are explained elsewhere in the document.

SOPs cite references and incorporate them in whole by reference rather than specifying details that duplicate these items. This is done so that changes in policy, guidelines or reference materials require no changes to SOPs themselves and the most current version of the referenced material is used. Examples are:

- Policy ABC-123, Administrative Policy for Electronic Records; Electronic Signatures
- Corporate, Plan for Compliance with 21 CFR Part 11, Electronic Records; Electronic Signatures
- SOP-456, Electronic Records; Electronic Signatures
- FDA, 21 CFR Part 11, Electronic Records; Electronic Signatures
- FDA, General Principals of Software Validation; Final Guidance for Industry and FDA Staff, January 11, 2002

Assumptions / Exclusions / Limitations

All assumptions, exclusions or limitations within the context of the SOP should be discussed. Specific issues should be directly addressed in this section of the SOP on SOPs. For example:

"Existing SOPs shall continue to be used. They shall be brought into conformance with this SOP upon their next scheduled periodic review. In the absence of a specified periodic review time, this document shall govern that process."

Glossary of Terms

A Glossary of Terms should be included that defines all of the terms and acronyms that are used in this document or that are specific to the operation being described and may not be generally known to the reader. The reader may also be referred to the FDA document, "Glossary of Computerized System and Software Development Terminology."

Roles and Responsibilities

The creation, implementation, distribution, control, maintenance and decommissioning of all Standard Operating Procedures should be, at a minimum, the joint effort of a Subject Matter Expert, the department responsible for the SOP on SOPs and a Reviewer with a member of QA(U) being the final signatory.

This section should clearly specify the participants by name, title, department or company affiliation and role in a table. At a minimum, they should include:

- Department Head or Designee
- Departmental SOP Coordinator
- Department Manager or Designee
- QA(U) SOP Administrator
- QA(U)
- User

The QA(U) should be responsible to verify that the details in the SOP are compliant with the applicable corporate practices and policies. They should also assure that the approach meets appropriate cGxP regulations and good business practices.

The qualifications of all key individuals involved in the project should be kept on file and made available as needed in accordance with cGxP.

Procedure

SOPs describe procedural events, not guidelines or policies. SOPs are needed for systems that gather, manipulate, transfer or store data within and between GxP environments. SOPs are directives and should be written as such. These should include how the systems are operated and maintained. The SOP on SOPs specifies the procedures on how to prepare all types of SOPs.

Content, Format and Structure

The content, format and structure of each Standard Operating Procedure should be in the form of the following specifications.

Cover Page

The Cover Page should contain the name and location of the company, the label "Standard Operating Procedure", SOP title, page numbering (e.g.- in the form "Page x of y"), a unique and sequential SOP number, revision number, department number, department name and effective date, the superseded documents' number and revision, effective date, system name and number (if applicable).

Note: An SOP Documentation Administrator should issue SOP numbers.

Approval Section

The Approval Section should contain a table with the Author and Approvers printed names and titles with a place to sign and date the SOP. The functional department (e.g., Production / Quality Assurance / Engineering) of the signatory should be placed below each printed name and title along with a statement associated with each approver indicating the person's role or qualification in the approval process. Indicate the significance of the associated signature (e.g. This document meets the requirements of Corporate Policy 9055, Electronic Records; Electronic Signatures, dated 22-Mar-2000).

The completion (signing) of this section indicates that the contents have been reviewed and approved by the listed individuals.

Page Headers

All page headers of the document should contain the name and location of the company, division or department, "Standard Operating Procedure", SOP title, page numbering in the form "Page x of y," SOP number, revision number, department number, department name and effective date.

Page Footers

The footer should contain the restrictive legend such as: "The information contained herein is proprietary to Example Pharmaceuticals Corporation and may not be used outside of the company without prior written permission."

Table of Contents

The SOP should contain a Table of Contents, which appears towards the beginning of the document. This Table of Contents provides a way to easily locate various topics contained in the document.

Date Format

To prevent confusion, the date format of DD-MMM-YYYY, as in "10-Mar-1999," should be used.

Body

The body of the SOP should begin on the page after the Table of Contents and contain, at a minimum, the following sections:

- Purpose
- Scope
- Equipment / Materials / Tools
- Warnings / Notes / General Information / Safety
- Policies / Guidelines / References
- Assumptions / Exclusions / Limitations
- Glossary of Terms
- Roles and Responsibilities
- Procedure
- Document Management and Storage
- Revision History
- Attachments

An entry should be placed in each section. If there are no entries for a section or it is not applicable to the SOP that is being written, then an entry of "None" or "Not Applicable" should be made in the section.

The required information for each section should include, but not be limited to, the following:

A. Purpose

This introduction to the SOP concisely describes the document's purpose in sufficient detail so that a non-technical reader can understand it. State the purpose of the SOP and an indication of the desired outcome, if appropriate. The areas, operations and procedures that are covered by the SOP should be defined. Identify:

- The process
- The procedures to be followed
- Workflow
- Equipment to be used
- Critical operational parameters
- Values to be measured and/or recorded

Additional information required to further provide a detailed description of the work to be performed

B. Scope

This section requires the identification of the limitations of the document with respect to the procedure to be performed. It should limit the scope to all GxP systems. It should apply to all Sites, Locations and Departments within the company. It should include electronic records, electronic signatures and handwritten signatures executed to electronic records. It should apply to all records that are created, modified, maintained, archived, retrieved or transmitted in electronic form under any requirements set forth in FDA regulations.

Require that the scope or range of control and authority of the SOP be specified. State who, what, where, when, why and how the SOP is to control the processes or procedures specified in the Purpose section.

C. Equipment / Materials / Tools

This section lists any required or recommended equipment, materials, tools, templates or guides. Provide the name and a brief description of the equipment, instruments, materials and/or tools to be used in the procedures or testing covered by the procedure or protocol. This is for reference purposes and requires the user to utilize the specific item or a designated equivalent. The purpose and use is explained elsewhere in the document. Substitution of comparable equipment may be made in accordance with procedures for Change Control.

If the information is included in a job aid or operating manual, a description of where the information may be found with a specific reference to the document number, title, date and revision number should be specified

D. Warnings / Notes / General Information / Safety

Warnings, notes, general or safety information applicable to the procedure or protocol are listed in this section.

E. Policies / Guidelines / References

This section should require the listing of any references to appropriate regulatory requirements, SOPs, policies, guidelines, or other relevant documents referred to in the SOP. It should require for the providing of the name and a brief description of applicable related SOPs, corporate policies, procedures and guidelines, Quality Control monographs, books, technical papers, publications, specific regulations, batch records, vendor or system manuals, and all other requirements of the SOP. This listing should be for reference purposes and require the user to use the specific item. The purpose and use should be required to be explained elsewhere in the document.

F. Assumptions / Exclusions / Limitations

This section should require the discussion of any assumptions, exclusions or limitations within the context of the SOP. Specific issues should be required to be directly addressed here.

G. Glossary of Terms

This section should require the listing of a Glossary of Terms with their definition in the context of the SOP. Define all of the terms and acronyms that are used in the document or that are specific to the operation being described and that may not be generally known to the reader. Also, the reader may be referred specific published glossaries that may be available, such as the FDA document, "Glossary of Computerized System and Software Development Terminology."

This section should require the definition of titles, terms, and definitions that are used in the corporate policies, procedures and guidelines.

H. Roles and Responsibilities

This section should require the clear identification of whom the SOP is directed to and the individual(s) responsible for its implementation. It should require the specification of the participants by name, title, department or company affiliation and responsibility in a table with specific for the procedures and activities delineated in the document.

I. Procedure

This section should describe the procedure to be used and followed by means of written instructions that describe who, what, when, where, why and how the procedure is to be performed. It should be written in a manner that is clearly understandable by the people qualified to perform the task and in enough detail that an individual not necessarily familiar with the requirements could perform the task with minimal training.

It should be presented in outline format with the heading of each section reflecting the activity described. The sections may vary with each type of procedure.

J. Document Management and Storage

This section should require that all documentation be stored and maintained by the Quality Assurance (Unit) Systems Administration. The location of the documentation should be in the corporate records management file room or other specific location based on the company policy and the SOP on Document Management and Storage. To comply with regulatory agency requirements, SOPs and their associated review, approval and decommission documents should be printed, signed, dated, distributed, managed and archived in accordance with a specific SOP on Document Management and Storage. This SOP should cover document change control and not be confused with computer, computerized system or software change control, which should be a separate SOP.

The hardcopy archived with Document Management should be required to be the official SOP. Electronic SOP versions should be allowed as working copies for reference purposes only.

K. Revision History

This section should require that a Revision History section be included in each SOP that includes the SOP Version Number and descriptive information about all Revisions made to the document.

The Revision History section should be provided by the Author and consist of a summary of what was changed and a brief statement explaining why the document was revised.

The Revision History section should include the numbers of all SOPs it replaces and be written so that the original SOP is not needed to understand the revisions or the revised document being issued.

A numbering system (see Number System later on in this chapter) should be specified that clearly indicates how revisions are to be identified. For example, if the SOP is new, enter "000" as the Revision Number and a statement such as "Initial Issue" in the comments section.

Regulatory or audit commitments that have been made regarding specific documents should be prevented from being inadvertently lost or modified in subsequent revisions. All entries or modifications made as a result of regulatory or audit commitments should be indicated as such in the Revision History and identify the organization and date, along with the identification of any other applicable references.

L. Attachments

This section should require that an SOP Attachment Index and Attachment section follow the text of an SOP. The top of each attachment page should be specified, such as "Attachment # x," where 'x' is the sequential number of each attachment. The second line of each attachment page might contain the attachment title. The number of pages for each attachment might be listed in the Attachment Index.

Attachments should be required to be placed directly after the Attachment Index and be sequentially numbered (e.g., Attachment # 1, Attachment # 2) with each page of an attachment containing the corresponding document number and page numbering with the total number of pages in the attachment (e.g., Page x of y). Each attachment should be consistently titled in the body of the SOP with the top of the attachment itself.

Attachments should be treated as part of the SOP. If an attachment needs to be changed, the entire SOP should be given a new (next consecutive) revision number and reissued. New attachments or revisions of existing attachments should be in accordance with a Document Management and Storage SOP.

Forms should be issued as attachments. If the form spans multiple SOPs then they should be included in the first SOP only and referenced in subsequent SOPs.

Numbering System

The SOP number should consist of the combination of a letter designation plus a three digit sequentially assigned number. Categories for the letter designation should be specified with the second part of the SOP alphanumeric designation being the sequential number for SOPs generated in that category (e.g.- B-002, B-001).

A three-digit number could be assigned as the Revision Number. It denotes the number of times that the document has been revised. "000" is an initial issue, "001" is a first revision, "002" is a second revision

The initial Revision Number for a new SOP should be required to be "000" whether or not it supersedes another SOP number. If SOP F-000-000 supersedes U000-003, then the Revision Number of the new SOP is –000, not –004.

Departmental codes should be assigned to SOPs. This allows tighter control and more clearly identifies where it may be appropriate for use. For example, SOPs with the department code 100 might be defined as Global SOPs that pertain to multiple departments. Global SOPs would reference "All" or list the specific department numbers under "Department." SOPs otherwise could be designated by a single three digit numerical Department Code and Name.

SOPs with departmental designation would be able to be used by another department contingent upon completion of the required training and with appropriate documentation tracking the use of the SOP. When several departments need the same procedure, list all department numbers using the originating department first.

Margins / Indentation / Font / Size

Page set up should be specified to assist in standardizing SOP form and format. An example might be:

Page margins:

Margin:

Top = 1.0"

Bottom = 1.0"

Left = 1.25"

Right = 1.25"

Gutter = 0.5"

From Edge:

Header = 0.5"

Footer = 0.75"

Gutter Position = Left

This is to allow the punching of holes in the SOPs without interfering with the text when they are placed in binders and allow for the printer margin at the bottom of the page.

Text format should be 12 point Times Roman or the equivalent. Point size may be varied in tables or charts, but should not be less than 8 point. Text should be left justified with 'single' line spacing.

Major headings may be added and should be 14 points, bold and left justified. Subheadings should be 12 points and left justified. Subsequent headings should be 12 points and indented 0.5 inches. Capitalize words in headings, but do not use all uppercase or underline.

Outline Level headings are presented in Outline Character Order as shown below:

1. Outline Level 1

1.1. Outline Level 2

1.1.1. Outline Level 3

1.1.1.1. Outline Level 4

1.1.1.1.1. Outline Level 5

Writing Style

Some indication of writing style may be desirable to aid the author in creating the document. Some example suggestions are:

- Write clear, concise descriptions of the processes and procedures
- Try to keep the number of steps as short as possible
- Make the sentence structure simple. Limit most sentences to one clause where possible
- Provide descriptions that do not require pictures or diagrams, or provide pictures with written descriptions. Do not use pictures or diagrams without associated narrative
- Avoid terms that are directional (left, right, up, down). Rather indicate an object or landmark, if possible
- Emphasize key information early in the narrative
- Use bullets, headings, underlining or bolding to emphasize or differentiate important points
- Short paragraphs are more easily understood
- Maintain a consistent tense
- Place references to attachments or form in parentheses
- Prepare documents in an approved company word processor

Attachments and Forms

If needed, an Attachments section is included. An Attachment Index and Attachment section follow the text of the SOP. The top of each attachment page indicates "Attachment # x," where 'x' is the sequential number of each attachment. The second line of each attachment page contains the attachment title. The number of pages and date of issue corresponding to each attachment are listed in the Attachment Index.

Attachments are placed directly after the Attachment Index and are sequentially numbered (e.g., Attachment # 1, Attachment # 2) with each page of an attachment containing the corresponding document number, page numbering, and total number of pages in the attachment (e.g., Page x of y). Each attachment is labeled consistent with the title of the attachment itself.

Attachments are treated as part of the SOP. If an attachment needs to be changed, the entire SOP requires a new revision number and is reissued. New attachments or revisions of existing attachments are in accordance with an established Change Control SOP and a Document Management and Storage SOP.

This section should require that the headers, bodies and footers of attachments and forms contain specific information to allow them to be easily identifiable. For example:

- Header

The header is the Attachment or Form Number.

- Body

The title of the form should be the first line of the body of the form or attachment. The body of the form is the actual attachment.

- Footer

The footer contains the restrictive legend such as, "The information contained herein is proprietary to Example Pharmaceuticals Corporation Name and may not be used outside of the company without prior written permission."

Triggers for SOP Initiation, Change or Decommission

This section should specify what triggers are to be put into place that enables control of the SOP during its life cycle. Here are some examples of circumstances that might trigger the need for a new or revised document or for the decommissioning of a document:

- Adjustment in Business Conditions or pertinent input from another location, such as that caused by regulatory activity
- Change in Management Philosophy or Corporate Policies
- Alteration in Organization or Responsibilities
- Modification to the Procedure or Process
- Change to an Attachment or Form
- Equipment Implementation, Change or Obsolescence

New or Revised SOP

When an existing SOP does not cover a process or procedure that is necessary, then an SOP needs to be created or revised to remediate such condition. The SOP on SOPs should specify how to remediate the situation. Some examples of what might be done follow.

Initiating a New or Revised SOP

When an SOP needs to be created or revised, the Originator shall:

Prepare the draft of the new or changed SOP after consulting with a Subject Matter Expert (SME), if necessary.

Prepare a Document Change Form (DCF) from the controlled system (e.g., computer network drive) by entering the following information:

- Originator's name and title
- Originator's department number and name
- SOP number and title
- Reason for the new or revised SOP
- Department / Documents affected
- Training requirements
- Originator's signature and date

The Originator creates a list of Reviewers and alternates, where possible. These include, at a minimum, the Originating Department Manager, System Owner or designee (who will be held to observing the procedure), a Subject Matter Expert (such as the Computer Validation Group) for technical correctness or impact on other items and the Quality Assurance (Unit), who is responsible for compliance with policies, procedures and regulations.

The Originator shall send the list of reviewers, the draft with the updated history revisions page and the DCF to the Departmental SOP Coordinator for document format review and completion.

SOP Review

For each SOP, the Departmental SOP Coordinator sets up and maintains an electronic folder that contains read-only copies of the following:

- Original SOP
- Revised SOPs
- Document Change Forms
- Revision History pages
- Reviewers' Comment Forms

The Departmental SOP Coordinator shall distribute the documentation set to the Reviewers for their review and comment indicating that the documentation must be reviewed by the close of business of a specific day. Not more than 5 business days should be allowed for completion.

Once each reviewer has completed reviewing the documentation, all comments and an indication of acceptance or rejection is entered on the Reviewer's Comment Form and returned to the Departmental SOP Coordinator.

The Originator of the new or revised SOP reviews the comments and incorporate them into the new or revised SOP. The Originator shall contact a reviewer directly to clarify any points of concern.

SOP Re-review

After the new or revised SOP has been modified to include the comments, it, along with the documents listed below, shall be distributed for a subsequent review as described in the SOP Approval Process

SOP Approval

The Departmental SOP Coordinator shall send the completed, reviewed new or revised SOP to the QA(U) SOP Administrator who shall perform the following:

- Log receipt of the SOP
- Create a circulation folder that includes:
- SOP Reviewer's Comments Form
- SOP Signature Routing Form
- Original SOP
- New or revised SOP
- Document Change Form
- History of Revisions
- Route the circulation folder for approval
- Track the circulation folder

When there is a required completion date, the process shall provide for sufficient lead-time so that all necessary Approval signatures are in place by that date.

The Approvers shall complete the approval of the documentation by the close of business of the specified date.

Approvers shall use the "Signature Routing Form" to route the circulation folder.

Note: Minor typographical or grammatical errors should be corrected but should not delay the approval process.

The Author of the new or revised SOP shall not be an Approver.

Representatives or designees of at least the following departments shall approve the SOP: Originating Department Manager, CVG and QA(U)

Quality Assurance (Unit) shall be the last department to approve the new or revised SOP.

After all Approvers have signed the documentation, it shall be returned to the QA(U) SOP Administrator who shall proceed to implement the SOP in accordance with the section on SOP Implementation.

Decommissioning an SOP

When an SOP no longer represents a process or procedure that is necessary or is currently performed, then that SOP should be decommissioned and no longer made available for use. An example of this might be achieved as follows.

Initiating the Decommissioning Process

The Originator of the request to decommission an SOP shall prepare a Document Change Form (DCF) and a Decommission Request Form (DRF) from the controlled system (e.g., computer network drive) by entering the following information:

- Originator's name and title
- Originator's department number and name

- SOP number and title
- Reason for decommissioning the SOP
- Department / Documents affected
- Originator's signature and date

The Originator shall create a list of Reviewers and alternates, where possible. These shall include, at a minimum, the manager of the department that originally issued the SOP, the System Owner or designee, if applicable, the Computer Validation Group or designee and the Quality Assurance (Unit) or designee.

The Originator shall send the list of reviewers, the draft with the updated history revisions page, the DCF and the DRF to the Departmental SOP Coordinator.

Decommissioning Approval Process

The Departmental SOP Coordinator forwards these to the QA(U) SOP Administrator who shall perform the following:

- Retrieve the original signed and dated SOP from the corporate records management file room
- Make a copy of SOP
- File 'original' SOP in file SOPs Routing for Deletion
- The SOP Database shall be updated to indicate that the SOP has been routed for decommission
- Log receipt of the SOP
- Create a circulation folder that includes:
- SOP Signature Routing Form
- A copy of the original SOP
- Revised SOPs
- Document Change Form
- Decommission Request Form
- History of Revisions
- Track the circulation folder

The Document Administrator distributes the documentation set to the Approvers for their review and approval indicating that the documentation must be reviewed by the close of business of a specific day. Not more than 5 business days should be allowed for completion.

When there is a required completion date, the process shall provide for sufficient lead-time so that all necessary Approval signatures are in place by that date.

The Approvers shall complete the approval of the documentation by the close of business of the specified date.

Representatives or designees of at least the following departments shall approve decommissioning the SOP: the SOP Author or designee, Cognizant Departmental Manager, Manager of all Departments affected by the SOP and QA(U)

The SOP Decommission Approvers shall approve the SOP decommissioning by signing the Deletion Request Form and routing the document package to the next approver using the "Signature Routing Form" to route the circulation folder.

Quality Assurance (Unit) shall be the last department to approve the decommissioning.

After all Approvers have signed the documentation, it shall be returned to the QA(U) SOP Administrator who shall retrieve the original signed/dated (hard copy) from the file SOPs Routing for Deletion.

Decommissioning

The Document Administrator is responsible for the decommissioning of the SOP.

All original "wet ink signature" documents shall have the following procedure performed:

Stamp the top of the cover page with the following in red ink:

DECOMMISSIONED	
Replaced by:	________________
SOP #:	________________
Version:	________________
Initials:	________________
Date:	________________

Figure 1 - Decommissioned Stamp

Fill in the stamp as appropriate.

Stamp all other pages at the top in red ink with the word DECOMMISSIONED.

The Document Administrator places the decommissioned SOP in a clear, side loading sheet protector and place it in a red interior file folder labeled with the deleted SOP number.

The Document Administrator places the file containing the SOP History file in the corporate records management file room.

On the workday (evening) prior to the SOP's decommissioned date, the Document Administrator places or moves the electronic record of the decommissioned SOP to the area on the network designated for decommissioned SOPs.

The Document Administrator sends a Notice of Decommissioning to each Departmental SOP Coordinator notifying the department of the date that the decommissioning is effective.

The Departmental SOP Coordinator is responsible for retrieving the issued copies of the decommissioned SOP and destroying each copy on or about the decommissioned date. No copy shall be in circulation or use after the decommissioned date.

SOP Maintenance

All SOPs should be required to follow specific maintenance procedures during their life cycle in order to assure that they are current and appropriate for their intended use. Examples of these procedures follow.

Periodic Review

SOPs shall be reviewed periodically by the Originating Department to ensure that the content is accurate. The periodic review period shall be that stated in the SOP and shall not exceed (2) years from the Effective Date of the SOP.

Approximately six months prior to the required review period completion date the Document Administrator begins the review process by sending a completed SOP ReviewForm to the appropriate Departmental SOP Coordinator.

Within 10 working days, the Departmental SOP Coordinator of the Originating Department shall return the completed SOP Review Form to the Document Administrator indicating the status of the review as either acceptable as written, in need of revision or to be decommissioned.

The Document Administrator files the completed ReviewForm in the applicable SOP file and proceed in accordance with the sections that discuss No Revision Required or Decommission as appropriate.

No Revision Required

The Document Administrator, upon receipt of the completed SOP ReviewForm, shall obtain agreement from QA(U). QA shall review and approved/disapprove the SOP Review Form.

If QA(U) approves the SOP ReviewForm, then the cover page shall be stamped as shown below. The Document Administrator re-issues the SOP with the stamped cover page of the SOP.

Periodic Review Completed
Content Accurate as Written

Date Completed: ____________________

Initials: ____________________

Date: ____________________

Figure 2 - Periodic Review Completed

Revision Required

The Originating Department, upon returning the completed SOP Review Form, shall initiate the revision process in accordance with the section that discusses New or Revised SOP.

Decommission Required

The Originating Department, upon returning the completed SOP ReviewForm, shall initiate the decommissioning in accordance with the section that discusses Decommissioning of an SOP.

SOP Implementation

All SOPs should be required to follow specific implementation procedures in order to assure that they are current, appropriate for their intended use and properly controlled. Examples of these procedures follow.

Each SOP shall have an effective date assigned by the QA(U) SOP Administrator in accordance with the following procedure. The Document Administrator verifies that:

All parties have approved the SOP by signing and dating their respective signature blocks on the Approval Page of the SOP

The cover page shall have a header and footer printed with "Official Copy" in red for official copies

Each official copy has been sequentially numbered

There is an Effective Date that is at least 30 days later than the last approval date or following approval by a Regulatory Agency unless specified otherwise by the Originating Department. This allows adequate time for training and distribution.

The Document Administrator updates the Master SOP Index and distribute the SOPs in accordance with the section that discusses Distribution and Tracking of Controlled Copies of an SOP.

The SOP shall be used on or after the Effective Date and after all training requirements have been satisfied.

Distribution and Tracking of Controlled Copies

Numbered copies shall be distributed based on the content of the SOP and the requirements of each department. The Document Administrator distributes SOPs based on requests from Departmental SOP Coordinators that have been authorized by the requesting Department's Manager. The Departmental SOP Coordinator shall track and distribute the SOPs.

The Departmental SOP Coordinator shall be responsible for the distribution of SOP updates, attachments, decommission notices for use in the areas where they are required.

Official SOPs shall not be copied except as follows:

SOPs may be copied for training purposes, provided that:

The first copy is made from an official SOP and the first page of that first copy is dated and stamped in red ink with the words: "For Training Purposes Only". This copy is to remain in the hands of the trainer.

Each succeeding copy made from the stamped copy, shall bear the copied statement "For Training Purposes Only" in the black photocopied format.

Distribution and destruction of all copies shall be under the supervision of the trainer. All copies shall be considered expired 30 days after the date-stamp.

SOPs may be copied for requests made by the FDA or other regulatory agencies. These copies shall be marked "Provided to the <regulatory agency name> on <date>" before issuance and recorded in the Quality Assurance (Unit) files.

SOPs may be copied for documentation purposes as for inclusion in Validation Protocols provided that they are identified for this purpose and marked "For Validation Use Only."

SOPs may be copied for the purpose of revisioning and are considered to be drafts (unapproved) and shall be marked "Not for Production Use," "Draft," or "For Review Purposes Only" as appropriate.

Superseded SOP copies and/or attachments shall be destroyed unless they are required to be archived for reference purposes, in which case they shall be marked OBSOLETE – FOR REFERENCE ONLY.

The Document Administrator issues a Destruction of SOP Notice to each Departmental SOP Coordinator for that SOP. The Destruction of SOP Notice will include the copy numbers of each copy issued to that department.

Each Departmental SOP Coordinator shall be responsible for the destruction of all copies in that department. Upon destroying the obsolete SOP, the Departmental SOP Coordinator shall sign and return the Destruction of SOP Notice to the QA SOP Coordinator, indicating the number of copies that have been destroyed.

The Document Administrator maintains a file of current, superseded and decommissioned SOPs.

Security

SOPs contain information that may not be generally known or available to the public. Therefore, employees, vendors or contractors, without express written consent of an authorized individual of the company, except in the performance of contractual services, shall not:

- Use or disclose this information outside the company
- Publish an article with respect thereto
- Remove or aid in the removal from the company any confidential information, property or material

Corrections and Dates

Manual changes or corrections to documentation are made in black or blue indelible ink by placing a single line through the entity to be corrected, entering any correction in a clear and concise manner, initialing and dating the correction. Do NOT obliterate any item.

Do *not* pre- or post-date any documents. If a date is in error or missed, enter the current date with initials and an explanation as to what transpired.

Document Management and Storage

Document Management and Storage defines the methodology that is to be followed for issue, revision replacement, control and management of this specific document. This section should be in accordance with a Document Management and Storage SOP.

Revision History

Changes to a document should be noted in its Revision History section. These changes should be in accordance with a Change Control Procedure SOP.

Attachments

This section should be used to list all documents that are attached. It might contain an index of those documents and may include such information as the Attachment Number, Document Number, Title, Revision Number, Date and number of pages.

Chapter 6 - The SOP on SOPs

Chapter 7 – Project Plan

NIST defines a project plan as "A management document describing the approach taken for a project. The plan typically describes work to be done, resources required, methods to be used, the configuration management and quality assurance procedures to be followed, the schedules to be met, the project organization Project in this context is a generic term. Some projects may also need integration plans, security plans, test plans, quality assurance plans" This chapter will provide procedures to be followed for creating project plans specific for validation work. It specifies the requirements for a plan that will determine how the project is rolled out and managed.

Best industry practices and FDA Guidance documents indicate that in order to demonstrate that a computer or computerized system performs as it purports and is intended to perform, a methodical plan is needed that details how this may be proven. It lays out what is to be done, how it is to be accomplished and the resources required for completion of each activity.

This SOP is used for all systems that create, modify, maintain, retrieve, transmit or archive records in electronic format. It is used in conjunction with an overlying corporate policy to provide sustainable compliance with the regulation.

Everyone involved in projects for a software application, computer or computerized system that controls equipment and/or produces electronic data or records in support of all regulated laboratory, manufacturing, clinical or distribution activities uses this SOP.

Title Block, Headers and Footers (SOP)

The Project Plan SOP should have a title block, title or cover page along with a header and/or footer on each page of the document that clearly identifies the document. The minimum information that should be displayed for the Project Plan SOP is as follows:

- Document Title
- Document Number
- Document Revision Number
- Project Name and Number
- Site/Location/Department Identification
- Date Document was issued, last revised or draft status
- Effective Date
- Pagination: page of total pages (e.g., Page x of y)

The Document Title should be in large type font that enables clear identification. It should clearly and concisely describe the contents and intent of the document. Since this document will be part of a set or package of documents within a project, it should be consistent with other documents used for the project. This information should be on every page of the document.

The Document Number should be in large type font that enables clear identification. Since the Project Plan SOP will be part of a set or package of documents within a project, it should be consistent with other documents used for the project. This information should appear on every page of the document.

The Document Revision Number, which identifies the revision of the document, should appear on each page of the document. Many documents will go through one or more changes over time, particularly during the draft stage and the review and approval cycle. It is important to maintain accurate revision control because even a

subtle change in a document can have an immense impact on the outcomes of activities. The SOP on SOPs should govern the numbering methodology.

See Chapter 6 – "The SOP on SOPs."

The Project Name and Number, if applicable, should be indicated on the cover or title page. This information identifies the overall project to which this document belongs.

The Site/Location/Department Identification should be indicated on the cover or title page. It identifies the Site/Location/Department that has responsibility for this document and its associated project.

The Date that the Document was issued, last revised or issued with draft status should be on each page of the document (e.g., 12-Dec-01). This will clearly identify the date on which the document was last modified.

The Effective Date of the Document should be on the title or cover page of the document (e.g., 12-Dec-01). This will clearly identify when to begin using the current version of the document.

Pagination, page of total pages (e.g., Page x of y), should be on each page of the document. This will help identify whether or not all of the pages of the document are present. Pages intentionally left blank should clearly be designated as such.

Approval Section *(SOP)*

The Approval Section should contain a table with the Author and Approvers printed names, titles and a place to sign and date the SOP. The SOP on SOPs should govern who the approvers of the document should be. The functional department (e.g., Production / Quality Assurance / Engineering) of the signatory should be placed below each printed name and title along with a statement associated with each approver indicating the person's role or qualification in the approval process. Indicate the significance of the associated signature (e.g. This document meets the requirements of Corporate Policy 9055, Electronic Records; Electronic Signatures, dated 22-Mar-2000).

The completion (signing) of this section indicates that the contents have been reviewed and approved by the listed individuals.

Table of Contents *(SOP)*

The SOP should contain a Table of Contents, which appears towards the beginning of the document. This Table of Contents provides a way to easily locate various topics contained in the document.

Purpose *(SOP)*

This introduction to the SOP concisely describes the document's purpose in sufficient detail so that a non-technical reader can understand it.

The Purpose section states the purpose of the SOP and an indication of the desired outcome, if appropriate. The areas, operations and procedures that are covered by the SOP are defined here.

The Purpose section should clearly state that the intention of the document is to provide a Standard Operating Procedure to be used to specify the requirements for the preparation of a Project Plan for a new or upgraded software application, computer or computerized system along with the required activities and deliverables.

These requirements are necessary for the Project Plan to be effective and compliant with company polices and government regulation.

As one of the primary governing documents of a validation project (the Validation Plan being the other), the Project Plan defines the scope of work, responsibilities and deliverables in order to demonstrate that a computer or computerized system meets its intended use in a reliable, consistent and compliant manner.

This SOP should not be a dissertation on project management as there is much information available on this subject.

Scope *(SOP)*

The Scope section identifies the limitations of the Project Plan SOP. It limits the scope to all GxP systems and is applicable to all computers, computerized systems, infrastructure, LANs and WANs including clinical, laboratory, manufacturing and distribution computer systems. It applies to all Sites, Locations and Departments within the company.

Equipment / Materials / Tools *(SOP)*

This section lists any required or recommended equipment, materials, tools, templates or guides. Provide the name and a brief description of the equipment, instruments, materials and/or tools to be used in the procedures or testing covered by the Project Plan SOP. This is for reference purposes and requires the user to utilize the specific item or a designated equivalent. The purpose and use is explained elsewhere in the document. Substitution of comparable equipment may be made in accordance with procedures for Change Control.

If the information is included in a job aid or operating manual, a description of where the information may be found with a specific reference to the document number, title, date and revision number should be specified.

Warnings / Notes / General Information / Safety *(SOP)*

Warnings, notes, general or safety information applicable to the Project Plan SOP are listed in this section.

Policies, Guidelines, References *(SOP)*

This section lists references to appropriate regulatory requirements, SOPs, policies, guidelines, or other relevant documents referred to in the Project Plan SOP. The name and a brief description of applicable related SOPs, corporate policies, procedures and guidelines, Quality Control monographs, books, technical papers, publications, specific regulations, batch records, vendor or system manuals, and all other relevant documents that are required by the Project Plan SOP should be included. This is for reference purposes and requires the user to use the specific item or a designated equivalent. The purpose and use are explained elsewhere in the document.

SOPs cite references and incorporate them in whole by reference rather than specifying details that duplicate these items. This is done so that changes in policy, guidelines or reference materials require no changes to SOPs themselves and the most current version of the referenced material is used. Examples are:

- Policy ABC-123, Administrative Policy for Electronic Records; Electronic Signatures
- Corporate, Plan for Compliance with 21 CFR Part 11, Electronic Records; Electronic Signatures
- SOP-456, Electronic Records; Electronic Signatures
- FDA, 21 CFR Part 11, Electronic Records; Electronic Signatures

- FDA, General Principals of Software Validation; Final Guidance for Industry and FDA Staff, January 11, 2002
- ANSI/PMI 99-001-2000, PMBOK® Guide 2000 Edition - A Guide to the Project Management Body of Knowledge
- Project Management, A Systems Approach to Planning, Scheduling, and Controlling, Sixth Edition by Harold Kerzner, Ph.D. published by John Wiley & Sons, Inc.

Assumptions / Exclusions / Limitations *(SOP)*

Any and all assumptions, exclusions or limitations within the context of the SOP should be discussed. Specific issues should be directly addressed in this section of the Project Plan SOP. For example:

> "The project manager and the approvers of the project plan shall determine the amount of detail and required documentation that is to be set forth in the Project Plan for the computer or computerized system. The determination of detail and deliverables shall be in accordance with SOP C-03, Overview of Computer and Computerized System Validation and SOP C-04, Roles and Responsibilities in Computer and Computerized System Validation."

Glossary of Terms *(SOP)*

A Glossary of Terms should be included that defines all of the terms and acronyms that are used in this document or that are specific to the operation being described and may not be generally known to the reader. The reader may also be referred to the FDA document, "Glossary of Computerized System and Software Development Terminology."

Roles and Responsibilities *(SOP)*

The Roles and Responsibilities of the Project Plan should be divided into two specific areas:

Creation and Maintenance

The System Owner of the computer or computerized system is responsible for the preparation and maintenance of the Project Plan.

It is common for a Project Plan to go through several revisions during the course of a project, as it is considered a living document, especially during the project's early stages. As a normal process, it helps to ensure that the validation activities are in compliance with cGxPs, FDA regulations, company policies and industry standards and guidelines.

The roles and responsibilities for preparing the Project Plan and other project documentation should be required to be addressed in accordance with established Standard Operating Procedures.

Use

The members of the Project Team must be familiar with the Project Plan in order to fulfill the project requirements. The Quality Assurance (Unit) and Project Team use the plan in concert with the other validation documentation as a contract that details the project and quality expectations.

Procedure

The SOP should contain instructions and procedures that enable the creation of a Project Plan. These include, but are not limited to, an overall description and requirements of the computer or computerized system validation process, a description of the overall system, and the deliverables, as well as the roles and responsibilities of the participants in its selection, design, delivery, validation and implementation.

The Project Plan should be prepared as a stand-alone document for all projects. It is not good practice to include it as part of other documents in an attempt to reduce the amount of workload or paperwork. This only causes more work over the length of the project resulting from misunderstandings and confusion. Clear, concise documents that address s pecific parts of any given project are desirable.

Project plans in the form of a narrative are preferable. Gant, Program Evaluation and Review Technique (PERT) and Critical Path Method (CPM) charts are desirable and may be used as an adjunct to narrative plans.

Purpose (protocol)

Clearly state the system with information as to why it is considered a regulated system and introduce planned activities in order to validate it. This short introduction should be concise and present details so that an unfamiliar reader can understand the purpose of the document with regard to the system.

Scope (protocol)

The Scope section identifies the limitations of a specific Project Plan in relationship to validating a specific software application, computer or computerized system. For example, if only a part of the computer system will be validated, then this section would be used to explain the reason for this decision.

Equipment / Materials / Tools (protocol)

This section lists any required or recommended equipment, materials, tools, templates or guides. Provide the name and a brief description of the equipment, instruments, materials and/or tools to be used in the procedures or testing covered by the Project Plan. This is for reference purposes and requires the user to utilize the specific item or a designated equivalent. The purpose and use is explained elsewhere in the document. Substitution of comparable equipment may be made in accordance with procedures for Change Control.

If the information is included in a job aid or operating manual, a description of where the information may be found with a specific reference to the document number, title, date and revision number should be specified

Warnings / Notes / General Information / Safety (protocol)

Warnings, notes, general or safety information applicable to the Project Plan are listed in this section.

Policies / Guidelines / References (protocol)

This section lists references to appropriate regulatory requirements, SOPs, policies, guidelines, or other relevant documents referred to in the Project Plan. The name and a brief description of applicable related SOPs, corporate policies, procedures and guidelines, Quality Control monographs, books, technical papers, publications, specific regulations, batch records, vendor or system manuals, and all other relevant documents that are required by the Project Plan should be included. This is for reference purposes and requires the user to use the specific item or a designated equivalent. The purpose and use are explained elsewhere in the document.

Assumptions / Exclusions / Limitations (protocol)

Discuss any assumptions, exclusions or limitations within the context of the plan. Specific issues should be directly addressed in this section of the Project Plan.

Glossary of Terms (protocol)

The Glossary of Terms should define all of the terms and acronyms that are used in the document or that are specific to the operation being described and may not be generally known to the reader. The reader may als o be referred to the FDA document, "Glossary of Computerized System and Software Development Terminology," or other specific glossaries that may be available.

Titles, terms, and definitions that are used in the corporate policies, procedures and guidelines should be defined.

Roles and Responsibilities (protocol)

Validation activities are the result of a cooperative effort involving many people, including the System Owners, End Users, the Computer Validation Group, Subject Matter Experts (SME) and Quality Assurance (Unit). This section should clearly identify the roles, responsibilities and individual(s) responsible for the implementation of the Project Plan. It should specify the participants by name, title, department or company affiliation and responsibility in a table. Responsibilities should be specific for the procedures and activities that are indicated in each specific document.

The System Owners should be required to be responsible for the development and understanding of the Project Plan. They should acquire and coordinate support from in-house and/or external consultants, the Computer Validation Group, Quality Assurance (Unit) and system developers.

The QA(U) is responsible to verify that the details in the Project Plan are compliant with applicable departmental, interdepartmental and corporate policies and procedures. The QA(U) must confirm that the approach is in accordance with cGxP regulations and good business practices, especially 21 CFR Part 11.

The validation efforts are planned and directed by the System Owner, Project Manager or designee. A Project Team is created and the respective expertise of each of the team members is documented in the Roles and Responsibilities section of the Project Plan. To assist in the selection of the team members, a list of qualified individuals, along with their relevant experience, is made available from the Human Resources Department and a list of qualified contractors from the Purchasing Department.

Appoint a Document Coordinator who maintains the project documentation in accordance with certain specifications of the SOP.

Procedure (protocol)

Project Overview

This section contains an overview of the business operations for the system including workflow, material flow and the control philosophy of what functions, calculations and processes are to be performed by the system. It also describes any interfaces that the system or application must be provided. It contains an outline of the activities required for commissioning the system.

Facilities

This section describes the facilities that will use the system and specific attributes of the site (e.g., manufacturing sites, analytical labs with associated instrumentation or equipment, clinical trial sites, distribution services, medical devices).

Deliverables - Work Breakdown Structures

The Project Plan provides a proposed list of all the associated deliverables, also called tasks or work breakdown structures (WBS), including documents that constitute the overall package. Each of the deliverables must clearly specify the required outcome of the work performed. This list provides detailed information on the work to be performed or the title of the document, document number, when available, document purpose and a listing of main sections or features to be included in the document. For a list of examples of the basic set of documents to be included in a Project Plan see Chapter 8 - Validation Plan - Procedure - Deliverables.

"A work breakdown structure is a deliverable-oriented grouping of project elements that organizes and defines the total scope of the project: work not in the WBS is outside the scope of the project."[6]

Acceptance Criteria

Acceptance Criteria is the criteria that must be met for a task, component or system to be accepted by the user, customer, or authority.

The Acceptance Criteria - Expected Results – are specified in clear and certain terms. The criteria define what must be accomplished in order for the work that was performed for each task required in the Project Plan to be acceptable to the Project Team. Each WBS with its associated criteria for acceptance is listed. Indicate any boundaries, exceptions and exclusions. As little room as possible should be left open for subjective interpretation of the requirements. The System Owner, CVG and QA(U), with assistance from an SME, should make the final determination of the acceptance criteria.

Integration Management

Integration Management is the process of coordinating the integration of multiple elements of a project. It involves making and prioritizing decisions that determine which tasks are to be performed, which ones are not and the order of performance. The Project Team needs to have a methodology that will assist them in making these decisions to have the work satisfactorily completed.

Scope Management

Managing the scope to prevent scope creep by clearly defined processes that can be followed by the project team will enable adds, changes and deletions to the project to be handled in an organized and effective manner. Specify the process by which changes are to be made. No work outside of the project Statement of Work - the project plan - can be permitted. Authorizing and enforcing these controls will assure the project completion on schedule and within budget.

---

[6] ANSI/PMI 99-001-2000, PMBOK® Guide 2000 Edition - *A Guide to the Project Management Body of Knowledge*

Time Management

Time checkpoints for critical tasks in the WBS should be specified. When a checkpoint is encountered, such as a specified interim period of elapsed time for a task, evaluate the status based on predetermined criteria. This will allow the Project Manager (PM) or team to determine if the task is on schedule, has enough resources and will complete on time. Based on this assessment, the PM will be able to manage the time and resources allocated for the task.

Cost Management

Cost checkpoints for critical tasks in the WBS should be specified. When a checkpoint is encountered, such as reaching a specified interim cost for a task, evaluate the status based on predetermined criteria. This will allow the PM to determine if the task is on target from a cost perspective. A possible cost overrun or underrun can be estimated, calculated and forecast. Based on this assessment in accordance with predefined guidelines, the PM will be able to manage the cost allocated for the task and transfer funds to or from other tasks as necessary.

A formal process for requesting additional project funding may be on order. However, it is outside the scope of this handbook.

Quality Management

Much has been written on the subject of quality management. The intent here is to briefly outline what needs to be covered in the context of computer and computerized system validation within a full SDLC methodology. Quality management includes the finalized and approved versions of the following documents, which comprise the key elements of SDLC:

- Letter of Notification / System Proposal
- Project Plan
- Validation Plan
- Software development plan, documentation, and programming tools
- User Requirements
- Functional Requirements
- Vendor Audit
- Design Specifications
- Review process
- Unit Testing Documentation
- Factory Acceptance Test Documents
- Test plan
- Testing procedures
- IQ Protocol
- IQ Summary Report
- OQ Protocol
- OQ Summary Report
- PQ Protocol
- PQ Summary Report
- Traceability Matrix Document
- Documentation - user manuals, administrator documentation, SOPs
- Document Management and Storage
- Training Documents
- Project Validation Final Report
- Configuration Management

- Change Control
- Periodic Review
- Quality management procedures
- SDLC definition
- Software audit program
- Training
- Decommissioning Plan & Report

Human Resource Management

Human resource staffing is the responsibility of the line manager or System Owner, not the Project Manager. The Project Manager has the responsibility to identify and request resources needed to fulfill project requirements and manage them in their performance of the specified tasks.

Many resources will be shared across departments, projects and, in the case of contractors and consultants, across companies. The use and time of the people staffing a project to effect a task completion - milestone achievement – is scheduled in a timely and cost effective manner. Most project scheduling software applications provide a tool for human resource management that includes time allotment, personnel allocation and workload leveling methods to foster an efficient means of scheduling staff and tracking actual performance against the objectives of the Project Plan.

Communication Management

Whether it is by meetings, e-mail correspondence, collaborative databases or some other medium of communication and collaboration, information should be exchanged in a prescribed scheduled manner. In this way everyone involved in the project will know when to expect updates and when they can provide input to the process. The more that is known about the goings on of the project, the better for everyone. A policy of "No surprises" means that planning may be accomplished on an on-going basis by requiring activities and actions that continually correct and improve the course and direction of the project. Management at every level will always be in the loop and know the current status of the project.

Risk Management

Risk is a measure of the probability and consequence of not achieving a defined project goal.[7]

The first part of project risk management identifies unpredicted or unwanted changes, determines the likelihood of those changes occurring and provides an assessment of the impact on the project before they can occur

Risk Management in action is the assessment of the unpredicted or unwanted changes when they occur, determining what corrective action is to be taken, obtaining approval from the affected parties, implementing the corrective action and communicating the activity to the stakeholders involved as the process progresses.

As much, too, has been written on risk management, the intent here is to identify what needs to be covered in the context of computer and computerized system validation within a full SDLC methodology. Risk management should include the key elements of SDLC as described in the Quality Management section of this chapter.

---

[7] Project Management, *A Systems Approach to Planning, Scheduling, and Controlling* Sixth Edition by Harold Kerzner, Ph.D. published by John Wiley & Sons, Inc.

Risk management in the context of an SDLC project should not be confused with a risk assessment that deals with business decisions impacting company operations such as whether or not to validate a system in compliance with regulatory requirements.

Risk management should be included in the every day activities of each team member. In continually assessing the progress of the project, anticipation, forecasting, and suitable preparation are accomplished. The earlier a risk is identified, the more easily and cost-effectively it can be handled.

It should be noted that in encouraging staff to report observations and assessments in regard to risk management, the Project Manager and management should be careful to not become critical of individuals for fulfilling this responsibility. Negative responses will discourage future reporting of such information and place the project at even greater risk of becoming problematic. Remember, the person on the front line sees it coming long before anyone else.

Procurement Management

A procurement plan worked out with the guidance and participation of the purchasing department should be implemented as early as possible in the project, preferably during the planning phase. This will ensure that the necessary outside resources - staff, materials, equipment, and licenses - will be available when they are needed. This will also allow negotiation of better pricing based on properly prepared statements of work and requests for quotation or invitations for bid resulting in competitive bidding by qualified contractors and vendors. By following a contracts and procurement methodology[8] that goes through the requirement, requisition, solicitation and award cycles, a detailed and well organized approach will be achieved.

Document Management and Storage (protocol)

A Document Management and Storage section is included in each Project Plan. It defines the Document Management and Storage methodology that is followed for issue, revision, replacement, control and management of each Project Plan. This methodology is in accordance with an established Change Control SOP and a Document Management and Storage SOP.

Revision History (protocol)

A Revision History section is included in each Project Plan. The Project Plan Version Number and all Revisions made to the document are recorded in this section. These changes are in accordance with a Change Control Procedure SOP.

The Revision History section consists of a summary of the changes, a brief statement explaining the reasons the document was revised, and the name and affiliation of the person responsible for the entries. It includes the numbers of all items the Project Plan replaces and is written so that the original Project Plan is not needed to understand the revisions or the revised document being issued.

When the Project Plan is a new document, a "000" as the Revision Number is entered with a statement such as "Initial Issue" in the comments section.

Regulatory or audit commitments require preservation from being inadvertently lost or modified in subsequent revisions. All entries or modifications made as a result of regulatory or audit commitments require indication

---

[8]See the chapter titled Contracts and Procurement in Project Management, *A Systems Approach to Planning, Scheduling, and Controlling*, Sixth Edition by Harold Kerzner, Ph.D. published by John Wiley & Sons, Inc.

as such in the Revision History and should indicate the organization, date and applicable references. A complete and accurate audit trail must be developed and maintained throughout the project lifecycle.

Attachments (protocol)

If needed, an Attachments section is included in each Project Plan in accordance with the SOP on SOPs. An Attachment Index and Attachment section follow the text of a Project Plan. The top of each attachment page indicates "Attachment # x," where 'x' is the sequential number of each attachment. The second line of each attachment page contains the attachment title. The number of pages and date of issue corresponding to each attachment are listed in the Attachment Index.

Attachments are placed directly after the Attachment Index and are sequentially numbered (e.g., Attachment # 1, Attachment # 2) with each page of an attachment containing the corresponding document number, page numbering, and total number of pages in the attachment (e.g., Page x of y). Each attachment is labeled consistent with the title of the attachment itself.

Attachments are treated as part of the Project Plan. If an attachment needs to be changed, the entire Project Plan requires a new revision number and is reissued. New attachments or revisions of existing attachments are in accordance with an established Change Control SOP and a Document Management and Storage SOP.

Document Management and Storage *(SOP)*

A Document Management and Storage section is included in the Project Plan SOP. It defines the Document Management and Storage methodology that is followed for issue, revision, replacement, control and management of this SOP. This methodology is in accordance with an established Change Control SOP and a Document Management and Storage SOP.

Revision History *(SOP)*

A Revision History section is included in the Project Plan SOP. The Project Plan SOP Version Number and all Revisions made to the document are recorded in this section. These changes are in accordance with a Change Control Procedure SOP.

The Revision History section consists of a summary of the changes, a brief statement explaining the reasons the document was revised, and the name and affiliation of the person responsible for the entries. It includes the numbers of all items the Project Plan SOP replaces and is written so that the original Project Plan SOP is not needed to understand the revisions or the revised document being issued.

When the Project Plan SOP is a new document, a "000" as the Revision Number is entered with a statement such as "Initial Issue" in the comments section.

Regulatory or audit commitments require preservation from being inadvertently lost or modified in subsequent revisions. All entries or modifications made as a result of regulatory or audit commitments require indication as such in the Revision History and should indicate the organization, date and applicable references. A complete and accurate audit trail must be developed and maintained throughout the project lifecycle.

Attachments *(SOP)*

If needed, an Attachments section is included in the Project Plan SOP in accordance with the SOP on SOPs. An Attachment Index and Attachment section follow the text of the Project Plan SOP. The top of each attachment page indicates "Attachment # x," where 'x' is the sequential number of each attachment. The second line of each attachment page contains the attachment title. The number of pages and date of issue corresponding to each attachment are listed in the Attachment Index.

Attachments are placed directly after the Attachment Index and are sequentially numbered (e.g., Attachment # 1, Attachment # 2) with each page of an attachment containing the corresponding document number, page numbering, and total number of pages in the attachment (e.g., Page x of y). Each attachment is labeled consistent with the title of the attachment itself.

Attachments are treated as part of the Project Plan SOP. If an attachment needs to be changed, the entire Project Plan SOP requires a new revision number and is reissued. New attachments or revisions of existing attachments are in accordance with an established Change Control SOP and a Document Management and Storage SOP.

Chapter 8 – Validation Plan

In this chapter we will discuss the Standard Operating Procedure (SOP) for creating a Validation Plan for a validation project.

Best industry practices and FDA Guidance documents indicate that a methodical plan is needed in order to demonstrate that a computer or computerized system performs as it purports and is intended to perform. The Validation Plan details how this may be proven.

This SOP is used for all systems that create, modify, maintain, retrieve, transmit or archive records in electronic format. The Validation Plan is used in conjunction with an overlying corporate policy to provide sustainable compliance with the regulations.

All staff members involved in the operation a software application, computer or computerized system that controls equipment and/or produces electronic data or records in support of all regulated laboratory, manufacturing, clinical or distribution activities use this SOP.

Title Block, Headers and Footers (SOP)

The Validation Plan SOP should have a title block, title or cover page along with a header and/or footer on each page of the document that clearly identifies the document. The minimum information that should be displayed for the Validation Plan is as follows:

- Document Title
- Document Number
- Document Revision Number
- Project Name and Number
- Site/Location/Department Identification
- Date Document was issued, last revised or draft status
- Effective Date
- Pagination: page of total pages (e.g., Page x of y)

The Document Title should be in large type font that enables clear identification. It should clearly and concisely describe the contents and intent of the document. Since this document will be part of a set or package of documents within a project, it should be consistent with other documents used for the project and specify that it is the Validation Plan SOP. This information should be on every page of the document.

The Document Number should be in large type font that enables clear identification. Since the Validation Plan SOP will be part of a set or package of documents within a project, it should be consistent with other documents used for the project. This information should appear on every page of the document.

The Document Revision Number, which identifies the revision of the document, should appear on each page of the document. Many documents will go through one or more changes over time, particularly during the draft stage and the review and approval cycle. It is important to maintain accurate revision control because even a subtle change in a document can have an immense impact on the outcomes of activities. The SOP on SOPs should govern the numbering methodology.

See Chapter 6 – "The SOP on SOPs."

The Project Name and Number, if applicable, should be indicated on the cover or title page. This information identifies the overall project to which this document belongs.

The Site/Location/Department Identification should be indicated on the cover or title page. It identifies the Site/Location/Department that has responsibility for this document and its associated project.

The Date that the Document was issued, last revised or issued with draft status should be on each page of the document (e.g., 12-Dec-01). This will clearly identify the date on which the document was last modified.

The Effective Date of the Document should be on the title or cover page of the document (e.g., 12-Dec-01). This will clearly identify when to begin using the current version of the document.

Pagination, page of total pages (e.g., Page x of y), should be on each page of the document. This will help identify whether or not all of the pages of the document are present. Pages intentionally left blank should clearly be designated as such.

Approval Section *(SOP)*

The Approval Section should contain a table with the Author and Approvers printed names, titles and a place to sign and date the SOP. The SOP on SOPs should govern who the approvers of the document should be. The functional department (e.g., Production / Quality Assurance / Engineering) of the signatory should be placed below each printed name and title along with a statement associated with each approver indicating the person's role or qualification in the approval process. Indicate the significance of the associated signature (e.g. This document meets the requirements of Corporate Policy 9055, Electronic Records; Electronic Signatures, dated 22-Mar-2000).

The completion (signing) of this section indicates that the contents have been reviewed and approved by the listed individuals.

Table of Contents *(SOP)*

The SOP should contain a Table of Contents, which appears towards the beginning of the document. This Table of Contents provides a way to easily locate various topics contained in the document.

Purpose *(SOP)*

This introduction to the SOP concisely describes the document's purpose in sufficient detail so that a non-technical reader can understand it.

The Purpose section states the purpose of the SOP and an indication of the desired outcome, if appropriate. The areas, operations and procedures that are covered by the SOP are defined here.

The Purpose section should clearly state that the intention of the document is to provide a Standard Operating Procedure to be used to specify the requirements for the preparation of a Validation Plan for a new or upgraded software application, computer or computerized system along with the required activities and deliverables. These requirements are necessary for the Validation Plan to be effective and compliant with company polices and government regulation.

As one of the primary governing documents of a validation project (the Project Plan being the other), the Validation Plan defines the scope of work, responsibilities and deliverables in order to demonstrate that a computer or computerized system meets its intended use in a reliable, consistent and compliant manner.

Scope (SOP)

The Scope section identifies the limitations of the Validation Plan SOP. It limits the scope to all GxP systems and is applicable to all computers, computerized systems, infrastructure, LANs and WANs including clinical, laboratory, manufacturing and distribution computer systems. It applies to all Sites, Locations and Departments within the company.

Equipment / Materials / Tools (SOP)

This section lists any required or recommended equipment, materials, tools, templates or guides. Provide the name and a brief description of the equipment, instruments, materials and/or tools to be used in the procedures or testing covered by the Validation Plan SOP. This is for reference purposes and requires the user to utilize the specific item or a designated equivalent. The purpose and use is explained elsewhere in the document. Substitution of comparable equipment may be made in accordance with procedures for Change Control.

If the information is included in a job aid or operating manual, a description of where the information may be found with a specific reference to the document number, title, date and revision number should be specified.

Warnings / Notes / General Information / Safety (SOP)

Warnings, notes, general or safety information applicable to the Validation Plan SOP are listed in this section.

Policies, Guidelines, References (SOP)

This section lists references to appropriate regulatory requirements, SOPs, policies, guidelines, or other relevant documents referred to in the Validation Plan SOP. The name and a brief description of applicable related SOPs, corporate policies, procedures and guidelines, Quality Control monographs, books, technical papers, publications, specific regulations, batch records, vendor or system manuals, and all other relevant documents that are required by the Validation Plan SOP should be included. This is for reference purposes and requires the user to use the specific item or a designated equivalent. The purpose and use are explained elsewhere in the document.

SOPs cite references and incorporate them in whole by reference rather than specifying details that duplicate these items. This is done so that changes in policy, guidelines or reference materials require no changes to SOPs themselves and the most current version of the referenced material is used. Examples are:

- Policy ABC-123, Administrative Policy for Electronic Records; Electronic Signatures
- Corporate, Plan for Compliance with 21 CFR Part 11, Electronic Records; Electronic Signatures
- SOP-456, Electronic Records; Electronic Signatures
- FDA, 21 CFR Part 11, Electronic Records; Electronic Signatures
- FDA, General Principals of Software Validation; Final Guidance for Industry and FDA Staff, January 11, 2002

Assumptions / Exclusions / Limitations (SOP)

Any and all assumptions, exclusions or limitations within the context of the SOP should be discussed. Specific issues should be directly addressed in this section of the Validation Plan SOP. For example:

"The project manager and the approvers of the plan shall determine the amount of detail and required documentation that is to be set forth in the Validation Plan for the computer or computerized system. The determination of detail and deliverables shall be in accordance with SOP C-03, *Overview of Computer and Computerized System Validation* and SOP C-04, *Roles and Responsibilities in Computer and Computerized System Validation.*"

Glossary of Terms *(SOP)*

A Glossary of Terms should be included that defines all of the terms and acronyms that are used in this document or that are specific to the operation being described and may not be generally known to the reader. The reader may also be referred to the FDA document, "Glossary of Computerized System and Software Development Terminology."

Roles and Responsibilities *(SOP)*

The Roles and Responsibilities of the Validation Plan should be divided into two specific areas:

Creation and Maintenance

The System Owner of the computer or computerized system should be required to be responsible for the preparation and maintenance of the Validation Plan.

It is common for a Validation Plan to go through several revisions during the course of a project, as it is considered a living document, especially during the project's early stages. As a normal process, it helps to ensure that the validation activities are in compliance with cGxPs, FDA regulations, company policies and industry standards and guidelines.

The roles and responsibilities for preparing the Validation Plan and other project documentation should be required to be addressed in accordance with established Standard Operating Procedures.

Use

The members of the Project Team must be familiar with the Validation Plan in order to fulfill the project requirements. The Quality Assurance (Unit) and Project Team use the plan in concert with the Project Plan as a contract that details the project and quality expectations.

Procedure

The SOP should contain instructions and procedures that enable the creation of a Validation Plan. These include, but are not limited to an overall description and requirements of the computer or computerized system validation process, a description of the overall system, the deliverables, the roles and responsibilities of the participants in its selection, design, delivery, validation and implementation.

The Validation Plan should be prepared as a stand-alone document for all projects. It is not good practice to include it as part of other documents in an attempt to reduce the amount of workload or paperwork. This only causes more work over the length of the project resulting from misunderstandings and confusion. Clear, concise documents that address specific parts of any given project are desirable.

Purpose (protocol)

Clearly state the system with information as to why it is considered a regulated system and introduce planned activities in order to validate it. This short introduction should be concise and present details so that an unfamiliar reader can understand the purpose of the document with regard to the system.

Scope (protocol)

The Scope section identifies the limitations of a specific Validation Plan in relationship to validating a specific software application, computer or computerized system. For example, if only a part of the computer system will be validated, then this section would be used to explain the reason for this decision.

Equipment / Materials / Tools (protocol)

This section lists any required or recommended equipment, materials, tools, templates or guides. Provide the name and a brief description of the equipment, instruments, materials and/or tools to be used in the procedures or testing covered by the Validation Plan. This is for reference purposes and requires the user to utilize the specific item or a designated equivalent. The purpose and use is explained elsewhere in the document. Substitution of comparable equipment may be made in accordance with procedures for Change Control.

If the information is included in a job aid or operating manual, a description of where the information may be found with a specific reference to the document number, title, date and revision number should be specified.

Warnings / Notes / General Information / Safety (protocol)

Warnings, notes, general or safety information applicable to the Validation Plan are listed in this section.

Policies / Guidelines / References (protocol)

This section lists references to appropriate regulatory requirements, SOPs, policies, guidelines, or other relevant documents referred to in the Validation Plan. The name and a brief description of applicable related SOPs, corporate policies, procedures and guidelines, Quality Control monographs, books, technical papers, publications, specific regulations, batch records, vendor or system manuals, and all other relevant documents that are required by the Validation Plan should be included. This is for reference purposes and requires the user to use the specific item or a designated equivalent. The purpose and use are explained elsewhere in the document.

Assumptions / Exclusions / Limitations (protocol)

Discuss any assumptions, exclusions or limitations within the context of the plan. Specific issues should be directly addressed in this section of the Validation Plan.

Glossary of Terms (protocol)

The Glossary of Terms should define all of the terms and acronyms that are used in the document or that are specific to the operation being described and may not be generally known to the reader. The reader may also be referred to the FDA document, "Glossary of Computerized System and Software Development Terminology," or other specific glossaries that may be available.

Roles and Responsibilities (protocol)

Validation activities are the result of a cooperative effort involving many people, including the System Owners, End Users, the Computer Validation Group, Subject Matter Experts (SME) and Quality Assurance (Unit). This section should clearly identify the roles, responsibilities and individual(s) responsible for the implementation of the Validation Plan. It should specify the participants by name, title, department or company affiliation and responsibility in a table. Responsibilities should be specific for the procedures and activities that are indicated in each specific document.

The System Owners should be required to be responsible for the development and understanding of the plan. They should acquire and coordinate support from in-house and/or external consultants, the Computer Validation Group, Quality Assurance (Unit) and system developers.

The QA(U) should be required to be responsible to verify that the details in the Validation Plan are compliant with applicable departmental, interdepartmental and corporate policies and procedures and that the approach is in accordance with cGxP regulations and good business practices, especially 21 CFR Part 11.

The validation efforts are planned and directed by the System Owner, Project Manager or designee. A Project Team is created and their respective expertise documented in the Roles and Responsibilities section of the Validation Plan. To assist in the selection of the team members, a list of qualified individuals is made available from the Human Resources Department and a list of qualified contractors from the Purchasing Department.

It might be appropriate to suggest the appointment of a Document Coordinator who should maintain the project documentation in accordance with certain specifications of the SOP.

Procedure (protocol)

System Overview and Process

A general description of the system that includes equipment, hardware, software, interfaces, workflows, processes and procedures should be included. This will assist the Project Team in understanding the overall perspective and expectations of the system.

Introduction

Provide an overview of the business operations for the system including workflow, material flow as well as the control philosophy of the functions, calculations and processes that are to be performed by the system, as well as interfaces that the system or application must be provided. It should outline all of the activities that are required for system commissioning.

Software Category

Because the software category determines the validation requirements, the Validation Plan requires the category listing as defined in an overview SOP or in GAMP4. The reasons for its selection are clearly stated in this section.

Operational Requirements / Critical Functions

This section of the Validation Plan contains the description of the operational requirements and critical functions that must be met in order for the system to function and be acceptable for use in the production environment. It should address automatic, manual, maintenance modes and operational interlock requirements of the system. Interlocks are hardware or software elements that inhibit or enable the use of features under specific conditions. They are used for data integrity, security or safety purposes.

This section provides detailed information, such as, circumstances that could directly impact the identity, strength, quality, or purity of a drug product, the operation of a medical device, or data that is subject to regulatory review or the safety of individuals who operate the system. This should summarize the critical items listed in the Functional Requirements document.

Description of Operations

Chapter 8 – Validation Plan

This section of the Validation Plan provides the description of the system operation as it is in its current state, whether manual or automated, and its desired future state. Both narrative and diagrammatic methods divided into modules should be used for large or complex systems. For small or simple systems, diagrams are optional, but are highly recommended. Diagrams should be accurate, concise and clearly labeled. Emphasis should be placed on those portions of the system that are critical.

Types of diagrams that could be used are:

- Hardware Configuration
- Network
- As Built
- Workflow
- Data Flow

Business and Operating Units

This section should require the description of the business and operating units involved in the project, such as:

- Business Unit / System Owner / End-User Representatives (each department affected)
- Management Representatives
- Quality Assurance (Unit)
- System Developer
- Interface Developer
- Vendors

The Products

This section, when applicable, provides the description of the products processed or controlled by the system with their specific require ments.

The Facilities

This section provides the description the facilities that will use the system and specific attributes of the site (e.g., manufacturing sites, analytical labs with associated instrumentation or equipment, clinical trial sites, distribution services, medical devices).

Application Software

This section requires the listing of the software application to be installed as part of the system. A description of the purpose along with the name of the manufacturer and version number should be provided. The anticipated number of concurrent users and total clients is specified.

It should require that sufficient information be provided about the software and its method validation. If the software was previously qualified (e.g. – the system was installed on an existing server), then the associated qualification documents are listed along with a description of the qualification activities and the identification and location of the validation documentation

The operating system that is to be used for the computer or computerized system should be specified in this section of the validation Plan.

Computer Hardware

This section should require listing of the primary hardware that will be utilized by the computer system. If the hardware was previously qualified (i.e., the system was installed on an existing server), then the associated qualification documents should be listed with a description of the qualification activities along with the identification and location of the validation documentation.

External Devices and Network Communications

This section should require the listing of external devices, such as interfaced instruments or manufacturing equipment and how they are to be qualified. External devices are elements that are not considered part of the computer system but can communicate with the computer system that is being validated. This category does not include client personal computer, which is dealt with separately. The external devices should require qualification prior to being validated with the computer system in question. This includes assuring current calibration of the devices, where appropriate. Interfaces to the network are included in this category.

Data Conversion / Migration

In computer systems where there is an upgrade to an existing system or the data is to be migrated and/or converted from a different computer system, the process and procedure for this migration requires specification. The details of how the data will be converted from the old to the new system and how it will be evaluated to ensure that the transferred data has maintained it's integrity is clearly described. In cases where data will originate from systems that are not validated, a plan is developed to ensure the accuracy after migration. All original data must be maintained in accordance with Predicate Rules.

Testing Procedures

Developing Test Scripts

All testing requires formal execution using approved protocols and test procedures. Testing that is required as part of an installation is specified in the IQ Protocol. In many cases, existing approved protocols and test scripts from previous implementations can be used. These protocols and plans need to be evaluated carefully and modified as needed to match the conditions of the current implementation. These test scripts require the same review and approval cycle as any other documentation.

All protocols and test scripts should be required to be traceable to the User and Functional Requirements for the system by means of a traceability matrix. Testing is required in accordance with written test procedures. Deviations should be resolved in accordance with established procedures.

Executed Test Script Review

A member of the Project Team not directly involved in the testing should be required to review each executed test procedure. If the test procedure is complete and meets its objectives, the reviewer should be required to sign and date it, thereby signifying approval. The review of any test outputs, evidence or attachments are included in this section of the Validation Plan.

Deviation Reporting

A requirement to address, on a case-by-case basis, all deviations from the expected results of the test protocol or failures identified during the testing process is included in this section of the Validation Plan. The Project Team is responsible for the evaluation and resolution of these issues. The System Owner or Project Manager and Quality Assurance (Unit) are responsible for assuring that evaluation and resolution of test deviations and failures are consistently and accurately carried out. Protocols having outstanding issues are considered incomplete.

Deviations / Failures are categorized in accordance with a defined scheme. An example of such a scheme:

- Critical

An issue that is critical to cGxP compliance; affects compliance with company SOPs or regulatory requirements; in any way compromises the integrity of the data; or that could directly impact the identity, strength, quality, or purity of a drug product, the operation of a medical device, data that is subject to regulatory review, or the safety of individuals who operate the system.

- Major

An issue for which there is an acceptable workaround solution.

- Minor

Errors in text fields that do not fall into the Critical or Major category and are unlikely to be misleading.

- Cosmetic

Trivial errors for which no action is required.

A report is required for Critical and Major categorizations. A report should be recommended for Minor or Cosmetic categorizations.

Upon review, Deviation / Failure Outcomes are handled according to a defined methodology. An example of such a methodology is as follows:

"Change in software configuration or code shall be in accordance with SOP C-027, Change Control and SOP C-028, Document Management and Storage.

Modification of the system requirements or specifications shall be in accordance with SOP C-27, Change Control and SOP C-028, Document Management and Storage. Major documentation changes shall require that the document be revised, the revision level incremented and the document sent for re-approval. Minor test procedure changes shall require a deviation report be generated. If such a change is required, the Project Team may elect to refer the problem to Example Pharmaceuticals Corporation management.

Develop a workaround procedure. Such workaround shall be incorporated into a user SOP.

Modification of the test procedure in accordance with SOP C-27, Change Control and SOP C-028, Document Management and Storage. If after evaluation, it is determined that the system is responding to a test as intended, but that the test procedure contains either the wrong instructions or an incorrect expected response, a change in the test procedure may be made. This shall require that the change in the test procedure be documented and the procedure re-executed after the change is approved."

Change Control

Change control is required for changes to the validated system in accordance with established Change Control and Document Management and Storage SOPs. The following are the minimum deliverables that require being secured and placed under formal version control once they are approved:

- Validation Plan
- User Requirements
- Functional Requirements
- IQ, OQ and PQ Protocols
- All Reports
- Traceability Matrix
- Job Aids
- SOPs

A system or document is subject to change control when the first activity is complete. Changes made to the System or to a version-controlled document during the project are handled as follows:

Major Changes

The document should be revised, the revision level incremented and the document sent for re-approval.

Minor Changes

A designated person maintains a marked-up copy of the document. If the number of changes is significant, the original document is updated at the end of the project and reissued for approval in accordance established Change Control and Document Management and Storage SOPs.

Changes to a document require notation in its Document History section. Documents that are affected by change control require listing in the Deliverables section of the Validation Plan. Software related problems identified during testing require handling in accordance with established Change Control and Document Management and Storage SOPs.

Summary Reports

A summary report is required for each executed IQ, OQ and PQ.

Deliverables

The Validation Plan requires the providing of a proposed list of all the associated documents that constitute the overall package. This list includes the title of the document, document number, availability, effective date, document purpose, and a listing of main sections or features to be included in the document. Examples of the basic set of documents to be included in a Validation Plan listing are as follows:

Letter of Notification or Change Control

A Letter of Notification or Change Control, as appropriate, is required to formally notify the responsible Quality Assurance (Unit) of the decision to validate a software application, computer or computerized system. The Manager or designee of the department that uses, or will use, a particular software application, computer or computerized system makes such a determination.

Validation Plan

Chapter 8 – Validation Plan

The Validation Plan is required to provide a clear description of the process by which the computer system will be validated. The plan includes a list of participants, required documents and an assignment of responsibilities.

User Requirements

A User Requirements document is required to provide a vehicle for the user to describe, in non-technical terms, what the system is supposed to do. The means of implementation should not be part of the description. The User Requirements document becomes the standard against which the system is built or procured, and tested. It is the guiding document that determines not only whether the system was built right, but also whether the right system was built.

Functional Requirements

A Functional Requirements document is required to provide a vehicle for communication between the User and the technology provider. It requires definition of the functional requirements of the system, interface projects, and the business needs that must be fulfilled. The Functional Requirements document defines, in specific terms, "what" the system will do and "how" it will do it.

System Specifications / Design

A System Specifications/Design document is required to provide the detailed requirements for the design, structure, and configuration of the system. The elements of the design are described. Items to consider including in this section are:

- Inputs
- Processes
- Outputs
- Constraints
- Workflow
- Hardware
- Firmware
- Software
- Data Structures
- Interfaces
- Performance
- Audit Trails
- Electronic Record Keeping
- Security
- Safety
- Maintenance

Installation Qualification (IQ)

An Installation Qualification test protocol is required to establish confidence that process equipment and ancillary systems are compliant with appropriate codes, approved design intentions, and that manufacturer's recommendations are suitably considered. The IQ Test protocol verifies that the hardware, software, and environmental specifications listed in the System Specification/Design documents have been correctly acquired or created, received, and installed. It addresses the elements of the system that are to be tested as part of the Installation Qualification process.

Each test procedure requires indication of the User and Functional Requirement that it tests.

A requirement that only qualified users can execute the IQ protocol is included.

An IQ Summary Report should be required.

The System Owner maintains the approved and executed copies of the IQ protocol along with supporting evidence and the IQ Summary Report. This information is archived in accordance with established Change Control and Document Management and Storage SOPs.

Operational Qualification (OQ)

An Operational Qualification test protocol is required to establish confidence that process equipment and sub-systems are capable of consistently operating within established limits and tolerances. The OQ protocol focuses on functionality and includes, but is not limited to, input, output, boundary, process and workflow testing, and error trapping.

Each test procedure requires indication of the User and Functional Requirement that it tests.

A requirement that only qualified users can execute the OQ protocol is included.

An OQ Summary Report should be required.

The System Owner maintains the approved and executed copies of the OQ protocol along with supporting evidence and the OQ Summary Report. This information is archived in accordance with established Change Control and Document Management and Storage SOPs.

Performance Qualification (PQ)

A Performance Qualification test protocol requires the establishment of confidence that the process is effective and reproducible and that through appropriate testing the finished product produced by a specified process meets all release requirements for functionality and safety. The PQ protocol may require that testing be site-specific due to unique workflow.

Each test procedure requires indication of the User and Functional Requirement that it tests.

A requirement that only qualified users can execute the PQ protocol is included.

A PQ Summary Report should be required.

The System Owner maintains the approved and executed copies of the PQ protocol along with supporting evidence and the PQ Summary Report. This information is archived in accordance with established Change Control and Document Management and Storage SOPs.

Traceablitiy Matrix

A Traceability Matrix document, which contains a matrix recording the relationship between the User Requirements, Functional Requirements and the design of the system, is required. The purpose of the Traceability Matrix is to check that all of the requirements of the User and Functional Requirements have been met in the System Specifications / Design documents and tested in the IQ, OQ and PQ protocols.

Requirements that are not met require an explanation as to why they have not been met, and a description of how any workaround will be accomplished.

Requirements that are to be "Phased In" during or after the project require identification as such. An explanation of the process to be followed in order to conduct and accomplish testing after project completion utilizing established Change Control procedures must be included.

Functional requirements not having a unique identification that associates them with a User or Functional Requirements document are referenced by their corresponding section number, name, and/or description

21 CFR Part 11 Assessment

A Regulatory Compliance Assessment based on 21 CFR Part 11 is required. An individual who is not assigned as a member of the Core Team should prepare such an assessment. This document addresses system compliance with applicable regulatory requirements along with an assessment of the project documentation with respect to the Project Plan. It should require compliance with an established SOP on 21 CFR Part 11 Assessment.

Acceptance Letter / Certification

At the completion and acceptance of the project, the System Owner sends an Acceptance Letter and Certification for Use to Quality Assurance (Unit) certifying that the system has been successfully tested and that all compliance related activities have been completed. This correspondence indicates that the system is certified for use.

Back Out Plan

Should the process defined in the Validation Plan fail, particularly in the case of modifications, a method must be defined to restore the system to its previous validated state. This entails a complete contingency plan that follows the requirements specified above for a full Validation Plan. The restoration of the system to its previous state must be validated.

System Maintenance

Specify the on-going activities needed in order to maintain the system in a state of sustainable compliance. Examples of areas to be addressed in this section are:

Change Control

This requirement specifies the methodology that will be used to manage changes to the system in accordance with an established SOP on Change Control.

Periodic Review

This requirement specifies the frequency and process that will be used to verify the compliance of the system with quality assurance standards in accordance with an established SOP on Periodic Review.

Document Management and Storage (protocol)

A Document Management and Storage section is included in each Validation Plan. It defines the Document Management and Storage methodology that is followed for issue, revision, replacement, control and management of each Validation Plan. This methodology is in accordance with an established Change Control SOP and a Document Management and Storage SOP.

Revision History (protocol)

A Revision History section is included in each Validation Plan. The Validation Plan Version Number and all Revisions made to the document are recorded in this section. These changes are in accordance with a Change Control Procedure SOP.

The Revision History section consists of a summary of the changes, a brief statement explaining the reasons the document was revised, and the name and affiliation of the person responsible for the entries. It includes the numbers of all items the Validation Plan replaces and is written so that the original Validation Plan is not needed to understand the revisions or the revis ed document being issued.

When the Validation Plan is a new document, a "000" as the Revision Number is entered with a statement such as "Initial Issue" in the comments section.

Regulatory or audit commitments require preservation from being inadvertently lost or modified in subsequent revisions. All entries or modifications made as a result of regulatory or audit commitments require indication as such in the Revision History and should indicate the organization, date and applicable references. A complete and accurate audit trail must be developed and maintained throughout the project lifecycle.

Attachments (protocol)

If needed, an Attachments section is included in each Validation Plan in accordance with the SOP on SOPs. An Attachment Index and Attachment section follow the text of a Validation Plan. The top of each attachment page indicates "Attachment # x," where 'x' is the sequential number of each attachment. The second line of each attachment page contains the attachment title. The number of pages and date of issue corresponding to each attachment are listed in the Attachment Index.

Attachments are placed directly after the Attachment Index and are sequentially numbered (e.g., Attachment # 1, Attachment # 2) with each page of an attachment containing the corresponding document number, page numbering, and total number of pages in the attachment (e.g., Page x of y). Each attachment is labeled consistent with the title of the attachment itself.

Attachments are treated as part of the Validation Plan. If an attachment needs to be changed, the entire Validation Plan requires a new revision number and is reissued. New attachments or revisions of existing attachments are in accordance with an established Change Control SOP and a Document Management and Storage SOP.

Document Management and Storage *(SOP)*

A Document Management and Storage section is included in the Validation Plan SOP. It defines the Document Management and Storage methodology that is followed for issue, revision, replacement, control and management of this SOP. This methodology is in accordance with an established Change Control SOP and a Document Management and Storage SOP.

Revision History *(SOP)*

A Revision History section is included in the Validation Plan SOP. The Validation Plan Version Number and all Revisions made to the document are recorded in this section. These changes are in accordance with a Change Control Procedure SOP.

The Revision History section consists of a summary of the changes, a brief statement explaining the reasons the document was revised, and the name and affiliation of the person responsible for the entries. It includes the numbers of all items the Validation Plan SOP replaces and is written so that the original Validation Plan SOP is not needed to understand the revisions or the revised document being issued.

When the Validation Plan SOP is a new document, a "000" as the Revision Number is entered with a statement such as "Initial Issue" in the comments section.

Regulatory or audit commitments require preservation from being inadvertently lost or modified in subsequent revisions. All entries or modifications made as a result of regulatory or audit commitments require indication as such in the Revision History and should indicate the organization, date and applicable references. A complete and accurate audit trail must be developed and maintained throughout the project lifecycle.

Attachments *(SOP)*

If needed, an Attachments section is included in the Validation Plan SOP in accordance with the SOP on SOPs. An Attachment Index and Attachment section follow the text of the Validation Plan SOP. The top of each attachment page indicates "Attachment # x," where 'x' is the sequential number of each attachment. The second line of each attachment page contains the attachment title. The number of pages and date of issue corresponding to each attachment are listed in the Attachment Index.

Attachments are placed directly after the Attachment Index and are sequentially numbered (e.g., Attachment # 1, Attachment # 2) with each page of an attachment containing the corresponding document number, page numbering, and total number of pages in the attachment (e.g., Page x of y). Each attachment is labeled consistent with the title of the attachment itself.

Attachments are treated as part of the Validation Plan SOP. If an attachment needs to be changed, the entire Validation Plan SOP requires a new revision number and is reissued. New attachments or revisions of existing attachments are in accordance with an established Change Control SOP and a Document Management and Storage SOP.

Chapter 9 – Design Qualification

In this chapter we are going to explore Design Qualification (DQ) elements, requirements, techniques and procedures that will be covered as specific cases in later Chapters. Design Qualification is an important part of the System Development Life Cycle (SDLC) process and is normally required as part of an overlying corporate policy. It ensures that the design meets the needs and requirements of the business and user. The elements comprising Design Qualification in SDLC are:

Stage	Associated Chapter
User Requirements with Intended Use	Chapter 10 – User Requirements
Functional Requirements	Chapter 11 – Functional Requirements
Vendor Audit and Selection	Chapter 12 – Vendor Audit
Software Design Specification	Chapter 13 – Software Design Specification
Software Module Design Specification	Chapter 13 – Software Design Specification
Hardware Design Specification	Chapter 14 – Hardware Design Specification
Summary Report	Chapter 25 – Summary Report
Traceability Matrix	Chapter 26 – Traceability Matrix
Manuals and SOPs	Chapter 31 – Manuals and SOPs
Training	Chapter 33 – Training
Ongoing Operation	Chapter 35 – Ongoing Operation

Table 3 - Design Qualification Elements

Best practices and GxP dictate the use of design controls that create and implement procedures to control the system design. This ensures that specified design requirements are met and fulfill the business and user needs and requirements for proper sustained compliance in accordance with company operations, policies and procedures, and current regulation. A review and approval process must be implemented in order to ensure the technical content and quality of the system design and documentation. This is accomplished prior to construction, installation and implementation. The level and complexity of the DQ process is determined by a project team, which is headed by the System Owner.

The result of DQ is the confirmation of proper use of SDLC methods – a system that is designed and verified as meeting the business and User needs and requirements for proper job performance and sustainable compliance with regulations. It also provides a venue for the System Owner to acquire assurance that internal resources and outside vendors are in agreement with the stated work requirements that need fulfillment. Finally, the DQ documents are used as a basis for qualification tests in the Testing phases of SDLC.

Also see Chapter 15 – Testing.

Deficiencies in DQ impact compliance with the Food, Drug and Cosmetic Act and its associated regulations with respect to drug safety, identity and strength, and the quality and purity characteristics that are purported or represented as possessing. Similar situations may arise for medical devices and data in support of GxP operations. They can also have a business impact, as well. Sufficient planning and resource allocation can aid in avoiding these problems.

The following items are included within the documents produced in the stages shown above:

- Flow Diagrams
- System Diagrams
- Operator and/or User Manuals
- System Administration Documentation
- Theory of Operation
- Design phase implementation and test requirements
- Drawings
- Material and equipment lists
- Quality plans
- User Requirements deviations and remedies
- Functional Requirements deviations and remedies
- Design Specification deviations and remedies
- Identification of critical functions
- Change control records
- Source code review

Requirements and Specifications

DQ requires the definition and specification of User Requirements, Functional Requirements and detail Design Specifications of a software application, computer system or computerized system. As a part of these stages, the User Requirements are specified in lay terms. An example user requirement is, "Provide for limited access to the networked system. Only authorized users shall be allowed access to the system with functionality based on their job function." The associated Functional Requirement that meets the User Requirement is then stated as, "User ID and password entry fields shall require entry of a valid User ID and password combination for permitted access and qualified level of use of the system."

Traceability Matrix

A Traceability Matrix is created that cross-references User Requirements, Functional Requirements, Design Specifications, Installation Qualification (IQ), Operational Qualification (OQ) and Performance Qualification (PQ) for a validation project. The requirements for preparing a Traceability Matrix that tracks and determines the compliance of a project with the business needs and requirements of the System Owner in support of a GxP and 21 CFR Part 11 project are specified.

This Traceability Matrix helps demonstrate that a computer or computerized system performs as it purports and is intended to perform. The Traceability Matrix is required to demonstrate, in a summarized format, that the system is being built in accordance with the user and business needs and requirements of the project. It aids in demonstrating that the correct system was built, that the system was built correctly, and properly implemented.

See Chapter 26 – Traceability Matrix.

Vendor Qualification

As part of the design qualification process, each vendor is audited and qualified prior to the contracting of goods and services. Vendor audit procedures provide requirements for auditing a vendor or contractor that may provide goods or services for or in support of a GxP project or Ongoing Operation. Only qualified vendors are used for GxP and 21 CFR Part 11 environments and operations.

See Chapter 12 – Vendor Audit

Manuals and SOPs

The procedures for obtaining or creating, using, managing, and storing system specific manuals and SOPs for software applications, computer or computerized systems that are used in a GxP and 21 CFR Part 11 environment or operation are specified in this section. This provides for the control of all documentation including SOPs and manuals.

See Chapter 31 – Manuals and SOPs.

Training

Training criteria and materials for computer or computerized systems that are validated systems are created, maintained, reviewed, and updated periodically. The requirements for the development, curricula, assessment, records, management and control of the documentation and procedures should be specified. This is necessary in order to perform and document sustainable compliance for all employee, consultant and contractor training and assessment in the GxP and 21 CFR Part 11 environment.

See Chapter 33 – Training.

Ongoing Operation Requirements

A key responsibility of the System Owner is the implementation of standard operating procedures that specify and require that all validated systems be maintained in a sustainedly compliant and validated state. All modifications and changes to a system must be documented in accordance with the SDLC methodology. This includes hardware, software, peripheral equipment, connectivity and communication equipment, training materials and all system documentation. The frequency of periodic review and training should be specified..

See Chapter 35 – Ongoing Operations.

Design Qualification Report

The System Owner produces a Design Qualification Summary Report and ensures that it is reviewed and approved by the System Owner, CVG and QA(U). In addition to those requirements detailed in Chapter 25 – Summary Report, the results of the DQ that are documented in the Design Qualification Summary Report include, but are not limited to:

- Overview of the Process
- Summary of tasks performed
- Documents produced with version numbers and dates
- Supporting documentation
- Discrepancies, Deviations and Remedies

See Chapter 25 – Summary Report.

21 CFR 820.30 Quality System Regulation: Design Controls

In keeping with the concept of Total Quality Management and when applied in a general manner, the following excerpt from 21 CFR Part 820, section 30, which deals with design controls, is an excellent source of information to guide in the process of design qualification.

General.

(1) Each manufacturer of any class III or class II device, and the class I devices listed in paragraph (a)(2)[9] of this section, shall establish and maintain procedures to control the design of the device in order to ensure that specified design requirements are met.

Design and development planning.

Each manufacturer shall establish and maintain plans that describe or reference the design and development activities and define responsibility for implementation. The plans shall identify and describe the interfaces with different groups or activities that provide, or result in, input to the design and developmentprocess. The plans shall be reviewed, updated, and approved as design and development evolves.

Design input.

Each manufacturer shall establish and maintain procedures to ensure that the design requirements relating to a device are appropriate and address the intended use of the device, including the needs of the user and patient. The procedures shall include a mechanism for addressing incomplete, ambiguous, or conflicting requirements. The design input requirements shall be documented and shall be reviewed and approved by a designated individual(s). The approval, including the date and signature of the individual(s) approving the requirements, shall be documented.

Design output.

Each manufacturer shall establish and maintain procedures for defining and documenting design output in terms that allow an adequate evaluation of conformance to design input requirements. Design output procedures shall contain or make reference to acceptance criteria and shall ensure that those design outputs that are essential for the proper functioning of the device are identified. Design output shall be documented, reviewed, and approved before release. The approval, including the date and signature of the individual(s) approving the output, shall be documented.

Design review.

Each manufacturer shall establish and maintain procedures to ensure that formal documented reviews of the design results are planned and conducted at appropriate stages of the device's design development. The procedures shall ensure that participants at each design review include representatives of all functions concerned with the design stage being reviewed and an individual(s) who does not have direct responsibility for the design stage being reviewed, as well as any specialists needed. The results of a design review, including identification of the design, the date, and the individual(s) performing the review, shall be documented in the design history file (the DHF).

Design verification.

Each manufacturer shall establish and maintain procedures date, and the individual(s) performing the verification, shall be documented in the DHF.

Design validation.

Each manufacturer shall establish and maintain procedures for validating the device design. Design validation shall be performed under defined operating conditions on initial production units, lots, or batches, or their equivalents. Design validation shall ensure that devices conform to defined user needs and intended uses and shall include testing of production units under actual or simulated use conditions. Design validation shall include software validation and risk analysis, where appropriate. The results of the design validation, including identification of the design, method(s), the date, and the individual(s) performing the validation, shall be documented in the DHF.

Design transfer.

[9] Paragraph (a)(2) lists the class I devices subject to design controls, which includes devices automated with computer controls.

Chapter 9 – Design Qualification

Each manufacturer shall establish and maintain procedures to ensure that the device design is correctly translated into production specifications.

Design changes.

Each manufacturer shall establish and maintain procedures for the identfication, documentation, validation or where appropriate verification, review, and approval of design changes before their implementation.

Design history file.

Each manufacturer shall establish and maintain a DHF for each type of device. The DHF shall contain or reference the records necessary to demonstrate that the design was developed in accordance with the approved design plan and the requirements of this part.

Chapter 10 – User Requirements

In this chapter we will discuss the Standard Operating Procedure (SOP) for creating user-defined requirements for a validation project. This SOP defines the method to be used for specifying, in non-technical terms, all of the functions that a system or system component must perform.

The User Requirements determine what will be built or procured and how it will operate and be used. Best industry practices and FDA Guidance documents indicate that in order to demonstrate that a computer or computerized system performs as it purports and is intended to perform, a methodical plan is needed that details how this may be proven. All testing will be based on THE User Requirements.

The SOP for User Requirements is used for all systems that create, modify, maintain, retrieve, transmit or archive records in electronic format. This document is used in conjunction with an overlying corporate policy to provide sustainable compliance with the regulations. The project team uses the SOP to create the User Requirements, from which, the Functional Requirements, Detail Design Specification, Hardware Design Specification, Factory Acceptance Testing and all validation qualification are derived. It is the procedure that is followed to create the foundation of the software application, computer or computerized system.

System Owners and End-Users use this SOP to create the User Requirements for a software application, computer or computerized system that controls equipment and/or produces electronic data or records in support of all regulated laboratory, manufacturing, clinical or distribution activities.

Title Block, Headers and Footers (SOP)

The User Requirements SOP should have a title block, title or cover page along with a header and/or footer on each page of the document that clearly identifies the document. The minimum information that should be displayed for the User Requirements SOP is as follows:

- Document Title
- Document Number
- Document Revision Number
- Project Name and Number
- Site/Location/Department Identification
- Date Document was issued, last revised or draft status
- Effective Date
- Pagination: page of total pages (e.g., Page x of y)

The Document Title should be in large type font that enables clear identification. It should clearly and concisely describe the contents and intent of the document. Since this document will be part of a set or package of documents within a project, it should be consistent with other documents used for the project and specify that it is the User Requirements SOP. This information should be on every page of the document.

The Document Number should be in large type font that enables clear identification. Since the User Requirements SOP will be part of a set or package of documents within a project, it should be consistent with other documents used for the project. This information should appear on every page of the document.

The Document Revision Number, which identifies the revision of the document, should appear on each page of the document. Many documents will go through one or more changes over time, particularly during the draft stage and the review and approval cycle. It is important to maintain accurate revision control because even a subtle change in a document can have an immense impact on the outcomes of activities. The SOP on SOPs should govern the numbering methodology.

See Chapter 6 – "The SOP on SOPs."

The Project Name and Number, if applicable, should be indicated on the cover or title page. This information identifies the overall project to which this document belongs.

The Site/Location/Department Identification should be indicated on the cover or title page. It identifies the Site/Location/Department that has responsibility for this document and its associated project.

The Date that the Document was issued, last revised or issued with draft status should be on each page of the document (e.g., 12-Dec-01). This will clearly identify the date on which the document was last modified.

The Effective Date of the Document should be on the title or cover page of the document (e.g., 12-Dec-01). This will clearly identify when to begin using the current version of the document.

Pagination, page of total pages (e.g., Page x of y), should be on each page of the document. This will help identify whether or not all of the pages of the document are present. Pages intentionally left blank should clearly be designated as such.

Approval Section *(SOP)*

The Approval Section should contain a table with the Author and Approvers printed names, titles and a place to sign and date the SOP. The SOP on SOPs should govern who the approvers of the document should be. The functional department (e.g., Production / Quality Assurance / Engineering) of the signatory should be placed below each printed name and title along with a statement associated with each approver indicating the person's role or qualification in the approval process. Indicate the significance of the associated signature (e.g. This document meets the requirements of Corporate Policy 9055, Electronic Records; Electronic Signatures, dated 22-Mar-2000).

The completion (signing) of this section indicates that the contents have been reviewed and approved by the listed individuals.

Table of Contents *(SOP)*

The SOP should contain a Table of Contents, which appears towards the beginning of the document. This Table of Contents provides a way to easily locate various topics contained in the document.

Purpose *(SOP))*

This introduction to the SOP concisely describes the document's purpose in sufficient detail so that a non-technical reader can understand its purpose.

The Purpose section should clearly state that the intention of the document is to provide a Standard Operating Procedure to be used to specify, in non-technical terms, the requirements for the preparation of User Requirements for a new or upgraded computer or computerized system (what the system is supposed to do) along with the required activities and deliverables. These requirements are necessary for the User Requirements to be effective and compliant with any company polices and government regulation. As one of the foundation of the software application, computer or computerized system, the User Requirements needs to define the scope of work, responsibilities and deliverables in order to demonstrate that a computer or computerized system meets its intended use in a reliable, consistent and compliant manner. The User Requirements document derived from this SOP will become the standard against which the system is tested. It shall be the guiding document that determines not only whether the system was built right, but also whether the right system was built.

Scope *(SOP)*

The Scope section identifies the limitations of the Design Qualification SOP. It limits the scope to all GxP systems and is applicable to all computers, computerized systems, infrastructure, LANs and WANs including clinical, laboratory, manufacturing and distribution computer systems. It applies to all Sites, Locations and Departments within the company.

Equipment / Materials / Tools *(SOP)*

This section lists any required or recommended equipment, materials, tools, templates or guides. Provide the name and a brief description of the equipment, instruments, materials and/or tools to be used in the procedures or testing covered by the User Requirements SOP. This is for reference purposes and requires the user to utilize the specific item or a designated equivalent. The purpose and use is explained elsewhere in the document. Substitution of comparable equipment may be made in accordance with procedures for Change Control.

If the information is included in a job aid or operating manual, a description of where the information may be found with a specific reference to the document number, title, date and revision number should be specified.

Warnings / Notes / General Information / Safety *(SOP)*

Warnings, notes, general or safety information applicable to the User Requirements SOP are listed in this section.

Policies, Guidelines, References *(SOP)*

This section lists references to appropriate regulatory requirements, SOPs, policies, guidelines, or other relevant documents referred to in the User Requirements SOP. The name and a brief description of applicable related SOPs, corporate policies, procedures and guidelines, Quality Control monographs, books, technical papers, publications, specific regulations, batch records, vendor or system manuals, and all other relevant documents that are required by the User Requirements SOP should be included. This is for reference purposes and requires the user to use the specific item or a designated equivalent. The purpose and use are explained elsewhere in the document.

SOPs cite references and incorporate them in whole by reference rather than specifying details that duplicate these items. This is done so that changes in policy, guidelines or reference materials require no changes to SOPs themselves and the most current version of the referenced material is used. Examples are:

- Policy ABC-123, Administrative Policy for Electronic Records; Electronic Signatures
- Corporate, Plan for Compliance with 21 CFR Part 11, Electronic Records; Electronic Signatures
- SOP-456, Electronic Records; Electronic Signatures
- FDA, 21 CFR Part 11, Electronic Records; Electronic Signatures
- FDA, General Principals of Software Validation; Final Guidance for Industry and FDA Staff, January 11, 2002

Assumptions / Exclusions / Limitations (SOP)

Any and all assumptions, exclusions or limitations within the context of the SOP should be discussed. Specific issues should be directly addressed in this section of the User Requirements SOP. For example:

"The means of implementation shall not be part of the description."

Glossary of Terms (SOP)

A Glossary of Terms should be included that defines all of the terms and acronyms that are used in this document or that are specific to the operation being described and may not be generally known to the reader. The reader may also be referred to the FDA document, "Glossary of Computerized System and Software Development Terminology."

Roles and Responsibilities (SOP)

The Roles and Responsibilities of the User Requirements are divided into two specific areas:

Creation and Maintenance

The System Owner of the computer or computerized system is required to be responsible for the preparation and maintenance of the User Requirements.

It is common for a User Requirements to go through several revisions during the course of a project, as it is considered a living document, especially during the project's early stages. As a normal process, it helps to ensure that the validation activities are in compliance with cGxPs, FDA regulations, company policies and industry standards and guidelines.

The roles and responsibilities for preparing the User Requirements and other project documentation are addressed in accordance with established Standard Operating Procedures.

Use

The members of the Project Team must be familiar with the User Requirements in order to fulfill the project requirements. The Quality Assurance (Unit) and Project Team use the User Requirements in concert with other project documentation as a contract that details the project and quality expectations.

The User Requirements are the governing document for all validation qualification and testing.

Procedure (SOP)

The SOP should contain instructions and procedures that enable the creation of a User Requirements protocol document. The User Requirements are needed in order to provide an overall non-technical description of the requirements of a software application, computer or computerized system.

The User Requirements are prepared as a stand-alone document for all projects. It is not good practice to include it as part of other documents in an attempt to reduce the amount of workload or paperwork. This only causes more work over the length of the project resulting from misunderstandings and confusion. Clear, concise documents that address specific parts of any given project are desirable.

Chapter 10 - User Requirements

Purpose (protocol)

Clearly state the system with information as to why it is considered a regulated system and introduce planned activities in order to validate it. This short introduction should be concise and present details so that an unfamiliar reader can understand the purpose of the document with regard to the system.

Scope (protocol)

The Scope section identifies the limitations of a specific User Requirements protocol in relationship to validating a specific software application, computer or computerized system. For example, if only a part of the computer system will be validated, then this section would be used to explain the reason for this decision.

Equipment / Materials / Tools (protocol)

This section lists any required or recommended equipment, materials, tools, templates or guides. Provide the name and a brief description of the equipment, instruments, materials and/or tools to be used in the procedures or testing covered by the User Requirements protocol. This is for reference purposes and requires the user to utilize the specific item or a designated equivalent. The purpose and use is explained elsewhere in the document. Substitution of comparable equipment may be made in accordance with procedures for Change Control.

If the information is included in a job aid or operating manual, a description of where the information may be found with a specific reference to the document number, title, date and revision number should be specified.

Warnings / Notes / General Information / Safety (protocol)

Warnings, notes, general or safety information applicable to the User Requirements protocol are listed in this section.

Policies / Guidelines / References (protocol)

This section lists references to appropriate regulatory requirements, SOPs, policies, guidelines, or other relevant documents referred to in the User Requirements protocol. The name and a brief description of applicable related SOPs, corporate policies, procedures and guidelines, Quality Control monographs, books, technical papers, publications, specific regulations, batch records, vendor or system manuals, and all other relevant documents that are required by the User Requirements protocol should be included. This is for reference purposes and requires the user to use the specific item or a designated equivalent. The purpose and use are explained elsewhere in the document.

Assumptions / Exclusions/ Limitations (protocol)

Discuss any assumptions, exclusions or limitations within the context of the User Requirements. Specific issues should be directly addressed in this section of the User Requirements protocol.

Glossary of Terms (protocol)

The Glossary of Terms should define all of the terms and acronyms that are used in the document or that are specific to the operation being described and may not be generally known to the reader. The reader may also be referred to the FDA document, "Glossary of Computerized System and Software Development Terminology," or other specific glossaries that may be available.

Titles, terms, and definitions that are used in the corporate policies, procedures and guidelines should be defined.

Roles and Responsibilities (protocol)

Validation activities are the result of a cooperative effort involving many people, including the System Owners, End Users, the Computer Validation Group, Subject Matter Experts (SME) and Quality Assurance (Unit).

Clearly identify the roles, responsibilities and individual(s) responsible for the creation and use of the User Requirements. Specify the participants by name, title, department or company affiliation and responsibility in a table. Responsibilities should be specific for the procedures and activities delineated in each specific document.

The System Owners are responsible for the development and understanding of the User Requirements. They acquire and coordinate support from in-house and/or external consultants, the Computer Validation Group, Quality Assurance (Unit) and system developers.

The QA(U) is responsible to verify that the details in the User Requirements are compliant with applicable departmental, interdepartmental and corporate policies and procedures and that the approach is in accordance with cGxP regulations and good business practices, especially 21 CFR Part 11.

Procedure - System Overview and Process (protocol)

This section requires a general non-technical description of the system that includes equipment, hardware, software, interfaces, workflows, processes and procedures.

Introduction (protocol)

This section contains a description of the intended use of the application, computer system or computerized system. It provides an overview of the business operations for the system including workflow, material flow, the control philosophy of what functions, calculations and processes are to be performed by the system; and information describing the interfaces that the system or application must be provided or created. It specifies an outline of the activities required for commissioning the system.

Software Category (protocol)

This section requires the software category be selected from that in an Overview of Computer and Computerized System Validation. The category chosen is clearly stated with the reason for its choice as the software category determines the validation requirements.

Operational Requirements / Critical Functions (protocol)

This section contains the description of the operational requirements and critical functions that must be met in order for the system to function and be acceptable for use.

It addresses automatic, manual, and maintenance modes, and operational interlock requirements of the system. Interlocks are hardware or software elements that inhibit or enable the use of features under specific conditions. They are used for data integrity, security or safety purposes.

Chapter 10 - User Requirements

Detailed information, such as, circumstances that could directly impact the identity, strength, quality, or purity of a drug product, the operation of a medical device, data subject to regulatory review or the safety of individuals who operate the system are provided in this section. Critical items listed in the User Requirements document are summarized in this section.

If it is not feasible to separate the critical functions from the rest of the requirements, then they should be identified as such.

All requirements should be indicated as being classified as one of the following:

- Mandatory - must be included
- Desirable - may be excluded

Description of Operations

This section describes the system operation as it is in its current state, be it manual or automated, and its desired future state. Both narrative and diagrammatic methods divided into modules should be used for large or complex systems. For small or simple systems, diagrams may be made optional, but they are highly recommended. Diagrams must be accurate, concise and clearly labeled with emphasis placed on those portions of the system that are critical.

Types of diagrams that can be used are:

- Hardware Configuration
- Network
- As Built
- Workflow
- Data Flow

Business and Operating Units (protocol)

This section describes the business and operating units involved in the project:

- Business Unit / System Owner / End-User Representatives (each department affected)
- Management Representatives
- Quality Assurance (Unit)
- System Developer
- Interface Developer
- Vendors

The Products (protocol)

The Products section, when applicable, describes the products processed or controlled by the system with their specific requirements.

The Facilities (protocol)

The Facilities section describes the facilities that will use the system and specific attributes of the site (e.g., manufacturing sites, analytical labs with associated instrumentation or equipment, clinical trial sites, distribution services, medical devices).

Application Software User Requirements (protocol)

The Application Software User Requirements section lists the non-technical user requirements that constitute the system. Functional entities, such as algorithms, menus, screens, fields, printing, forms, reports, audit trails, electronic signatures, electronic records, storage and retrieval, system administration, regulatory, security, support, system are addressed here.

This section lists the software application, if applicable, to be installed as part of the system. Provide a description of the purpose along with the name of the manufacturer and version number. Specify the anticipated number of concurrent users and total clients.

The Application Software User Requirements should provide sufficient information about the software application, computer or computerized system and its method of validation. If the software was previously qualified (i.e., system was installed on an existing server), then the associated qualification documents are listed with a description of the qualification activities and the identification and location of the validation documentation.

The operating system that is to be used for the computer or computerized system requires specification, as well.

Computer Hardware (protocol)

This section may require the listing the primary hardware that is to be utilized by the system. It may not be feasible or desirable at the user requirements level to specify hardware.

If the specified hardware was previously qualified (i.e., system was installed on an existing server), then the associated qualification documents are listed with a description of the qualification activities and the identification and location of the validation documentation. This may be the case if an existing server is available and has the required space and performance for running a new application.

Environmental, Mechanical, Electrical (protocol)

This section specifies, in non-technical terms, as applicable, environmental, mechanical and electrical requirements in which the hardware is expected to operate.

External Devices and Network Communications (protocol)

This section lists any external devices, such as interfaced instruments or manufacturing equipment, and any Network Interfaces.

Data Conversion / Migration (protocol)

In computer systems where there is an upgrade to an existing system or in situations where the data is to be migrated and/or converted from a different computer system, the process and procedure for this migration is specified. The details of how the data will be converted from the old to the new system and how the information will be evaluated to ensure that the transferred data has maintained it's integrity requires clear description in non-technical terms. In cases where data will originate from systems that are not validated, a requirement should be require to be specified to ensure the accuracy after migration.

Predicate rules must be followed to ensure proper storage and integrity of the old data.

Document Management and Storage (protocol)

A Document Management and Storage section is included in each User Requirements protocol. It defines the Document Management and Storage methodology that is followed for issue, revision, replacement, control and management of each User Requirements protocol. This methodology is in accordance with an established Change Control SOP and a Document Management and Storage SOP.

Revision History (protocol)

A Revision History section is included in each User Requirements protocol. The User Requirements protocol Version Number and all Revisions made to the document are recorded in this section. These changes are in accordance with a Change Control Procedure SOP.

The Revision History section consists of a summary of the changes, a brief statement explaining the reasons the document was revised, and the name and affiliation of the person responsible for the entries. It includes the numbers of all items the User Requirements protocol replaces and is written so that the original User Requirements protocol is not needed to understand the revisions or the revised document being issued.

When the User Requirements protocol is a new document, a "000" as the Revision Number is entered with a statement such as "Initial Issue" in the comments section.

Regulatory or audit commitments require preservation from being inadvertently lost or modified in subsequent revisions. All entries or modifications made as a result of regulatory or audit commitments require indication as such in the Revision History and should indicate the organization, date and applicable references. A complete and accurate audit trail must be developed and maintained throughout the project lifecycle.

Attachments (protocol)

If needed, an Attachments section is included in each User Requirements protocol in accordance with the SOP on SOPs. An Attachment Index and Attachment section follow the text of a User Requirements protocol. The top of each attachment page indicates "Attachment # x," where 'x' is the sequential number of each attachment. The second line of each attachment page contains the attachment title. The number of pages and date of issue corresponding to each attachment are listed in the Attachment Index.

Attachments are placed directly after the Attachment Index and are sequentially numbered (e.g., Attachment # 1, Attachment # 2) with each page of an attachment containing the corresponding document number, page numbering, and total number of pages in the attachment (e.g., Page x of y). Each attachment is labeled consistent with the title of the attachment itself.

Attachments are treated as part of the User Requirements protocol. If an attachment needs to be changed, the entire User Requirements protocol requires a new revision number and is reissued. New attachments or revisions of existing attachments are in accordance with an established Change Control SOP and a Document Management and Storage SOP.

***Document Management and Storage** (SOP)*

A Document Management and Storage section is included in the User Requirements SOP. It defines the Document Management and Storage methodology that is followed for issue, revision, replacement, control and management of this SOP. This methodology is in accordance with an established Change Control SOP and a Document Management and Storage SOP.

Revision History *(SOP)*

A Revision History section is included in the User Requirements SOP. The User Requirements SOP Version Number and all Revisions made to the document are recorded in this section. These changes are in accordance with a Change Control Procedure SOP.

The Revision History section consists of a summary of the changes, a brief statement explaining the reasons the document was revised, and the name and affiliation of the person responsible for the entries. It includes the numbers of all items the User Requirements SOP replaces and is written so that the original User Requirements SOP is not needed to understand the revisions or the revised document being issued.

When the User Requirements SOP is a new document, a "000" as the Revision Number is entered with a statement such as "Initial Issue" in the comments section.

Regulatory or audit commitments require preservation from being inadvertently lost or modified in subsequent revisions. All entries or modifications made as a result of regulatory or audit commitments require indication as such in the Revision History and should indicate the organization, date and applicable references. A complete and accurate audit trail must be developed and maintained throughout the project lifecycle.

Attachments *(SOP)*

If needed, an Attachments section is included in the User Requirements SOP in accordance with the SOP on SOPs. An Attachment Index and Attachment section follow the text of the User Requirements SOP. The top of each attachment page indicates "Attachment # x," where 'x' is the sequential number of each attachment. The second line of each attachment page contains the attachment title. The number of pages and date of issue corresponding to each attachment are listed in the Attachment Index.

Attachments are placed directly after the Attachment Index and are sequentially numbered (e.g., Attachment # 1, Attachment # 2) with each page of an attachment containing the corresponding document number, page numbering, and total number of pages in the attachment (e.g., Page x of y). Each attachment is labeled consistent with the title of the attachment itself.

Attachments are treated as part of the User Requirements SOP. If an attachment needs to be changed, the entire User Requirements SOP requires a new revision number and is reissued. New attachments or revisions of existing attachments are in accordance with an established Change Control SOP and a Document Management and Storage SOP.

Chapter 11 – Functional Requirements

In this chapter we will discuss the Standard Operating Procedure (SOP) for creating Functional Requirements for a validation project. This SOP defines the method to be used for specifying the functions that a system or system component must perform.

Best industry practices and FDA Guidance documents indicate that in order to demonstrate that a computer or computerized system performs as it purports and is intended to perform, a methodical plan is needed that details how this may be proven. The Functional Requirements determine what will be built or procured and how it will operate and be used. In concert with the User Requirements, the Detail Design Specification, and subsequent procurement or creation of the application software, the computer system and/or computerized system will based on the protocol created along with all testing.

This SOP is used for all systems that create, modify, maintain, retrieve, transmit or archive records in electronic format. It is used in conjunction with an overlying corporate policy to provide sustainable compliance with the regulation. The project team uses it to create the Functional Requirements, from which, the Detail Design Specification, Hardware Design Specification, Factory Acceptance Testing and all validation qualification is derived. It is the procedure that is followed to create the foundation of the software application, computer or computerized system.

System Owners, End-Users, and Subject Matter Experts (SME) use this SOP to create the Functional Requirements for a software application, computer or computerized system that controls equipment and/or produces electronic data or records in support of all regulated laboratory, manufacturing, clinical or distribution activities. The protocols produced in accordance with this SOP are used by procurement to acquire the specified items and by system developers to create Detail Design Specifications that are used to build the specified system.

Title Block, Headers and Footers (SOP)

The Functional Requirements SOP should have a title block, title or cover page along with a header and/or footer on each page of the document that clearly identifies the document. The minimum information that should be displayed for the Functional Requirements SOP is as follows:

- Document Title
- Document Number
- Document Revision Number
- Project Name and Number
- Site/Location/Department Identification
- Date Document was issued, last revised or draft status
- Effective Date
- Pagination: page of total pages (e.g., Page x of y)

The Document Title should be in large type font that enables clear identification. It should clearly and concisely describe the contents and intent of the document. Since this document will be part of a set or package of documents within a project, it should be consistent with other documents used for the project and specify that it is the Functional Requirements SOP. This information should be on every page of the document.

The Document Number should be in large type font that enables clear identification. Since the Functional Requirements SOP will be part of a set or package of documents within a project, it should be consistent with other documents used for the project. This information should appear on every page of the document.

The Document Revision Number, which identifies the revision of the document, should appear on each page of the document. Many documents will go through one or more changes over time, particularly during the draft stage and the review and approval cycle. It is important to maintain accurate revision control because even a subtle change in a document can have an immense impact on the outcomes of activities. The SOP on SOPs should govern the numbering methodology.

See Chapter 6 – The SOP on SOPs.

The Project Name and Number, if applicable, should be indicated on the cover or title page. This information identifies the overall project to which this document belongs.

The Site/Location/Department Identification should be indicated on the cover or title page. It identifies the Site/Location/Department that has responsibility for this document and its associated project.

The Date that the Document was issued, last revised or issued with draft status should be on each page of the document (e.g., 12-Dec-01). This will clearly identify the date on which the document was last modified.

The Effective Date of the Document should be on the title or cover page of the document (e.g., 12-Dec-01). This will clearly identify when to begin using the current version of the document.

Pagination, page of total pages (e.g., Page x of y), should be on each page of the document. This will help identify whether or not all of the pages of the document are present. Pages intentionally left blank should clearly be designated as such.

Approval Section *(SOP)*

The Approval Section should contain a table with the Author and Approvers printed names, titles and a place to sign and date the SOP. The SOP on SOPs should govern who the approvers of the document should be. The functional department (e.g., Production / Quality Assurance / Engineering) of the signatory should be placed below each printed name and title along with a statement associated with each approver indicating the person's role or qualification in the approval process. Indicate the significance of the associated signature (e.g. This document meets the requirements of Corporate Policy 9055, Electronic Records; Electronic Signatures, dated 22-Mar-2000).

The completion (signing) of this section indicates that the contents have been reviewed and approved by the listed individuals.

Table of Contents *(SOP)*

The SOP should contain a Table of Contents, which appears towards the beginning of the document. This Table of Contents provides a way to easily locate various topics contained in the document.

Purpose *(SOP)*

This introduction to the SOP concisely describes the document's purpose in sufficient detail so that a non-technical reader can understand its purpose.

The Purpose section should clearly state that the intention of the document is to provide a Standard Operating Procedure to be used to specify, in non-technical terms, the requirements for the preparation of Functional Requirements for a new or upgraded computer or computerized system (what the system is supposed to do) along with the required activities and deliverables. These requirements are necessary for the Functional

Requirements to be effective and compliant with any company polices and government regulation. As one of the foundation of the software application, computer or computerized system, the Functional Requirements needs to define the scope of work, responsibilities and deliverables in order to demonstrate that a computer or computerized system meets its intended use in a reliable, consistent and compliant manner. The Functional Requirements document derived from this SOP will become the standard against which the system is tested. It shall be the guiding document that determines not only whether the system was built right, but also whether the right system was built.

Scope *(SOP)*

The Scope section identifies the limitations of the Functional Requirements SOP. It limits the scope to all GxP systems and is applicable to all computers, computerized systems, infrastructure, LANs and WANs including clinical, laboratory, manufacturing and distribution computer systems. It applies to all Sites, Locations and Departments within the company.

Equipment / Materials / Tools *(SOP)*

This section lists any required or recommended equipment, materials, tools, templates or guides. Provide the name and a brief description of the equipment, instruments, materials and/or tools to be used in the procedures or testing covered by the Functional Requirements SOP. This is for reference purposes and requires the user to utilize the specific item or a designated equivalent. The purpose and use is explained elsewhere in the document. Substitution of comparable equipment may be made in accordance with procedures for Change Control.

If the information is included in a job aid or operating manual, a description of where the information may be found with a specific reference to the document number, title, date and revision number should be specified.

Warnings / Notes / General Information / Safety *(SOP)*

Warnings, notes, general or safety information applicable to the Functional Requirements SOP are listed in this section.

Policies, Guidelines, References *(SOP)*

This section lists references to appropriate regulatory requirements, SOPs, policies, guidelines, or other relevant documents referred to in the Functional Requirements SOP. The name and a brief description of applicable related SOPs, corporate policies, procedures and guidelines, Quality Control monographs, books, technical papers, publications, specific regulations, batch records, vendor or system manuals, and all other relevant documents that are required by the Functional Requirements SOP should be included. This is for reference purposes and requires the user to use the specific item or a designated equivalent. The purpose and use are explained elsewhere in the document.

SOPs cite references and incorporate them in whole by reference rather than specifying details that duplicate these items. This is done so that changes in policy, guidelines or reference materials require no changes to SOPs themselves and the most current version of the referenced material is used. Examples are:

- Policy ABC-123, Administrative Policy for Electronic Records; Electronic Signatures
- Corporate, Plan for Compliance with 21 CFR Part 11, Electronic Records; Electronic Signatures
- SOP-456, Electronic Records; Electronic Signatures
- FDA, 21 CFR Part 11, Electronic Records; Electronic Signatures

- FDA, General Principals of Software Validation; Final Guidance for Industry and FDA Staff, January 11, 2002

Assumptions / Exclusions / Limitations *(SOP)*

Any and all assumptions, exclusions or limitations within the context of the SOP should be discussed. Specific issues should be directly addressed in this section of the Functional Requirements SOP. For example:

"The means of implementation shall not be part of the description."

Glossary of Terms *(SOP)*

A Glossary of Terms should be included that defines all of the terms and acronyms that are used in this document or that are specific to the operation being described and may not be generally known to the reader. The reader may also be referred to the FDA document, "Glossary of Computerized System and Software Development Terminology."

Roles and Responsibilities *(SOP)*

The Roles and Responsibilities of the Functional Requirements are divided into two specific areas:

Creation and Maintenance

The System Owner of the computer or computerized system is responsible for the preparation and maintenance of the Functional Requirements.

It is common for the Functional Requirements to go through several revisions during the course of a project, as it is considered a living document, especially during the project's early stages. As a normal process, it helps to ensure that the validation activities are in compliance with cGxPs, FDA regulations, company policies and industry standards and guidelines.

The roles and responsibilities for preparing the Functional Requirements and other project documentation are addressed in accordance with established Standard Operating Procedures.

Use

Each of the members of the Project Team must be familiar with the Functional Requirements in order to fulfill the project requirements. The Quality Assurance (Unit) and Project Team use the Functional Requirements in conjunction with other project documentation as a contract that details the project and quality expectations.

The Functional Requirements is one of the governing documents for validation qualification and testing.

Procedure *(SOP)*

The SOP should contain instructions and procedures that enable the creation of a Functional Requirements protocol document. The Functional Requirements provide an overall description of the requirements of a software application, computer or computerized system that can be used by a procurement group to acquire the specified system or can be used by a system developer to create the Detail Design Specifications from which the system can be built.

Chapter 11 - Functional Requirements

The Functional Requirements are prepared as a stand-alone document for all projects. It is not good practice to include it as part of other documents in an attempt to reduce the amount of workload or paperwork. This only creates a need for more work over the length of the project resulting from misunderstandings and confusion. Clear, concise documents that address specific parts of any given project are necessary.

Purpose (protocol)

Clearly state the system with information as to why it is considered a regulated system and introducing planned activities in order to validate it in this section. This short introduction should be concise and present details so that an unfamiliar reader can understand the purpose of the document with regard to the system.

Scope (protocol)

The Scope section identifies the limitations of a specific Functional Requirements protocol in relationship to validating a specific software application, computer or computerized system. For example, if only a part of the computer system will be validated, then this section would be used to explain the reason for this decision.

Equipment / Materials / Tools (protocol)

This section lists any required or recommended equipment, materials, tools, templates or guides. Provide the name and a brief description of the equipment, instruments, materials and/or tools to be used in the procedures or testing covered by the Functional Requirements protocol. This is for reference purposes and requires the user to utilize the specific item or a designated equivalent. The purpose and use is explained elsewhere in the document. Substitution of comparable equipment may be made in accordance with procedures for Change Control.

If the information is included in a job aid or operating manual, a description of where the information may be found with a specific reference to the document number, title, date and revision number should be specified.

Warnings / Notes / General Information / Safety (protocol)

Warnings, notes, general or safety information applicable to the Functional Requirements protocol are listed in this section.

Policies / Guidelines / References (protocol)

This section lists references to appropriate regulatory requirements, SOPs, policies, guidelines, or other relevant documents referred to in the Functional Requirements protocol. The name and a brief description of applicable related SOPs, corporate policies, procedures and guidelines, Quality Control monographs, books, technical papers, publications, specific regulations, batch records, vendor or system manuals, and all other relevant documents that are required by the Functional Requirements protocol should be included. This is for reference purposes and requires the user to use the specific item or a designated equivalent. The purpose and use are explained elsewhere in the document.

Assumptions / Exclusions / Limitations (protocol)

Discuss any assumptions, exclusions or limitations within the context of the Functional Requirements. Specific issues should be directly addressed in this section of the Functional Requirements protocol.

Glossary of Terms (protocol)

The Glossary of Terms should define all of the terms and acronyms that are used in the document or that are specific to the operation being described and may not be generally known to the reader. The reader may also be referred to the FDA document, "Glossary of Computerized System and Software Development Terminology," or other specific glossaries that may be available.

Titles, terms, and definitions that are used in the corporate policies, procedures and guidelines should be defined.

Roles and Responsibilities (protocol)

Validation activities are the result of a cooperative effort involving many people, including the System Owners, End Users, the Computer Validation Group, Subject Matter Experts (SME) and Quality Assurance (Unit).

The roles, responsibilities and individual(s) responsible for the creation and use of the Functional Requirements should be clearly identified. A table should be included that specifies all of the participants by name, title, department or company affiliation and responsibility. Responsibilities should be specific for the procedures and activities delineated in each specific document.

The System Owners are responsible for the development and understanding of the Functional Requirements. They acquire and coordinate support from in-house and/or external consultants, the Computer Validation Group, Quality Assurance (Unit) and system developers.

The QA(U) is responsible for verifying that the details in the Functional Requirements are compliant with applicable departmental, interdepartmental and corporate policies and procedures and that the approach is in accordance with cGxP regulations and good business practices, especially 21 CFR Part 11.

Procedure - System Overview and Process (protocol)

This section describes and documents the operations, workflow, and expectations, as specified in the User Requirements, of a desired system in order to meet the business needs and requirements of the System Owner and End-User. This should reflect the End-User's existing and future procedures, workflows, and methods. This enables the System Developer or the Purchasing Department to fulfill their respective responsibilities in supporting the System Owner and End-User.

The requirements of the following areas should be clearly defined:

- Inputs
- Processes
- Outputs
- Constraints
- Workflow
- Hardware
- Firmware
- Software
- Data Structures
- Interfaces
- Performance
- Audit Trails

- Electronic Record Keeping
- Security
- Safety
- Maintenance

The Functional Requirements clearly defines what the system is to do and how it is to do it. It defines the functions to perform, the data on which to operate, equipment to operate and the operating environment. The required functionality, not the method of implementation, is indicated. Items such as time and cost constraints and deliverables to be supplied are also included.

The Functional Requirements should:

- Indicate whether functionality is mandatory/regulatory or desirable
- Be able to be tested
- Not duplicate one another
- Not contradict one another
- Not be design solutions

The Functional Requirements document is one of the governing documents in the process. Hardware, software, equipment, systems and/or vendors are selected after the approval of the Functional Requirements.

The Functional Requirements document is prepared as a stand-alone document for all projects.

Introduction (protocol)

The Introduction section contains an overview of the business operations for the system including workflow, material flow, and the control philosophy of what functions, calculations, and processes are to be performed by the system. It also describes interfaces that the system or application must be provided. The Introduction section outlines the activities required for commissioning the system.

Software Category (protocol)

This section lists the software category selected from that in an Overview of Computer and Computerized System Validation. Clearly state the category chosen and the reason for its choice as the software category determines the validation requirements.

The Software Category section lists the software category selection that is made from a group of categories in an Overview of Computer and Computerized System Validation. The category chosen and the reason for its choice are clearly stated, since the software category determines the validation requirements.

Operational Requirements / Critical Functions (protocol)

The Operational Requirements / CriticalFunctions section describes the operational requirements and critical functions that must be met in order for the system to function and be acceptable for use.

This section addresses automatic, manual, and maintenance modes, as well as operational interlock requirements of the system. These interlocks are hardware or software elements that inhibit or enable the use of features under specific conditions. They are used for data integrity, security or safety purposes.

This section provides detailed information, such as, circumstances that could directly impact the identity, strength, quality, or purity of a drug product, the operation of a medical device, data subject to regulatory review or the safety of individuals who operate the system. Critical items listed in the Functional Requirements document are summarized in this section.

If it is not feasible to separate the critical functions from the rest of the requirements, then they should be identified as such.

All requirements should be indicated as being classified as one of the following:

- Mandatory - must be included
- Desirable - may be excluded

Description of Operations

The Description of Operations section describes the system operation in its current state, be it manual or automated, and its desired future state. Both narrative and diagrammatic methods divided into modules should be used for large or complex systems. For small or simple systems, diagrams may be made optional, but they are highly recommended. Diagrams should be required to be accurate, concise and clearly labeled with emphasis placed on those portions of the system that are critical. The Description of Operations should conform to the User Requirements protocol.

Types of diagrams that could be included in this section are:

- Hardware Configuration
- Network
- As Built
- Workflow
- Data Flow

Business and Operating Units (protocol)

The Business and Operating Units section describes the business and operating units that are involved in the project:

- Business Unit / System Owner / End-User Representatives (each department affected)
- Management Representatives
- Quality Assurance (Unit)
- System Developer
- Interface Developer
- Vendors

The Products (protocol)

The Products section, when applicable, describes the products that are processed or controlled by the system, along with their specific requirements.

The Facilities (protocol)

The Facilities section describes the facilities that will use the system and specific attributes of the site (e.g., manufacturing sites, analytical labs with associated instrumentation or equipment, clinical trial sites, distribution services, medical devices).

Chapter 11 - Functional Requirements

Application Software Functional Requirements (protocol)

The Application Softwa re Functional Requirements section lists all of the functional requirements that constitute the system. Functional entities, such as algorithms, menus, screens, fields, printing, forms, reports, audit trails, electronic signatures, electronic records, storage and retrieval, system administration, regulatory, security, support, system are addressed.

This section lists the software application, if applicable, to be installed as part of the system. A description of the purpose along with the name of the manufacturer and version number is provided. The anticipated number of concurrent users and total clients is specified.

Sufficient information about the software application, computer or computerized system and its method of validation should be provided. If the software was previously qualified (i.e., system was installed on an existing server), then the associated qualification documents are listed in this section, along with a description of the qualification activities and the identification and location of the validation documentation.

The operating system that is to be used for the computer or computerized system requires specification, as well in this section.

Computer Hardware (protocol)

This section lists the primary hardware that is to be utilized by the system. If the hardware was previously qualified (i.e., system was installed on an existing server), then the associated qualification documents are listed with a description of the qualification activities and the identification and location of the validation documentation

Environmental, Mechanical, Electrical (protocol)

This section specifies as applicable, all of the environmental, mechanical and electrical requirements in which the hardware is expected to operate.

External Devices and Network Communications (protocol)

This section lists any external devices, such as interfaced instruments or manufacturing equipment and any Network Interfaces.

Data Conversion / Migration (protocol)

In computer systems where there is an upgrade to an existing system or the data is to be migrated and/or converted from a different computer system, the process and procedure for this migration is specified in this section. The details of how the data will be converted from the old to the new system and how it will be evaluated to ensure that the transferred data has maintained its integrity requires clear description in non-technical terms. In cases where data will originate from systems that are not validated, a requirement should be require to be specified to ensure the accuracy after migration.

Predicate rules must be followed to ensure proper storage and integrity of the old data.

Document Management and Storage (protocol)

A Document Management and Storage section is included in each Functional Requirements protocol. It defines the Document Management and Storage methodology that is followed for issue, revision, replacement, control and management of each Functional Requirements protocol. This methodology is in accordance with an established Change Control SOP and a Document Management and Storage SOP.

Revision History (protocol)

A Revision History section is included in each Functional Requirements. The Functional Requirements Version Number and all Revisions made to the document are recorded in this section. These changes are in accordance with a Change Control Procedure SOP.

The Revision History section consists of a summary of the changes, a brief statement explaining the reasons the document was revised, and the name and affiliation of the person responsible for the entries. It includes the numbers of all items the Functional Requirements replaces and is written so that the original Functional Requirements is not needed to understand the revisions or the revised document being issued.

When the Functional Requirements is a new document, a "000" as the Revision Number is entered with a statement such as "Initial Issue" in the comments section.

Regulatory or audit commitments require preservation from being inadvertently lost or modified in subsequent revisions. All entries or modifications made as a result of regulatory or audit commitments require indication as such in the Revision History and should indicate the organization, date and applicable references. A complete and accurate audit trail must be developed and maintained throughout the project lifecycle.

Attachments (protocol)

If needed, an Attachments section is included in each Functional Requirements protocol in accordance with the SOP on SOPs. An Attachment Index and Attachment section follow the text of a Functional Requirements protocol. The top of each attachment page indicates "Attachment # x," where 'x' is the sequential number of each attachment. The second line of each attachment page contains the attachment title. The number of pages and date of issue corresponding to each attachment are listed in the Attachment Index.

Attachments are placed directly after the Attachment Index and are sequentially numbered (e.g., Attachment # 1, Attachment # 2) with each page of an attachment containing the corresponding document number, page numbering, and total number of pages in the attachment (e.g., Page x of y). Each attachment is labeled consistent with the title of the attachment itself.

Attachments are treated as part of the Functional Requirements protocol. If an attachment needs to be changed, the entire Functional Requirements protocol requires a new revision number and is reissued. New attachments or revisions of existing attachments are in accordance with an established Change Control SOP and a Document Management and Storage SOP.

Document Management and Storage *(SOP)*

A Document Management and Storage section is included in the Functional Requirements SOP. It defines the Document Management and Storage methodology that is followed for issue, revision, replacement, control and management of this SOP. This methodology is in accordance with an established Change Control SOP and a Document Management and Storage SOP.

Revision History *(SOP)*

A Revision History section is included in the Functional Requirements SOP. The Functional Requirements SOP Version Number and all Revisions made to the document are recorded in this section. These changes are in accordance with a Change Control Procedure SOP.

The Revision History section consists of a summary of the changes, a brief statement explaining the reasons the document was revised, and the name and affiliation of the person responsible for the entries. It includes the numbers of all items the Functional Requirements SOP replaces and is written so that the original Functional Requirements SOP is not needed to understand the revisions or the revised document being issued.

When the Functional Requirements SOP is a new document, a "000" as the Revision Number is entered with a statement such as "Initial Issue" in the comments section.

Regulatory or audit commitments require preservation from being inadvertently lost or modified in subsequent revisions. All entries or modifications made as a result of regulatory or audit commitments require indication as such in the Revision History and should indicate the organization, date and applicable references. A complete and accurate audit trail must be developed and maintained throughout the project lifecycle.

Attachments *(SOP)*

If needed, an Attachments section is included in the Functional Requirements SOP in accordance with the SOP on SOPs. An Attachment Index and Attachment section follow the text of the Functional Requirements SOP. The top of each attachment page indicates "Attachment # x," where 'x' is the sequential number of each attachment. The second line of each attachment page contains the attachment title. The number of pages and date of issue corresponding to each attachment are listed in the Attachment Index.

Attachments are placed directly after the Attachment Index and are sequentially numbered (e.g., Attachment # 1, Attachment # 2) with each page of an attachment containing the corresponding document number, page numbering, and total number of pages in the attachment (e.g., Page x of y). Each attachment is labeled consistent with the title of the attachment itself.

Attachments are treated as part of the Functional Requirements SOP. If an attachment needs to be changed, the entire Functional Requirements SOP requires a new revision number and is reissued. New attachments or revisions of existing attachments are in accordance with an established Change Control SOP and a Document Management and Storage SOP.

Chapter 12 – Vendor Audit

In this chapter we will discuss the Standard Operating Procedure (SOP) for performing a Vendor Audit for a validation project. It specifies the requirements for auditing a vendor or contractor that is desired to provide goods or services for or in support of a Good Clinical, Laboratory, Manufacturing or Distribution Practices (GxP) project or on-going operation.

Best industry practices and FDA Guidance documents indicate that in order to demonstrate that a computer or computerized system performs as it purports and is intended to perform, vendors and contractors providing goods and/or services in a GxP environment must be qualified.

This SOP is used for all systems that create, modify, maintain, retrieve, transmit or archive records in electronic format. It is used in conjunction with an overlying corporate policy to provide sustainable compliance with the regulations.

This SOP is used prior to contracting goods and/or services by everyone involved in projects for a software application, computer or computerized system that controls equipment and/or produces electronic data or records in support of all regulated laboratory, manufacturing, clinical or distribution activities.

Title Block, Headers and Footers (SOP)

The Vendor Audit SOP should have a title block, title or cover page along with a header and/or footer on each page of the document that clearly identifies the document. The minimum information that should be displayed for the Vendor Audit SOP is as follows:

- Document Title
- Document Number
- Document Revision Number
- Project Name and Number
- Site/Location/Department Identification
- Date Document was issued, last revised or draft status
- Effective Date
- Pagination: page of total pages (e.g., Page x of y)

The Document Title should be in large type font that enables clear identification. It should clearly and concisely describe the contents and intent of the document. Since this document will be part of a set or package of documents within a project, it should be consistent with other documents used for the project and specify that it is the Vendor Audit SOP. This information should be on every page of the document.

The Document Number should be in large type font that enables clear identification. Since the Vendor Audit SOP will be part of a set or package of documents within a project, it should be consistent with other documents used for the project. This information should appear on every page of the document.

The Document Revision Number, which identifies the revision of the document, should appear on each page of the document. Many documents will go through one or more changes over time, particularly during the draft stage and the review and approval cycle. It is important to maintain accurate revision control because even a subtle change in a document can have an immense impact on the outcomes of activities. The SOP on SOPs should govern the numbering methodology.

See Chapter 6 – The SOP on SOPs.

The Project Name and Number, if applicable, should be indicated on the cover or title page. This information identifies the overall project to which this document belongs.

The Site/Location/Department Identification should be indicated on the cover or title page. It identifies the Site/Location/Department that has responsibility for this document and its associated project.

The Date that the Document was issued, last revised or issued with draft status should be on each page of the document (e.g., 12-Dec-01). This will clearly identify the date on which the document was last modified.

The Effective Date of the Document should be on the title or cover page of the document (e.g., 12-Dec-01). This will clearly identify when to begin using the current version of the document.

Pagination, page of total pages (e.g., Page x of y), should be on each page of the document. This will help identify whether or not all of the pages of the document are present. Pages intentionally left blank should clearly be designated as such.

Approval Section *(SOP)*

The Approval Section should contain a table with the Author and Approvers printed names, titles and a place to sign and date the SOP. The SOP on SOPs should govern who the approvers of the document should be. The functional department (e.g., Production / Quality Assurance / Engineering) of the signatory should be placed below each printed name and title along with a statement associated with each approver indicating the person's role or qualification in the approval process. Indicate the significance of the associated signature (e.g. This document meets the requirements of Corporate Policy 9055, Electronic Records; Electronic Signatures, dated 22-Mar-2000).

The completion (signing) of this section indicates that the contents have been reviewed and approved by the listed individuals.

Table of Contents *(SOP)*

The SOP should contain a Table of Contents, which appears towards the beginning of the document. This Table of Contents provides a way to easily locate various topics contained in the document.

Purpose *(SOP)*

The Purpose section should clearly state that the intention of the document is to provide a Standard Operating Procedure to be used to specify the requirements for the preparation of a Vendor Audit for a new or upgraded computer or computerized system along with the required activities and deliverables. These requirements are necessary for the Vendor Audit to be effective and compliant with any company polices and government regulation. The Vendor Audit SOP defines the scope of work, responsibilities and deliverables of a Vendor Audit Plan in order to demonstrate that a computer or computerized system meets its intended use in a reliable, consistent and compliant manner.

This introduction to the SOP concisely describes the document's purpose in sufficient detail so that a non-technical reader can understand its purpose.

Scope *(SOP)*

The Scope section identifies the limitations of the Vendor Audit SOP. It limits the scope to all GxP systems and is applicable to all computers, computerized systems, infrastructure, LANs and WANs including clinical, laboratory, manufacturing and distribution computer systems. It applies to all Sites, Locations and Departments within the company.

Equipment / Materials / Tools *(SOP)*

This section lists any required or recommended equipment, materials, tools, templates or guides. Provide the name and a brief description of the equipment, instruments, materials and/or tools to be used in the procedures or testing covered by the Vendor Audit SOP. This is for reference purposes and requires the user to utilize the specific item or a designated equivalent. The purpose and use is explained elsewhere in the document. Substitution of comparable equipment may be made in accordance with procedures for Change Control.

If the information is included in a job aid or operating manual, a description of where the information may be found with a specific reference to the document number, title, date and revision number should be specified.

Warnings / Notes / General Information / Safety *(SOP)*

Warnings, notes, general or safety information applicable to the Vendor Audit SOP are listed in this section.

Policies, Guidelines, References *(SOP)*

This section lists references to appropriate regulatory requirements, SOPs, policies, guidelines, or other relevant documents referred to in the Vendor Audit SOP. The name and a brief description of applicable related SOPs, corporate policies, procedures and guidelines, Quality Control monographs, books, technical papers, publications, specific regulations, batch records, vendor or system manuals, and all other relevant documents that are required by the Vendor Audit SOP should be included. This is for reference purposes and requires the user to use the specific item or a designated equivalent. The purpose and use are explained elsewhere in the document.

SOPs cite references and incorporate them in whole by reference rather than specifying details that duplicate these items. This is done so that changes in policy, guidelines or reference materials require no changes to SOPs themselves and the most current version of the referenced material is used. Examples are:

- Policy ABC-123, Administrative Policy for Electronic Records; Electronic Signatures
- Corporate, Plan for Compliance with 21 CFR Part 11, Electronic Records; Electronic Signatures
- SOP-456, Electronic Records; Electronic Signatures
- FDA, 21 CFR Part 11, Electronic Records; Electronic Signatures
- FDA, General Principals of Software Validation; Final Guidance for Industry and FDA Staff, January 11, 2002
- GAMP, Supplier Guide for Validation of Automated Systems in Pharmaceutical Manufacture

Assumptions / Exclusions / Limitations *(SOP)*

Any and all assumptions, exclusions or limitations within the context of the SOP should be discussed. Specific issues should be directly addressed in this section of the Vendor Audit SOP.

Glossary of Terms *(SOP)*

A Glossary of Terms should be included that defines all of the terms and acronyms that are used in this document or that are specific to the operation being described and may not be generally known to the reader.

The reader may also be referred to the FDA document, "Glossary of Computerized System and Software Development Terminology."

Roles and Responsibilities *(SOP)*

The Roles and Responsibilities of the Vendor Audit are divided into two specific areas:

Creation and Maintenance

The corporate purchasing department and QA(U) are responsible for the preparation and execution of all the steps included in the Vendor Audit process.

The roles and responsibilities for preparing the Vendor Audit and other project documentation are addressed in accordance with established Standard Operating Procedures.

Use

The members of the Project Team must be familiar with the Vendor Audit in order to fulfill the project requirements. The Quality Assurance (Unit) and Project Team use the Audit Plan to qualify prospective vendors and contractors.

Procedure *(SOP)*

The SOP should contain instructions and procedures that enable the creation of a Vendor Audit plan and the performance of an audit in accordance with its specifications. These include, but are not limited to, an overall description and requirements of the roles and responsibilities of both the vendor or contractor and the participants in its selection.

The Vendor Audit is prepared as a stand-alone document for all projects. It is not good practice to include it as part of other documents in an attempt to reduce the amount of workload or paperwork. This only causes more work over the length of the project resulting from misunderstandings and confusion. Clear, concise documents that address specific parts of any given project are desirable.

The GAMP guide "Supplier Guide for Validation of Automated Systems in Pharmaceutical Manufacture" is an excellent source of information for performing vendor and supplier audits. The Audit Team might use it in determining additional audit requirements. It may also be desirable to have the Vendor Audit SOP require its use.

Purpose (protocol)

Clearly state that the purpose of the audit is to satisfy requirements for the qualification of a particular vendor or contractor. This short introduction should be concise and present details so that an unfamiliar reader can understand the purpose of the document with regard to its use. It addresses the specific reason that this type of vendor or contractor is being sought.

Scope (protocol)

The Scope section identifies the limitations of a specific Vendor Audit in relationship to the purpose of the audit. The objective of an audit is to determine the ability of a vendor or contractor to fulfill a specific project requirement including financial capability, performance, delivery, and quality assurance standards.

Equipment / Materials / Tools (protocol)

This section lists any required or recommended equipment, materials, tools, templates or guides. Provide the name and a brief description of the equipment, instruments, materials and/or tools to be used in the procedures or testing covered by the Vendor Audit protocol. This is for reference purposes and requires the user to utilize the specific item or a designated equivalent. The purpose and use is explained elsewhere in the document. Substitution of comparable equipment may be made in accordance with procedures for Change Control.

If the information is included in a job aid or operating manual, a description of where the information may be found with a specific reference to the document number, title, date and revision number should be specified.

Warnings / Notes / General Information / Safety (protocol)

Warnings, notes, general or safety information applicable to the Vendor Audit protocol are listed in this section.

Policies / Guidelines / References (protocol)

This section lists references to appropriate regulatory requirements, SOPs, policies, guidelines, or other relevant documents referred to in the Vendor Audit protocol. The name and a brief description of applicable related SOPs, corporate policies, procedures and guidelines, Quality Control monographs, books, technical papers, publications, specific regulations, batch records, vendor or system manuals, and all other relevant documents that are required by the Vendor Audit protocol should be included. This is for reference purposes and requires the user to use the specific item or a designated equivalent. The purpose and use are explained elsewhere in the document.

Assumptions / Exclusions / Limitations (protocol)

Discuss any assumptions, exclusions or limitations within the context of the plan. Specific issues should be directly addressed in this section of the Vendor Audit protocol.

Glossary of Terms (protocol)

The Glossary of Terms should define all of the terms and acronyms that are used in the document or that are specific to the operation being described and may not be generally known to the reader. The reader may also be referred to the FDA document, "Glossary of Computerized System and Software Development Terminology," or other specific glossaries that may be available.

Titles, terms, and definitions that are used in the corporate policies, procedures and guidelines should be defined.

Roles and Responsibilities (protocol)

Validation activities are the result of a cooperative effort involving many people, including the System Owners, End Users, the Computer Validation Group, Subject Matter Experts (SME) and Quality Assurance (Unit). This section identifies the roles, responsibilities and individual(s) of the Audit Team responsible for the Vendor Audit. It specifies the participants by name, title, department or company affiliation and responsibility in a table. Responsibilities should be specific for the procedures and activities that are indicated in each specific document.

The company purchasing department and QA(U) are responsible for the development and understanding of the plan. They acquire and coordinate support from in-house and/or external consultants, the Computer Validation Group and Project Team members.

The QA(U) verifies that the details in the Vendor Audit are compliant with applicable departmental, interdepartmental and corporate policies and procedures and that the approach is in accordance with cGxP regulations and good business practices.

The System Owner, Project Manager or designee coordinates this effort. A Project Team is created and their respective expertise documented in the Roles and Responsibilities section of the Vendor Audit. To assist in the selection of the team members, a list of qualified individuals is made available from the Human Resources Department and a list of qualified contractors is provided by the Purchasing Department.

Procedure (protocol)

Introduction

Provide a brief explanation of the intended outcome of the audit and the vendor or contractor's anticipated relationship to the project.

Schedule Audit

The audit should be scheduled so that there is enough time for the team to perform the audit and for the vendor or contractor being audited to respond to questions and provide requested information. A minimum of one complete business day should be allocated for the performance of the audit.

Visit Confirmation

A letter confirming the scheduled visit for the purpose of performing an audit is sent at least 2 weeks in advance to the prospective vendor or supplier. This letter may be in the form of a formal paper document posted through the mail or it may be an electronic communication. The members of the Audit Team, the purpose and scope of the audit, the date(s), time(s), location and any other pertinent information are included in the confirmation letter. A copy of the agenda is usually enclosed with it, although it may be sent under separate cover.

Audit Agenda

An agenda is prepared well in advance and reviewed by everyone involved. This keeps the audit under control and within its scope. The Audit Agenda allows the vendor or supplier to properly prepare for the audit and provide the information required to the Audit Team in a timely and organized manner.

Audit Items

Name of Company Under Audit

The contact information, including the name, address at which the audit is to take place, telephone number, FAX number, and principal contact of the company to be audited, are specified.

Organizational Information

The business structure of the company and how it is organized are indicated. Organizational charts or listings of department managers and their responsibilities should be provided.

Organizational Capabilities

The functional capabilities or products supplied by the company that is being audited are evaluated

Personnel/Staffing

The skills and qualifications of the staff that will perform the work required, as well as the suitability of their training records and the effectiveness of the training itself, are evaluated.

Quality Assurance Program

The quality assurance practices and procedures, management participation and general Quality Assurance activities are evaluated.

Software Development Life Cycle

If applicable, the software development life cycle policies, practices and procedures surrounding software development are evaluated, with a focus on sustainable compliance with 21 CFR Part 11. Key items to consider are:

- Software Documentation
- Design Management
- Quality Assurance
- Coding Practices
- Testing at all levels
- Change Control and Configuration Management
- Release policies and procedures
- Changes, upgrades, enhancements and fixes
- Validation Maintenance, Periodic Review and Re-Validation
- Policies on obsolescence
- Support
- Escrow of code

Documentation Management

The Document Management and Storage methodology that is followed for issue, revision, replacement, control and management of documents is evaluated. This methodology should be in accordance with a Document Management and Storage SOP.

See Chapter 28 – Document Management and Storage for more information on requirements.

Code Management

If applicable, the software coding practices and management are evaluated. The following areas require major focus:

- Configuration Management and Change Control

- Conventions
- Deviation reporting, tracking and resolution
- Procedures
- Security
- Validation
- Virus protection

Testing

If applicable, evaluate the software testing practices and management. The following areas require major focus:

- 21 CFR Part 11
- Acceptance criteria
- Administration controls
- Alterations
- Boundary testing
- Error testing
- Regression testing
- Reports
- Review and Approval process
- Specified test cases
- Stress Testing
- Test procedures

Change Control

The overall policy and Standard Operating Procedures for Change Control are evaluated.

See Chapter 27 – *Change Control* for more information on requirements.

Security and Safety

The prospective vendor's or contractor's policies and procedures for security should be evaluated. The following areas of security require major focus:

- Archives
- Computer and computerized systems
- Documentation
- Physical location
- Software systems
- Source code

The site should be evaluated for safety, fire detection and suppression systems, OSHA compliance and local code compliance.

Chapter 12 – Vendor Audit

Audit Worksheets

Audit worksheets that include, all required, diagrams , checklists, tables, worksheets, and flowcharts are prepared and reviewed well in advance by the Audit Team. This ensures that all of the required information is obtained and properly documented.

Audit Report

The Audit Team issues an Audit Report to the Project Manager and the vendor or supplier. This report details the findings of the Audit Team with its recommendation of approval or denial. The Audit Report is issued in a timely manner to facilitate the project schedule.

Document Management and Storage (protocol)

A Document Management and Storage section is included in each Vendor Audit protocol. It defines the Document Management and Storage methodology that is followed for issue, revision, replacement, control and management of each Vendor Audit protocol. This methodology is in accordance with an established Change Control SOP and a Document Management and Storage SOP.

Revision History (protocol)

A Revision History section is included in each Vendor Audit. The Vendor Audit Version Number and all Revisions made to the document are recorded in this section. These changes are in accordance with a Change Control Procedure SOP.

The Revision History section consists of a summary of the changes, a brief statement explaining the reasons the document was revised, and the name and affiliation of the person responsible for the entries. It includes the numbers of all items the Vendor Audit replaces and is written so that the original Vendor Audit is not needed to understand the revisions or the revised document being issued.

When the Vendor Audit is a new document, a "000" as the Revision Number is entered with a statement such as "Initial Issue" in the comments section.

Regulatory or audit commitments require preservation from being inadvertently lost or modified in subsequent revisions. All entries or modifications made as a result of regulatory or audit commitments require indication as such in the Revision History and should indicate the organization, date and applicable references. A complete and accurate audit trail must be developed and maintained throughout the project lifecycle.

Attachments (protocol)

If needed, an Attachments section is included in each Vendor Audit protocol in accordance with the SOP on SOPs. An Attachment Index and Attachment section follow the text of a Vendor Audit protocol. The top of each attachment page indicates "Attachment # x," where 'x' is the sequential number of each attachment. The second line of each attachment page contains the attachment title. The number of pages and date of issue corresponding to each attachment are listed in the Attachment Index.

Attachments are placed directly after the Attachment Index and are sequentially numbered (e.g., Attachment # 1, Attachment # 2) with each page of an attachment containing the corresponding document number, page numbering, and total number of pages in the attachment (e.g., Page x of y). Each attachment is labeled consistent with the title of the attachment itself.

Attachments are treated as part of the Vendor Audit protocol. If an attachment needs to be changed, the entire Vendor Audit protocol requires a new revision number and is reissued. New attachments or revisions of existing attachments are in accordance with an established Change Control SOP and a Document Management and Storage SOP.

Document Management and Storage *(SOP)*

A Document Management and Storage section is included in the Vendor Audit SOP. It defines the Document Management and Storage methodology that is followed for issue, revision, replacement, control and management of this SOP. This methodology is in accordance with an established Change Control SOP and a Document Management and Storage SOP.

Revision History *(SOP)*

A Revision History section is included in the Vendor Audit SOP. The Vendor Audit SOP Version Number and all Revisions made to the document are recorded in this section. These changes are in accordance with a Change Control Procedure SOP.

The Revision History section consists of a summary of the changes, a brief statement explaining the reasons the document was revised, and the name and affiliation of the person responsible for the entries. It includes the numbers of all items the Vendor Audit SOP replaces and is written so that the original Vendor Audit SOP is not needed to understand the revisions or the revised document being issued.

When the Vendor Audit SOP is a new document, a "000" as the Revision Number is entered with a statement such as "Initial Issue" in the comments section.

Regulatory or audit commitments require preservation from being inadvertently lost or modified in subsequent revisions. All entries or modifications made as a result of regulatory or audit commitments require indication as such in the Revision History and should indicate the organization, date and applicable references. A complete and accurate audit trail must be developed and maintained throughout the project lifecycle.

Attachments *(SOP)*

If needed, an Attachments section is included in the Vendor Audit SOP in accordance with the SOP on SOPs. An Attachment Index and Attachment section follow the text of the Vendor Audit SOP. The top of each attachment page indicates "Attachment # x," where 'x' is the sequential number of each attachment. The second line of each attachment page contains the attachment title. The number of pages and date of issue corresponding to each attachment are listed in the Attachment Index.

Attachments are placed directly after the Attachment Index and are sequentially numbered (e.g., Attachment # 1, Attachment # 2) with each page of an attachment containing the corresponding document number, page numbering, and total number of pages in the attachment (e.g., Page x of y). Each attachment is labeled consistent with the title of the attachment itself.

Attachments are treated as part of the Vendor Audit SOP. If an attachment needs to be changed, the entire Vendor Audit SOP requires a new revision number and is reissued. New attachments or revisions of existing attachments are in accordance with an established Change Control SOP and a Document Management and Storage SOP.

Chapter 13 – Software Design Specification

In this chapter we will discuss the Standard Operating Procedure (SOP) for creating a Software Design Specification for a validation project. This SOP defines the method to be used for specifying the design that a system developer will use in creating a software application and/or system.

Best industry practices and FDA Guidance documents indicate that in order to demo nstrate that a computer or computerized system performs as it purports and is intended to perform, a methodical plan is needed that details how this may be proven. The Software Design Specification defines the specifications that are to be used in building the software and/or system specified in the User and Functional Requirements protocols.

This SOP is used for all systems that create, modify, maintain, retrieve, transmit or archive records in electronic format. It is used in conjunction with an overlying corporate policy to provide sustainable compliance with the regulations. The project team uses it to create the Software Design Specification, from which, the software application and/or system are created. It is the specification that is used to create the software application, computer and/or computerized system.

System Owners, End-Users, and Subject Matter Experts (SME) use this SOP to create the Software Design Specification protocol for a software application, computer or computerized system that controls equipment and/or produces electronic data or records in support of all regulated laboratory, manufacturing, clinical or distribution activities. The protocols produced in accordance with this SOP are used by procurement to acquire the specified items and by system developers to create the software application, computer or computerized system specified.

Title Block, Headers and Footers (SOP)

The Software Design Specification SOP should have a title block, title or cover page along with a header and/or footer on each page of the document that clearly identifies the document. The minimum information that should be displayed for the Software Design Specification SOP is as follows:

- Document Title
- Document Number
- Document Revision Number
- Project Name and Number
- Site/Location/Department Identification
- Date Document was issued, last revised or draft status
- Effective Date
- Pagination: page of total pages (e.g., Page x of y)

The Document Title should be in large type font that enables clear identification. It should clearly and concisely describe the contents and intent of the document. Since this document will be part of a set or package of documents within a project, it should be consistent with other documents used for the project and specify that it is the Software Design Specification SOP. This information should be on every page of the document.

The Document Number should be in large type font that enables clear identification. Since the Software Design Specification SOP will be part of a set or package of documents within a project, it should be consistent with other documents used for the project. This information should appear on every page of the document.

The Document Revision Number, which identifies the revision of the document, should appear on each page of the document. Many documents will go through one or more changes over time, particularly during the draft stage and the review and approval cycle. It is important to maintain accurate revision control because even a subtle change in a document can have an immense impact on the outcomes of activities. The SOP on SOPs should govern the numbering methodology.

See Chapter 6 – The SOP on SOPs.

The Project Name and Number, if applicable, should be indicated on the cover or title page. This information identifies the overall project to which this document belongs.

The Site/Location/Department Identification should be indicated on the cover or title page. It identifies the Site/Location/Department that has responsibility for this document and its associated project.

The Date that the Document was issued, last revised or issued with draft status should be on each page of the document (e.g., 12-Dec-01). This will clearly identify the date on which the document was last modified.

The Effective Date of the Document should be on the title or cover page of the document (e.g., 12-Dec-01). This will clearly identify when to begin using the current version of the document.

Pagination, page of total pages (e.g., Page x of y), should be on each page of the document. This will help identify whether or not all of the pages of the document are present. Pages intentionally left blank should clearly be designated as such.

Approval Section *(SOP)*

The Approval Section should contain a table with the Author and Approvers printed names, titles and a place to sign and date the SOP. The SOP on SOPs should govern who the approvers of the document should be. The functional department (e.g., Production / Quality Assurance / Engineering) of the signatory should be placed below each printed name and title along with a statement associated with each approver indicating the person's role or qualification in the approval process. Indicate the significance of the associated signature (e.g. This document meets the requirements of Corporate Policy 9055, Electronic Records; Electronic Signatures, dated 22-Mar-2000).

The completion (signing) of this section indicates that the contents have been reviewed and approved by the listed individuals.

Table of Contents *(SOP)*

The SOP should contain a Table of Contents, which appears towards the beginning of the document. This Table of Contents provides a way to easily locate various topics contained in the document.

Purpose *(SOP))*

This introduction to the SOP concisely describes the document's purpose in sufficient detail so that a non-technical reader can understand its purpose.

The Purpose section should clearly state that the intention of the document is to provide a Standard Operating Procedure to be used to specify the requirements for the preparation of Software Design Specification for a new or upgraded computer or computerized system (what the system is supposed to do) along with the required activities and deliverables. These requirements are necessary for the Software Design Specification to be effective and compliant with any company polices and government regulation. As one of the foundations of the software application, computer or computerized system, the Software Design Specification needs to define the scope of work, responsibilities and deliverables in order to create the software application, computer and/or computerized system and demonstrate that it meets its intended use in a reliable, consistent and compliant manner. The Software Design Specification document derived from this SOP will become the specification against which the system is created and tested.

Scope (SOP)

The Scope section should identify the limitations of the document with respect to the Software Design Specification. It should limit the scope to all GxP systems and is applicable to all computers, computerized systems, infrastructure, LANs and WANs including clinical, laboratory, manufacturing and distribution computer systems. It should apply to all Sites, Locations and Departments within the company.

Equipment / Materials / Tools (SOP)

This section lists any required or recommended equipment, materials, tools, templates or guides. Provide the name and a brief description of the equipment, instruments, materials and/or tools to be used in the procedures or testing covered by the Software Design Specification SOP. This is for reference purposes and requires the user to utilize the specific item or a designated equivalent. The purpose and use is explained elsewhere in the document. Substitution of comparable equipment may be made in accordance with procedures for Change Control.

If the information is included in a job aid or operating manual, a description of where the information may be found with a specific reference to the document number, title, date and revision number should be specified.

Warnings / Notes / General Information / Safety (SOP)

Warnings, notes, general or safety information applicable to the Software Design Specification SOP are listed in this section.

Policies, Guidelines, References (SOP)

This section lists references to appropriate regulatory requirements, SOPs, policies, guidelines, or other relevant documents referred to in the Software Design Specification SOP. The name and a brief description of applicable related SOPs, corporate policies, procedures and guidelines, Quality Control monographs, books, technical papers, publications, specific regulations, batch records, vendor or system manuals, and all other relevant documents that are required by the Software Design Specification SOP should be included. This is for reference purposes and requires the user to use the specific item or a designated equivalent. The purpose and use are explained elsewhere in the document.

SOPs cite references and incorporate them in whole by reference rather than specifying details that duplicate these items. This is done so that changes in policy, guidelines or reference materials require no changes to SOPs themselves and the most current version of the referenced material is used. Examples are:

- Policy ABC-123, Administrative Policy for Electronic Records; Electronic Signatures
- Corporate, Plan for Compliance with 21 CFR Part 11, Electronic Records; Electronic Signatures
- SOP-456, Electronic Records; Electronic Signatures
- FDA, 21 CFR Part 11, Electronic Records; Electronic Signatures
- FDA, General Principals of Software Validation; Final Guidance for Industry and FDA Staff, January 11, 2002

Assumptions / Exclusions / Limitations *(SOP)*

Any and all assumptions, exclusions or limitations within the context of the SOP should be discussed. Specific issues should be directly addressed in this section of the Software Design Specification SOP. For example:

"It is assumed that the means of implementation will be part of the specification."

Glossary of Terms *(SOP)*

A Glossary of Terms should be included that defines all of the terms and acronyms that are used in this document or that are specific to the operation being described and may not be generally known to the reader. The reader may als o be referred to the FDA document, "Glossary of Computerized System and Software Development Terminology."

Roles and Responsibilities *(SOP)*

The Roles and Responsibilities of the Software Design Specification should be divided into two specific areas:

Creation and Maintenance

The System Owner of the computer or computerized system is responsible for the preparation and maintenance of the Software Design Specification.

It is common for a Software Design Specification to go through several revisions during the course of a project, as it is considered a living document, especially during the project's early stages. As a normal process, it helps to ensure that the validation activities are in compliance with cGxPs, FDA regulations, company policies and industry standards and guidelines.

The roles and responsibilities for preparing the Software Design Specification protocol and other project documentation are addressed in accordance with established Standard Operating Procedures.

Use

The members of the Project Team must be familiar with the Software Design Specification in order to fulfill the project requirements. The Quality Assurance (Unit) and Project Team use the Software Design Specification in concert with other project documentation as a contract that details the project and quality expectations.

The Software Design Specification is one of the governing documents for validation qualification and testing.

Procedure *(SOP)*

The SOP should contain instructions and procedures that enable the creation of a Software Design Specification protocol document. The Software Design Specification should be required to provide the specifications for the creation of the software applicatio n, computer and/or computerized system that can be used by a procurement group to acquire the services needed to create the specified system or by a system developer to create the actual system.

The Software Design Specification should be prepared as a stand-alone document for all projects. It is not good practice to include it as part of other documents in an attempt to reduce the amount of workload or paperwork. This only causes more work over the length of the project resulting from misunderstandings and confusion. Clear, concise documents that address specific parts of any given project are desirable.

Purpose (protocol)

This section should clearly state the system with information as to why it is considered a regulated system and introducing planned activities in order to create it. This short introduction should be concise and present details so that an unfamiliar reader can understand the purpose of the document with regard to the system.

Scope (protocol)

The Scope section identifies the limitations of a specific Software Design Specification protocol in relationship to validating a specific software application, computer or computerized system. For example, if part of the system consists of a Commercial Off The Shelf software application and another part consists of manufacturing or laboratory equipment, then the part of the computerized system that is equipment specific may be excluded by stating that it is being addressed elsewhere.

Equipment / Materials / Tools (protocol)

This section lists any required or recommended equipment, materials, tools, templates or guides. Provide the name and a brief description of the equipment, instruments, materials and/or tools to be used in the procedures or testing covered by the Software Design Specification. This is for reference purposes and requires the user to utilize the specific item or a designated equivalent. The purpose and use is explained elsewhere in the document. Substitution of comparable equipment may be made in accordance with procedures for Change Control.

If the information is included in a job aid or operating manual, a description of where the information may be found with a specific reference to the document number, title, date and revision number should be specified.

Warnings / Notes / General Information / Safety (protocol)

Warnings, notes, general or safety information applicable to the Software Design Specification protocol are listed in this section.

Policies / Guidelines / References (protocol)

This section lists references to appropriate regulatory requirements, SOPs, policies, guidelines, or other relevant documents referred to in the Software Design Specification. The name and a brief description of applicable related SOPs, corporate policies, procedures and guidelines, Quality Control monographs, books, technical papers, publications, specific regulations, batch records, vendor or system manuals, and all other relevant documents that are required by the Software Design Specification should be included. This is for reference purposes and requires the user to use the specific item or a designated equivalent. The purpose and use are explained elsewhere in the document.

Assumptions / Exclusions / Limitations (protocol)

Discuss any assumptions, exclusions or limitations within the context of the Software Design Specification Specific issues should be directly addressed in this section of the Software Design Specification.

Glossary of Terms (protocol)

The Glossary of Terms should define all of the terms and acronyms that are used in the document or that are specific to the operation being described and may not be generally known to the reader. The reader may also be referred to the FDA document, "Glossary of Computerized System and Software Development Terminology," or other specific glossaries that may be available.

Titles, terms, and definitions that are used in the corporate policies, procedures and guidelines should be defined.

Roles and Responsibilities (protocol)

Validation activities are the result of a cooperative effort involving many people, including the System Owners, End Users, the Computer Validation Group, Subject Matter Experts (SME) and Quality Assurance (Unit).

Clearly identify the roles, responsibilities, and individual(s) responsible for the creation and use of the Software Design Specification. Specify the participants by name, title, department or company affiliation and responsibility in a table. Responsibilities should be specific for the procedures and activities delineated in each specific document.

The System Owner is responsible for the development and understanding of the Software Design Specification. They acquire and coordinate support from in-house and/or external consultants, the Computer Validation Group, Quality Assurance (Unit) and system developers.

The QA(U) is responsible to verify that the details in the Software Design Specification are compliant with applicable departmental, interdepartmental and corporate policies and procedures and that the approach is in accordance with cGxP regulations and good business practices, especially 21 CFR Part 11.

The System Developers are responsible for understanding the Computer System Design Specifications document and working with the System Owner/End-Users to develop a Design Specification document for the system that will satisfy the business needs as indicated in the Functional Requirements.

Procedure - System Overview and Process (protocol)

This section describes and documents the operations, workflow, and expectations, as specified in the User Requirements and Functional Requirements, of a desired system in order to meet the business needs and requirements of the System Owner and End-User. It reflects the End-User's existing and future procedures, workflows, and methods. This enables the System Developer to fulfill the required responsibilities in supporting the System Owner and End-User.

The requirements of the following areas should be clearly defined:

- Inputs
- Firmware
- Electronic Record Keeping
- Processes
- Software
- Security
- Outputs
- Data Structures

- Safety
- Constraints
- Interfaces
- Maintenance
- Workflow
- Performance
- Hardware

Audit Trails

The Software Design Specification clearly defines what the system is to do and how it is to do it. It defines the functions to perform, the data on which to operate, equipment to operate, and the operating environment. The required functionality and the method of implementation is indicated.

The Software Design Specification should:

- Indicate whether mandatory/regulatory or desirable
- Be able to be tested
- Not duplicate one another
- Not contradict one another

Hardware, software, equipment, systems, and/or vendors are selected after the approval of the Software Design Specification, using it as one of the governing documents in the process.

The Software Design Specification document is prepared as a stand-alone document for all projects.

Introduction (protocol)

This section contains an overview of the business operations for the system including workflow, material flow, and the control philosophy of what functions, calculations, and processes are to be performed by the system, as well as interfaces that the system or application must be provided. It outlines the activities required for commissioning the system.

Software Category (protocol)

This section contains the software category that is selected from an Overview of Computer and Computerized System Validation. Clearly state the category chosen and the reason for its choice as the software category determines the validation requirements.

Operational Requirements / Critical Functions (protocol)

This section describes the operational requirements and critical functions that must be met in order for the system to function and be acceptable for use.

It addresses automatic, manual, and maintenance modes, and operational interlock requirements of the system. Interlocks are hard ware or software elements that inhibit or enable the use of features under specific conditions. They are used for data integrity, security or safety purposes.

It provides detailed information, such as circumstances that could directly impact the identity, strength, quality, or purity of a drug product, the operation of a medical device, data subject to regulatory review or the safety of individuals who operate the system. The critical items listed in the Software Design Specification document are summarized in this section.

If it is not feasible to separate the critical functions from the rest of the requirements, then they should be identified as such.

All requirements should be indicated as being classified as one of the following:

- Mandatory - must be included
- Desirable - may be excluded.

Description of Operations

This section describes the system operation in its current state, be it manual or automated, and its desired future state. Both narrative and diagrammatic methods divided into modules should be used for large or complex systems. For small or simple systems, diagrams may be made optional, but they are highly recommended. Diagrams should be accurate, concise and clearly labeled with emphasis placed on those portions of the system that are critical. It should conform to the User Requirements protocol.

Types of diagrams that could be used are:

- Hardware Configuration
- Network
- As Built
- Workflow
- Data Flow

Business and Operating Units (protocol)

This section describes the business and operating units involved in the project:

- Business Unit / System Owner / End-User Representatives (each department affected)
- Management Representatives
- Quality Assurance (Unit)
- System Developer
- Interface Developer
- Vendors

The Products (protocol)

This section, when applicable, describes the products processed or controlled by the system with their specific requirements.

The Facilities (protocol)

This section describes the facilities that will use the system and specific attributes of the site (e.g., manufacturing sites, analytical labs with associated instrumentation or equipment, clinical trial sites, distribution services, medical devices).

Chapter 13 – Software Design Specification

Development Methods

This section describes the method used for this design. Where necessary, include a reference to published methods used. Explain all methods considered and why they were or were not chosen for use.

Architectural and Infrastructural Strategies

This section describes the design methods and decisions affecting the overall system. The strategies should provide insight into the key areas and methods used in the architecture of the system. Indicate the logic used to decide how each key area is handled. Include an explanation of compromised areas and why they had to be compromised.

At a minimum, consider the following areas:

- Communication
- Databases (internal and external)
- Products
- Compilers
- Error processing
- Programming languages
- Concurrency
- Error trapping
- Resource Management
- Control
- Future enhancement
- Software
- Data control
- Interfaces (hardware and software)
- Synchronization
- Data management
- Libraries
- Data storage
- Memory management

System Architecture

This section provides an overview of the system composition. Include components and modules, explaining how they are organized and function in achieving fulfillment of the System Owner and End-User's business needs and requirements.

Include diagrams, flowcharts, scenarios and use-cases of the system behavior and structure.

Subsystem Architecture

Where appropriate for clarity, an overview of the subsystem composition should be provided.

Platform Architecture

This section provides an overview of the platform composition. Include components and modules, explaining how they are organized and function in achieving fulfillment of the System Owner and End-User's business requirements.

Include diagrams , flowcharts, scenarios and use-cases of the system behavior and structure.

System Utilization Analysis

This section provides an overview of the system utilization. Explain what system components and services will be utilized in the operation of the system.

System Environments

This section provides an overview of the system environment. Explain what environments the system requires and how they will be used.

Outage and Recovery

This section specifies the requirements for compliance with corporate policies and procedures, SOPs, cGxPs, 21 CFR Part 11 and applicable statutes and regulations along with the User and Functional Requirements.

Mandatory and desired items, as indicated in the User and Functional Requirements require individual specification and identification.

Specify the details of the automatic, manual, maintenance, interlock and security mode requirements of the system.

Identify items that may impact the identity, strength, quality, or purity of a drug product, medical devices and data subject to regulatory review or operator safety.

Identify application software, if applicable, used as part of the system. Provide a description of the purpose along with the name of the manufacturer and version number.

Indicate the anticipated number of concurrent users and total clients of the system.

This section should also provide information on qualification/validation. If the software was previously qualified (i.e., system was installed on an existing server), then the associated qualification documentation is referenced with a brief description of the qualification activities.

The components described in the System Architecture section above require a detailed discussion. Other, lower-level components and sub-components should also be described. At a minimum, the following software component attributes should require description for each entity specified:

Detailed System Design

Associated Functional Requirement

This section identifies the functional requirement fulfilled in the Functional Requirements protocol.

Purpose / Scope

This section describes the purpose and scope of the component referencing specific items in the Functional Requirements protocol. It should describe the functions and/or activities of the component – what it is required to accomplish; role(s) it performs; and services it provides to its clients.

Chapter 13 – Software Design Specification

Classification

This section describes the classification of the component with an explanation of the classification. For example: subsystem, module, class, package, function, form, macro, interface, library or file.

Assumptions / Exclusions / Limitations

This section describes any assumptions, exclusions and/or limitations of the component. It should include such information as:

- Component state
- Constraints on values
- Data access
- Data formats
- Exceptions
- Invariants
- Local or global values
- Post conditions
- Preconditions
- Security
- Storage
- Synchronization
- Timing

Composition

This section describes any sub-components that are a part of this component.

Uses / Interactions

This section lists the components and/or interfaces with which this component interacts and describes the interaction. Describe the effects it has on the system as a whole. Object-oriented designs should include a description of sub-classes, super-classes, and meta-classes.

Resources

This section lists and describes the resources that are managed, affected and/or required by this component. For example: memory, processors, printers, databases and software libraries. This should include a discussion of resource contention issues and their resolution.

Processing

This section describes how the component functions to meet its requirements. For example, the services that it provides (e.g., resources, data, types, constants, subroutines and exceptions). It should provide a description of items such as:

- Actions
- Exit Action
- Post Close
- Algorithms
- Field Title
- Post Open
- Buttons
- Fields
- Pull-Downs
- Change Modes
- Forms
- Query Close
- Cleanup
- Initialization
- Query Open
- Creation Methods
- Input Translation
- Query Save
- Data Type
- Input Validation
- Recalculation
- Dialog Boxes
- Message Boxes
- State Changes
- Entry Action
- Mouse Moves
- Termination
- Exception Handling
- Population

Interfaces / Imports / Exports

This section describes user, software, hardware, and communications interfaces utilized in the performance of the requirements. Provide sufficient detail that the developer can effectively create the required interaction.

It also describes import and export functions in sufficient detail that the developer can effectively create the interaction required.

The definition or declaration of each entity are specified, along with a description of the meanings of values with their parameters

Detailed Subsystem Design

In addition to the information in the Detailed System Design section, this section contains a detailed description of this subsystem component. Include diagrams showing the details of subsystem component structure, behavior, information flow and control.

Computer Hardware , Peripherals, Operating Systems and Network Operating Systems.

This section lists primary hardware that is utilized by the computer system, referencing the associated Hardware Design Specification.

Hardware, Peripherals, Operating Systems and/or Network Operating Systems design specifications must be in accordance with Data Center SOPs and are covered in separate documentation.

Document Management and Storage (protocol)

A Document Management and Storage section is included in each Software Design Specification. It defines the Document Management and Storage methodology that is followed for issue, revision, replacement, control and management of each Software Design Specification. This methodology is in accordance with an established Change Control SOP and a Document Management and Storage SOP.

Revision History (protocol)

A Revision History section is included in each Software Design Specification. The Software Design Specification Version Number and all Revisions made to the document are recorded in this section. These changes are in accordance with a Change Control Procedure SOP.

The Revision History section consists of a summary of the changes, a brief statement explaining the reasons the document was revised, and the name and affiliation of the person responsible for the entries. It includes the numbers of all items the Software Design Specification replaces and is written so that the original Software Design Specification is not needed to understand the revisions or the revised document being issued.

When the Software Design Specification is a new document, a "000" as the Revision Number is entered with a statement such as "Initial Issue" in the comments section.

Regulatory or audit commitments require preservation from being inadvertently lost or modified in subsequent revisions. All entries or modifications made as a result of regulatory or audit commitments require indication as such in the Revision History and should indicate the organization, date and applicable references. A complete and accurate audit trail must be developed and maintained throughout the project lifecycle.

Attachments (protocol)

If needed, an Attachments section is included in each Software Design Specification protocol in accordance with the SOP on SOPs. An Attachment Index and Attachment section follow the text of a Software Design Specification protocol. The top of each attachment page indicates "Attachment # x," where 'x' is the sequential number of each attachment. The second line of each attachment page contains the attachment title. The number of pages and date of issue corresponding to each attachment are listed in the Attachment Index.

Attachments are placed directly after the Attachment Index and are sequentially numbered (e.g., Attachment # 1, Attachment # 2) with each page of an attachment containing the corresponding document number, page numbering, and total number of pages in the attachment (e.g., Page x of y). Each attachment is labeled consistent with the title of the attachment itself.

Attachments are treated as part of the Software Design Specification protocol. If an attachment needs to be changed, the entire Software Design Specification protocol requires a new revision number and is reissued. New attachments or revisions of existing attachments are in accordance with an established Change Control SOP and a Document Management and Storage SOP.

Document Management and Storage (SOP)

A Document Management and Storage section is included in the Software Design Specification SOP. It defines the Document Management and Storage methodology that is followed for issue, revision, replacement, control and management of this SOP. This methodology is in accordance with an established Change Control SOP and a Document Management and Storage SOP.

Revision History *(SOP)*

A Revision History section is included in the Software Design Specification SOP. The Software Design Specification SOP Version Number and all Revisions made to the document are recorded in this section. These changes are in accordance with a Change Control Procedure SOP.

The Revision History section consists of a summary of the changes, a brief statement explaining the reasons the document was revised, and the name and affiliation of the person responsible for the entries. It includes the numbers of all items the Software Design Specification SOP replaces and is written so that the original Software Design Specification SOP is not needed to understand the revisions or the revised document being issued.

When the Software Design Specification SOP is a new document, a "000" as the Revision Number is entered with a statement such as "Initial Issue" in the comments section.

Regulatory or audit commitments require preservation from being inadvertently lost or modified in subsequent revisions. All entries or modifications made as a result of regulatory or audit commitments require indication as such in the Revision History and should indicate the organization, date and applicable references. A complete and accurate audit trail must be developed and maintained throughout the project lifecycle.

Attachments *(SOP)*

If needed, an Attachments section is included in the Software Design Specification SOP in accordance with the SOP on SOPs. An Attachment Index and Attachment section follow the text of the Software Design Specification SOP. The top of each attachment page indicates "Attachment # x," where 'x' is the sequential number of each attachment. The second line of each attachment page contains the attachment title. The number of pages and date of issue corresponding to each attachment are listed in the Attachment Index.

Attachments are placed directly after the Attachment Index and are sequentially numbered (e.g., Attachment # 1, Attachment # 2) with each page of an attachment containing the corresponding document number, page numbering, and total number of pages in the attachment (e.g., Page x of y). Each attachment is labeled consistent with the title of the attachment itself.

Attachments are treated as part of the Software Design Specification SOP. If an attachment needs to be changed, the entire Software Design Specification SOP requires a new revision number and is reissued. New attachments or revisions of existing attachments are in accordance with an established Change Control SOP and a Document Management and Storage SOP.

Chapter 14 – Hardware Design Specification

In this chapter we will discuss the Standard Operating Procedure (SOP) for creating a Hardware Design Specification for a validation project. This SOP defines the method to be used for specifying the design that a computer operations and a procurement group will use in acquiring, configuring and implementing computer and computer related hardware.

Best industry practices and FDA Guidance documents indicate that in order to demonstrate that a computer or computerized system performs as it purports and is intended to perform, a methodical plan is needed that details how this may be proven. The Hardware Design Specification defines the specifications that are to be used in acquiring, configuring and implementing computer and computer related hardware in fulfillment of the requirements of the User and Functional Requirements protocols.

This SOP is used for all systems that create, modify, maintain, retrieve, transmit or archive records in electronic format. It is used in conjunction with an overlying corporate policy to provide sustainable compliance with the regulation. The project team uses it to create the Hardware Design Specification, from which, the computer operations and procurement group acquire, configure and implement it. It is the specification that is used to define a piece of computer or computer related hardware.

System Owners, End-Users, and Subject Matter Experts (SME) use this SOP to create the Hardware Design Specification protocol for computer or computer related hardware that controls equipment and/or produces electronic data or records in support of all regulated laboratory, manufacturing, clinical or distribution activities. The protocols produced in accordance with this SOP are used by the procurement group to acquire the specified items and by a computer operations group to configure and implement them.

Title Block, Headers and Footers (SOP)

The Hardware Design Specification SOP should have a title block, title or cover page along with a header and/or footer on each page of the document that clearly identifies the document. The minimum information that should be displayed for the Hardware Design Specification SOP is as follows:

- Document Title
- Document Number
- Document Revision Number
- Project Name and Number
- Site/Location/Department Identification
- Date Document was issued, last revised or draft status
- Effective Date
- Pagination: page of total pages (e.g., Page x of y)

The Document Title should be in large type font that enables clear identification. It should clearly and concisely describe the contents and intent of the document. Since this document will be part of a set or package of documents within a project, it should be consistent with other documents used for the project and specify that it is the Hardware Design Specification SOP. This information should be on every page of the document.

The Document Number should be in large type font that enables clear identification. Since the Hardware Design Specification SOP will be part of a set or package of documents within a project, it should be consistent with other documents used for the project. This information should appear on every page of the document.

The Document Revision Number, which identifies the revision of the document, should appear on each page of the document. Many documents will go through one or more changes over time, particularly during the draft stage and the review and approval cycle. It is important to maintain accurate revision control because even a subtle change in a document can have an immense impact on the outcomes of activities. The SOP on SOPs should govern the numbering methodology.

See Chapter 6 – The SOP on SOPs.

The Project Name and Number, if applicable, should be indicated on the cover or title page. This information identifies the overall project to which this document belongs.

The Site/Location/Department Identification should be indicated on the cover or title page. It identifies the Site/Location/Department that has responsibility for this document and its associated project.

The Date that the Document was issued, last revised or issued with draft status should be on each page of the document (e.g., 12-Dec-01). This will clearly identify the date on which the document was last modified.

The Effective Date of the Document should be on the title or cover page of the document (e.g., 12-Dec-01). This will clearly identify when to begin using the current version of the document.

Pagination, page of total pages (e.g., Page x of y), should be on each page of the document. This will help identify whether or not all of the pages of the document are present. Pages intentionally left blank should clearly be designated as such.

Approval Section *(SOP)*

The Approval Section should contain a table with the Author and Approvers printed names, titles and a place to sign and date the SOP. The SOP on SOPs should govern who the approvers of the document should be. The functional department (e.g., Production / Quality Assurance / Engineering) of the signatory should be placed below each printed name and title along with a statement associated with each approver indicating the person's role or qualification in the approval process. Indicate the significance of the associated signature (e.g. This document meets the requirements of Corporate Policy 9055, Electronic Records; Electronic Signatures, dated 22-Mar-2000).

The completion (signing) of this section indicates that the contents have been reviewed and approved by the listed individuals.

Table of Contents *(SOP)*

The SOP should contain a Table of Contents, which appears towards the beginning of the document. This Table of Contents provides a way to easily locate various topics contained in the document.

Purpose *(SOP)*

This introduction to the SOP concisely describes the document's purpose in sufficient detail so that a non-technical reader can understand its purpose.

The Purpose section should clearly state that the intention of the document is to provide a Standard Operating Procedure to be used to specify the requirements for the preparation of Hardware Design Specification for a new or upgraded computer or computerized system (what the system is supposed to do) along with the required activities and deliverables. These requirements are necessary for the Hardware Design Specification to be effective and compliant with any company polices and government regulation. As one of the foundations of the software application, computer or computerized system, the Hardware Design Specification needs to define the scope of work, responsibilities and deliverables in order to create the software application, computer and/or computerized system and demonstrate that it meets its intended use in a reliable, consistent and compliant manner. The Hardware Design Specification document derived from this SOP will become the specification against which the system is created and tested.

Scope *(SOP)*

The Scope section identifies the limitations of the Hardware Design Specification SOP. It limits the scope to all GxP systems and is applicable to all computers, computerized systems, infrastructure, LANs and WANs including clinical, laboratory, manufacturing and distribution computer systems. It applies to all Sites, Locations and Departments within the company.

Equipment / Materials / Tools *(SOP)*

This section lists any required or recommended equipment, materials, tools, templates or guides. Provide the name and a brief description of the equipment, instruments, materials and/or tools to be used in the procedures or testing covered by the Hardware Design Specification SOP. This is for reference purposes and requires the user to utilize the specific item or a designated equivalent. The purpose and use is explained elsewhere in the document. Substitution of comparable equipment may be made in accordance with procedures for Change Control.

If the information is included in a job aid or operating manual, a description of where the information may be found with a specific reference to the document number, title, date and revision number should be specified.

Warnings / Notes / General Information / Safety *(SOP)*

Warnings, notes, general or safety information applicable to the Hardware Design Specification SOP are listed in this section.

Policies, Guidelines, References *(SOP)*

This section lists references to appropriate regulatory requirements, SOPs, policies, guidelines, or other relevant documents referred to in the Hardware Design Specification SOP. The name and a brief description of applicable related SOPs, corporate policies, procedures and guidelines, Quality Control monographs, books, technical papers, publications, specific regulations, batch records, vendor or system manuals, and all other relevant documents that are required by the Hardware Design Specification SOP should be included. This is for reference purposes and requires the user to use the specific item or a designated equivalent. The purpose and use are explained elsewhere in the document.

SOPs cite references and incorporate them in whole by reference rather than specifying details that duplicate these items. This is done so that changes in policy, guidelines or reference materials require no changes to SOPs themselves and the most current version of the referenced material is used. Examples are:

- Policy ABC-123, Administrative Policy for Electronic Records; Electronic Signatures
- Corporate, Plan for Compliance with 21 CFR Part 11, Electronic Records; Electronic Signatures
- SOP-456, Electronic Records; Electronic Signatures
- FDA, 21 CFR Part 11, Electronic Records; Electronic Signatures
- FDA, General Principals of Software Validation; Final Guidance for Industry and FDA Staff, January 11, 2002

Assumptions / Exclusions / Limitations *(SOP)*

Any and all assumptions, exclusions or limitations within the context of the SOP should be discussed. Specific issues should be directly addressed in this section of the Hardware Design Specification SOP. For example:

"It is assumed that the means of implementation will be part of the specification."

Glossary of Terms *(SOP)*

A Glossary of Terms should be included that defines all of the terms and acronyms that are used in this document or that are specific to the operation being described and may not be generally known to the reader. The reader may also be referred to the FDA document, "Glossary of Computerized System and Software Development Terminology."

Roles and Responsibilities *(SOP)*

The Roles and Responsibilities of the Hardware Design Specification should be divided into two specific areas:

Creation and Maintenance

The System Owner of the computer or computerized system and Computer Operations is responsible for the preparation and maintenance of the Hardware Design Specification.

It is common for a Hardware Design Specification to go through several revisions during the course of a project, as it is considered a living document, especially during the project's early stages. As a normal process, it helps to ensure that the validation activities are in compliance with cGxPs, FDA regulations, company policies and industry standards and guidelines.

The roles and responsibilities for preparing the Hardware Design Specification protocol and other project documentation are addressed in accordance with established Standard Operating Procedures.

Use

The members of the Project Team must be familiar with the Hardware Design Specification in order to fulfill the project requirements. The Quality Assurance (Unit), Project Team and Computer Operations use the Hardware Design Specification in concert with other project documentation as a contract that details the project and quality expectations.

The Hardware Design Specification is one of the governing documents for the qualification and testing of the hardware it specifies.

Procedure *(SOP)*

The SOP should contain instructions and procedures that enable the creation of a Hardware Design Specification protocol document. The Hardware Design Specification is required in order to provide the specifications for the required hardware that can be used by a procurement group to acquire the equipment.

The Hardware Design Specification is prepared as a stand-alone document for all projects. It is not good practice to include it as part of other documents in an attempt to reduce the amount of workload or paperwork.

This only causes more work over the length of the project resulting from misunderstandings and confusion. Clear, concise documents that address specific parts of any given project are desirable.

Purpose (protocol)

Clearly state the equipment with information as to why it is considered a regulated item and introducing planned activities in order to acquire it. This short introduction should be concise and present details so that an unfamiliar reader can understand the purpose of the document with regard to the equipment.

Scope (protocol)

The Scope section identifies the limitations of a specific Hardware Design Specification protocol in relationship to validating a specific software application, computer or computerized system. For example, if part of the hardware consists of standard catalog items and another part consists of manufacturing or laboratory equipment, then the part of the hardware that is a standard catalog item may be specifically excluded.

Equipment / Materials / Tools (protocol)

This section lists any required or recommended equipment, materials, tools, templates or guides. Provide the name and a brief description of the equipment, instruments, materials and/or tools to be used in the procedures or testing covered by the Hardware Design Specification. This is for reference purposes and requires the user to utilize the specific item or a designated equivalent. The purpose and use is explained elsewhere in the document. Substitution of comparable equipment may be made in accordance with procedures for Change Control.

If the information is included in a job aid or operating manual, a description of where the information may be found with a specific reference to the document number, title, date and revision number should be specified.

Warnings / Notes / General Information / Safety (protocol)

Warnings, notes, general or safety information applicable to the Hardware Design Specification protocol are listed in this section.

Policies / Guidelines / References (protocol)

This section lists references to appropriate regulatory requirements, SOPs, policies, guidelines, or other relevant documents referred to in the Hardware Design Specification. The name and a brief description of applicable related SOPs, corporate policies, procedures and guidelines, Quality Control monographs, books, technical papers, publications, specific regulations, batch records, vendor or system manuals, and all other relevant documents that are required by the SOP should be included. This is for reference purposes and requires the user to use the specific item or a designated equivalent. The purpose and use are explained elsewhere in the document.

Assumptions / Exclusions / Limitations (protocol)

Discuss any assumptions, exclusions or limitations within the context of the Hardware Design Specification. Specific issues should be directly addressed in this section of the Hardware Design Specification.

Glossary of Terms (protocol)

The Glossary of Terms should define all of the terms and acronyms that are used in the document or that are specific to the operation being described and may not be generally known to the reader. The reader may also be referred to the FDA document, "Glossary of Computerized System and Software Development Terminology," or other specific glossaries that may be available.

Titles, terms, and definitions that are used in the corporate policies, procedures and guidelines should be defined.

Roles and Responsibilities (protocol)

Validation activities are the result of a cooperative effort involving many people, including the System Owners, End Users, the Computer Validation Group, Subject Matter Experts (SME) and Quality Assurance (Unit).

This section requires clear identification of the roles, responsibilities and individual(s) responsible for the creation and use of the Hardware Design Specification. It should specify the participants by name, title, department or company affiliation and responsibility in a table. Responsibilities should be specific for the procedures and activities delineated in each specific document.

The System Owner is responsible for the development and understanding of the Hardware Design Specification. They acquire and coordinate support from in-house and/or external consultants, the Computer Validation Group, Quality Assurance (Unit) and system developers.

The QA(U) is responsible to verify that the details in the Hardware Design Specification are compliant with applicable departmental, interdepartmental and corporate policies and procedures and that the approach is in accordance with cGxP regulations and good business practices, especially 21 CFR Part 11.

Computer Operations is responsible for understanding the Computer System Design Specifications document and working with the System Owner/End-Users to develop a Design Specification document for the hardware that will satisfy the business needs as indicated in the User and Functional Requirements and the Software Design Specification.

Procedure (protocol)

The Hardware Design Specification clearly defines what the equipment is to do and how it is to do it. It defines the functions to perform and the operating environment. The required functionality and the method of implementation is indicated.

The Hardware Design Specification should:

- Indicate whether mandatory/regulatory or desirable
- Be able to be tested
- Not duplicate one another
- Not contradict one another

Vendors are selected after the approval of the Hardware Design Specification, using it as one of the governing documents in the process.

The Hardware Design Specification document is prepared as a stand-alone document for all projects.

Chapter 14 – Hardware Design Specification

Introduction (protocol)

This section contains an overview of the business operations in which the hardware is to function including workflow, material flow and the control philosophy of what functions, calculations and processes are to be performed by the hardware, as well as interfaces that the equipment must be provided.

Operational Requirements / Critical Functions (protocol)

This section describes the operational requirements and critical functions that must be met in order for the hardware to function and be acceptable for use.

It addresses automatic, manual, and maintenance modes, and operational interlock requirements of the system. Interlocks are hardware or software elements that inhibit or enable the use of features under specific conditions. They are used for data integrity, security or safety purposes.

It provides detailed information, such as circumstances that could directly impact the identity, strength, quality, or purity of a drug product, the operation of a medical device, data subject to regulatory review or the safety of individuals who operate the system. Critical items listed in the Hardware Design Specification document are summarized in this section.

If it is not feasible to separate the critical functions from the rest of the requirements, then they are identified as such.

All requirements are indicated as being classified as one of the following:

- Mandatory - must be included
- Desirable - may be excluded.

Description of Operations

This section describes the hardware in its intended future state. Both narrative and diagrammatic methods divided into modules should be used for large or complex equipment. For small or simple equipment, diagrams may be made optional, but they are highly recommended. Diagrams are required to be accurate, concise and clearly labeled with emphasis placed on those portions of the equipment that are critical. The methods should conform to the User and Functional Requirements protocol.

Types of diagrams that could be used are:

- Hardware Configuration
- Network
- Wiring Diagrams
- Environment
- Construction

Business and Operating Units (protocol)

This section describes the business and operating units involved in the project:

- Business Unit / System Owner / End-User Representatives (each department affected)
- Management Representatives
- Quality Assurance (Unit)

- Computer Operations
- Interface Developer
- Vendors

The Products (protocol)

This section, when applicable, describes the products processed or controlled by the hardware with their specific requirements.

The Facilities (protocol)

This section describes the facilities that will use the hardware and specific attributes of the site (e.g., manufacturing sites, analytical labs with associated instrumentation or equipment, clinical trial sites, distribution services, medical devices).

Detailed Hardware Design

This section specifies the requirements needed for compliance with corporate policies and procedures, SOPs, cGxPs, 21 CFR Part 11 and applicable statutes and regulations along with the User and Functional Requirements.

Mandatory and desired items, as indicated in the User and Functional Requirements should be specified and identified individually.

The details of the automatic, manual, maintenance, interlock and security mode requirements of the system should be specified.

Items that may impact the identity, strength, quality, or purity of a drug product, medical devices and data subject to regulatory review or operator safety should be identified.

Application software, if applicable, used as part of the hardware operation, should be identified. A description of the purpose along with the name of the manufacturer and version number should be provided/included.

The anticipated number of concurrent users and total clients of the hardware should be indicated.

This section also provides information on qualification. If the hardware was previously qualified (i.e., hardware was installed on another existing system), then the associated qualification documentation should be referenced with a brief description of the qualification activities.

The anticipated number of concurrent users and total clients of the hardware should be indicated.

At a minimum, the following areas should be addressed:

- Application
- Hardware Specification
- Performance Sizing
- Hardware Components
- Operating System and Other Components and Descriptions

Chapter 14 – Hardware Design Specification

Configuration and Installation

This section describes the requirements for configuration and installation of the specified hardware. It should include, but not be limited to, the following:

- Hardware
- Software
- System Power-up
- Startup
- Connectivity

Environment and Facility Specification

This section details the requirements needed for the system to function as specified in the environment it is intended to operate. Items, such as, but not limited to, the following are addressed:

Ambient Air Temperature & Humidity

Specify the operational temperature range that the hardware is required to operate in.

Specify the operational relative humidity (non-condensing) range that the hardware is required to operate in.

Electromagnetic Interference

All equipment should comply with FCC rules and regulations, Part 15, as a class A digital devices.

No equipment should be capable of being subjected to any electromagnetic interference outside of the equipment manufacturer's specifications.

Radio Frequency Interference

All equipment should comply with FCC rules and regulations, Part 15, as a class A digital device.

No equipment should be capable of being subjected to any radio frequency interference outside of the equipment manufacturer's specifications.

Power

Specify the electrical power supply for each piece of equipment including voltage range, type of current (AC or DC), and power line frequency, if AC.

Specify the circuit ampacity, protection, and number of phases along with the NEMA connector type for connectivity.

Physical Security

If applicable, in order to protect confidential data, preserve data integrity and avoid system failure, keep hardware in a physically secure location and accessed only by authorized individuals. The room in which the hardware is located should be equipped with a locking mechanism on all points of access to prevent unauthorized entry. Authorized personnel must be properly qualified and screened to determine satisfactory grounds for authorization to access the hardware or equipment.

Logical Security

If applicable, only authorized individuals have logical access to hardware.

Document Management and Storage (protocol)

A Document Management and Storage section is included in each Hardware Design Specification. It defines the Document Management and Storage methodology that is followed for issue, revision, replacement, control and management of each Hardware Design Specification. This methodology is in accordance with an established Change Control SOP and a Document Management and Storage SOP.

Revision History (protocol)

A Revision History section is included in each Hardware Design Specification. The Hardware Design Specification Version Number and all Revisions made to the document are recorded in this section. These changes are in accordance with a Change Control Procedure SOP.

The Revision History section consists of a summary of the changes, a brief statement explaining the reasons the document was revised, and the name and affiliation of the person responsible for the entries. It includes the numbers of all items the Hardware Design Specification replaces and is written so that the original Hardware Design Specification is not needed to understand the revisions or the revised document being issued.

When the Hardware Design Specification is a new document, a "000" as the Revision Number is entered with a statement such as "Initial Issue" in the comments section.

Regulatory or audit commitments require preservation from being inadvertently lost or modified in subsequent revisions. All entries or modifications made as a result of regulatory or audit commitments require indication as such in the Revision History and should indicate the organization, date and applicable references. A complete and accurate audit trail must be developed and maintained throughout the project lifecycle.

Attachments (protocol)

If needed, an Attachments section is included in each Hardware Design Specification protocol in accordance with the SOP on SOPs. An Attachment Index and Attachment section follow the text of a Hardware Design Specification protocol. The top of each attachment page indicates "Attachment # x," where 'x' is the sequential number of each attachment. The second line of each attachment page contains the attachment title. The number of pages and date of issue corresponding to each attachment are listed in the Attachment Index.

Attachments are placed directly after the Attachment Index and are sequentially numbered (e.g., Attachment # 1, Attachment # 2) with each page of an attachment containing the corresponding document number, page numbering, and total number of pages in the attachment (e.g., Page x of y). Each attachment is labeled consistent with the title of the attachment itself.

Attachments are treated as part of the Hardware Design Specification protocol. If an attachment needs to be changed, the entire Hardware Design Specification protocol requires a new revision number and is reissued. New attachments or revisions of existing attachments are in accordance with an established Change Control SOP and a Document Management and Storage SOP.

***Document Management and Storage** (SOP)*

A Document Management and Storage section is included in the Hardware Design Specification SOP. It defines the Document Management and Storage methodology that is followed for issue, revision, replacement, control and management of this SOP. This methodology is in accordance with an established Change Control SOP and a Document Management and Storage SOP.

***Revision History** (SOP)*

A Revision History section is included in the Hardware Design Specification SOP. The Hardware Design Specification SOP Version Number and all Revisions made to the document are recorded in this section. These changes are in accordance with a Change Control Procedure SOP.

The Revision History section consists of a summary of the changes, a brief statement explaining the reasons the document was revised, and the name and affiliation of the person responsible for the entries. It includes the numbers of all items the Hardware Design Specification SOP replaces and is written so that the original Hardware Design Specification SOP is not needed to understand the revisions or the revised document being issued.

When the Hardware Design Specification SOP is a new document, a "000" as the Revision Nu mber is entered with a statement such as "Initial Issue" in the comments section.

Regulatory or audit commitments require preservation from being inadvertently lost or modified in subsequent revisions. All entries or modifications made as a result of regulatory or audit commitments require indication as such in the Revision History and should indicate the organization, date and applicable references. A complete and accurate audit trail must be developed and maintained throughout the project lifecycle.

***Attachments** (SOP)*

If needed, an Attachments section is included in the Hardware Design Specification SOP in accordance with the SOP on SOPs. An Attachment Index and Attachment section follow the text of the Hardware Design Specification SOP. The top of each attachment page indicates "Attachment # x," where 'x' is the sequential number of each attachment. The second line of each attachment page contains the attachment title. The number of pages and date of issue corresponding to each attachment are listed in the Attachment Index.

Attachments are placed directly after the Attachment Index and are sequentially numbered (e.g., Attachment # 1, Attachment # 2) with each page of an attachment containing the corresponding document number, page numbering, and total number of pages in the attachment (e.g., Page x of y). Each attachment is labeled consistent with the title of the attachment itself.

Attachments are treated as part of the Hardware Design Specification SOP. If an attachment needs to be changed, the entire Hardware Design Specification SOP requires a new revision number and is reissued. New attachments or revisions of existing attachments are in accordance with an established Change Control SOP and a Document Management and Storage SOP.

Chapter 15 – Testing

In this chapter we are going to investigate overall testing types and requirements, techniques and procedures that will be covered as specific cases in Chapters 16 through 22. These chapters provide the framework for the SOPs that govern the creation of protocols in support of qualification efforts. This chapter is provided to assist the Technical Test Writer in understanding and writing the qualification tasks. The types of testing in the System Development Life Cycle (SDLC) are:

- Unit or Module Testing
- Factory Acceptance Testing (FAT)
- End-User Testing
- Installation Qualification (IQ) Testing
- Operational Qualification (OQ) Testing
- Performance Qualification (PQ) Testing

Unit Testing

As discussed in Chapter 3 – Overview of Computer and Computerized System Validation, Unit Testing, also known as Module Testing, involves the basic detailed examination and testing of a new or changed program at the programming level. It is the lowest level of testing and is normally the responsibility of the developer. Its purpose is to verify that an application program is structurally sound and conforms to design specifications and appropriate programming standards.

Unit Testing is generally performed for all code produced under GAMP4 Categories 4 and 5.

See Chapter 16 – "Unit Testing."

Factory Acceptance Testing

As discussed in Chapter 3 – Overview of Computer and Computerized System Validation, Factory Acceptance Testing (FAT) combines units that were previously tested. Although the responsibility of the system developer, the client (i.e., System Owner or End-User) may be present to assure that the computer system being developed meets the requirements. FAT, often called system testing, normally takes place at the vendor facility. Simulators are often used for testing on larger systems, rather than involving actual hardware. The purpose of Factory Acceptance Testing is to demonstrate that the combined units and modules operate correctly together. FAT is performed in accordance with an approved protocol (normally approved by vendor management) and executed in accordance with good documentation practices.

See Chapter 17 – Factory Acceptance Testing.

End-User Testing

End-User Testing, also known as qualification or validation testing, begins once development of the computer system is complete and the system is installed and operating in a qualified environment. The System Owner is responsible for the validation of the computer system. Test protocols are written specific to the computer or computerized system and the appropriate End-User groups prior to system use and are reviewed and approved. The protocols are then executed, after which, QA(U) reviews the results. This review consists of two parts, verification that the tests were completed in accordance with the procedure and that the entire protocol was completed; and that all deviations were properly documented and remediated.

End-User testing consists of Installation, Operational, and Performance Qualification testing. Installation Qualification (IQ) provides documented verification of the installation of the software and hardware components and that these components are in accordance with the approved design. Operational Qualification (OQ) provides documented verification that each unit or subsystem operates within the defined operating range. Performance Qualification (PQ) provides documented verification that the integrated system, as tested by users, performs in its normal operating environment.

See Chapters 18 through 22.

Qualification Tasks

One of the most difficult parts of the validation process is the writing of the qualification tasks. These are the test steps in the protocols that are executed in the performance of Unit Testing, Installation Qualification, Operational Qualification and Performance Qualification. They are different and generally unique for each type of qualification and situation. These include, but are not limited to, software, hardware, systems such as Supervisory Control and Data Acquisition (SCADA), Programmable Logic Controller (PLC), Distributed Control System (DCS), Manufacturing Execution Systems (MES), Instrumentation and Process Control (IPC), Laboratory Information Management System (LIMS) Manufacturing Resource Planning (MRP), Office Systems like Local Area Networks (LAN), Enterprise Resource Planning (ERP) and Enterprise Resource Management (ERP). A Technical Test Writer conjures up test procedures or steps that indicate both an understanding of what is required for qualification and a demonstration that what is being tested establishes documented evidence that a system does what it is intended and purports to do.

The term conjure is used here because there are times when test writers go into a state of meditation and simply hope that a miracle manifests itself to enlighten them on how to test some of the work at hand.

It is not possible to test every possible permutation and combination of scenarios in a series of test steps. The order of magnitude can reach into the millions and sometimes trillions. Choosing a reasonable combination of tests is, in itself, a daunting task.

The most important objective is that the tests be reasonable and practical without being overwhelming. They must clearly indicate fulfillment of the User Requirements primarily, as well as the Functional Requirements and Design Specifications.

It is a good idea for the Technical Test Writer to begin work on the test script requirements early in the process so that testing may be constructed as the system evolves. This avoids misunderstanding and confusion.

Analysis

In order to write reasonable tests in support of the validation process, the Technical Test Writer must have an understanding of the User Requirements, how the system is desired to function, and how it actually works. They are seldom the same. In writing the tests, each test step must be tied back to a User Requirement first, then Functional Requirements and Design Specifications, thereafter. Each individual function, algorithm, process, routine, subroutine, action and outcome must be made accountable and proved to do what it is purported to do.

In cases where the Technical Test Writer is privy to the design, the test steps follow a path that is a combination of the design and operation to provide a logical flow for testing. Where the design is not available, as in Commercial-Off-The-Shelf situations, a reverse engineering process must be performed first. This entails taking apart what is to be tested and breaking it down into basic components. These components are then tested.

For example, an injector on a High Throughput Liquid Chromatography (HTLC) system requires verification that 100 cc ± .001% of a liquid is placed into a vial, then the actual steps performed by the operator in setting up the injector screen on the computer, the physical operation of the HTLC, the measured amount of liquid by a qualified method, field input values, system and error messages, pull down menu items and so forth, must all be tested.

In order to accomplish this, the Technical Test Writer must fully understand how everything involved works. If there are boundary limitations on values to be entered into fields on the computer screen, then values within, on and outside the required parameter must be used during the testing. This ensures that the field input validation code functions correctly. If a tolerance is permitted on the outcome of the process, i.e.- 100 cc ± .001%, a determination must be made if it is a computer analyzed tolerance or an equipment limitation. Further, is it permissible to have a cumulative tolerance that is a combination of both the equipment and software control or is it the tolerance for the actual outcome of the filling process? The tests for each of these two scenarios are different.

In another example, the checking of patient information collected during clinical trial visits requires a slightly different analysis. There are a series of visits across which an algorithm must verify parameters entered by the clinician. A truth table is set up with the rows representing the criteria and the columns representing the states or combination sets of those criteria. Entered in each column is the permutation and combination of conditions that are possible for each state or situation. This will provide the test steps as well as indicate the feasibility of the conditions themselves. Sometimes the truth table will show that there are no possible correct answer sets even though the criteria have been in use for many years. This is because previously a Clinical Research Associate performed an analysis and manually qualified the final entries. Computers cannot do that.

Construction

Each test is constructed in accordance with a Standard Operating Procedure governing the writing of Unit Testing, Installation Qualification, Operational Qualification, and Performance Qualification protocols.

See the appropriate chapter for the specific SOP requirements.

Each Test Step or Test Step Series contains a brief statement as to its purpose. This simple explanation describes what is to be accomplished. It may range from something as simple as "Password Security" to "Hardware Inventory Checklist and Microsoft Windows 2000 Server Pre-Installation Worksheet for the ABC-23A01 Archive Server."

Execution procedures, if appropriate, are the next instruction for the tester. This provides rules specific to this Test Step or series that guide the tester through the steps or to perform an initialization that does not require step-by-step elaboration. For example:

Install and configure the Microsoft Windows 2000 Server OS Software components in accordance with the System Design Specification document Revision 0 dated 22-Jun-2002. Retain copies of all documents used to install and configure the system software. The title page of each document and any check box items within it must be signed and dated by the installer.

Reference the Functional Requirement and Design Specification being tested. This may be one or more parts of the referenced item being tested, parts of multiple items or a combination. It must be clear from the information provided or the Traceability Matrix that all of the items are actually tested.

The procedure to be performed should be clear and concise. If there is predefined data to be used, it may be included here or in a referenced area located elsewhere in the protocol. It is not advisable to have predefined data in a separate document unless the document has gone through formal review and approval cycles or is a standard publication.

Simple steps that are reasonably easy for a qualified tester to understand and follow are the best way accomplish the goal of qualification. For example:

> "Fill the maximum number of plates that can be used in the plate stack with deionized water.
>
> Program the autosampler to sequentially inject 3.0 μL from each well in each plate: Top of left corner, top of right corner, center well (well D6 or 42), bottom of left corner and bottom of right corner."

The expected results or outcome of the test must be clear and unambiguous. The Tester and subsequent Reviewer must be able to conclude whether the step was completed properly, passed or failed, depending on the activity. If multiple outcomes are possible, all cases must be stated with their conditions of permissibility as being acceptable. For example:

> "The autosampler arm moves to the sample wells programmed into the system in accordance with the sequence specified and injects 3.0 μL from each well in each plate."

Avoid expected result statements that merely state "The autosampler works as expected." This gives no clear indication of what is supposed to occur.

Chapter 16 – Unit Testing

In this chapter we will discuss the Standard Operating Procedure (SOP) for creating a Unit Testing protocol for a validation project. This SOP defines the method to be used for specifying the Unit Testing that a System Developer will use in testing software application code at the unit or module level.

Best industry practices and FDA Guidance documents indicate that in order to demonstrate that a computer or computerized system performs as it purports and is intended to perform, a methodical plan is needed that details how this may be proven. The Unit Testing protocol defines the tests that are to be used in qualifying software application code at the unit or module level.

This SOP is used for all systems that create, modify, maintain, retrieve, transmit or archive records in electronic format. It is used in conjunction with an overlying corporate policy to provide sustainable compliance with the regulation. The System Developer uses it to create and execute the Unit Testing protocol from which the unit or module level software application code is qualified for use.

System Developers use this SOP to create and execute the Unit Testing protocol for software application code that controls equipment and/or produces electronic data or records in support of all regulated laboratory, manufacturing, clinical or distribution activities.

Title Block, Headers and Footers (SOP)

The Unit Testing SOP should have a title block, title or cover page along with a header and/or footer on each page of the document that clearly identifies the document. The minimum information that should be displayed for the Unit Testing SOP is as follows:

- Document Title
- Document Number
- Document Revision Number
- Project Name and Number
- Site/Location/Department Identification
- Date Document was issued, last revised or draft status
- Effective Date
- Pagination: page of total pages (e.g., Page x of y)

The Document Title should be in large type font that enables clear identification. It should clearly and concisely describe the contents and intent of the document. Since this document will be part of a set or package of documents within a project, it should be consistent with other documents used for the project and specify that it is the Unit Testing SOP. This information should be on every page of the document.

The Document Number should be in large type font that enables clear identification. Since the Unit Testing SOP will be part of a set or package of documents within a project, it should be consistent with other documents used for the project. This information should appear on every page of the document.

The Document Revision Number, which identifies the revision of the document, should appear on each page of the document. Many documents will go through one or more changes over time, particularly during the draft stage and the review and approval cycle. It is important to maintain accurate revision control because even a subtle change in a document can have an immense impact on the outcomes of activities. The SOP on SOPs should govern the numbering methodology.

See Chapter 6 – The SOP on SOPs.

The Project Name and Number, if applicable, should be indicated on the cover or title page. This information identifies the overall project to which this document belongs.

The Site/Location/Department Identification should be indicated on the cover or title page. It identifies the Site/Location/Department that has responsibility for this document and its associated project.

The Date that the Document was issued, last revised or issued with draft status should be on each page of the document (e.g., 12-Dec-01). This will clearly identify the date on which the document was last modified.

The Effective Date of the Document should be on the title or cover page of the document (e.g., 12-Dec-01). This will clearly identify when to begin using the current version of the document.

Pagination, page of total pages (e.g., Page x of y), should be on each page of the document. This will help identify whether or not all of the pages of the document are present. Pages intentionally left blank should clearly be designated as such.

Approval Section *(SOP)*

The Approval Section should contain a table with the Author and Approvers printed names, titles and a place to sign and date the SOP. The SOP on SOPs should govern who the approvers of the document should be. The functional department (e.g., Production / Quality Assurance / Engineering) of the signatory should be placed below each printed name and title along with a statement associated with each approver indicating the person's role or qualification in the approval process. Indicate the significance of the associated signature (e.g. This document meets the requirements of Corporate Policy 9055, Electronic Records; Electronic Signatures, dated 22-Mar-2000).

The completion (signing) of this section indicates that the contents have been reviewed and approved by the listed individuals.

Table of Contents *(SOP)*

The SOP should contain a Table of Contents, which appears towards the beginning of the document. This Table of Contents provides a way to easily locate various topics contained in the document.

Purpose *(SOP)*

This introduction to the SOP concisely describes the document's purpose in sufficient detail so that a non-technical reader can understand its purpose.

The Purpose section should clearly state that the intention of the document is to provide a Standard Operating Procedure to be used to specify the requirements for the preparation of Unit Testing protocol for a new or upgraded computer or computerized system (how the hardware is supposed to be installed) along with the required activities and deliverables. These requirements are necessary for the Unit Testing protocol to be effective and compliant with company polices and government regulation. As one of the foundations of the software application, computer or computerized system, the Unit Testing protocol needs to define the scope of work, responsibilities and deliverables in order to create the software application, computer and/or computerized system and demonstrate that it meets its intended use in a reliable, consistent and compliant manner. The Unit Testing protocol document derived from this SOP will become the standard that is used to qualify specific software code.

Scope (SOP)

The Scope section identifies the limitations of the Unit Testing SOP. It limits the scope to all GxP systems and is applicable to all computers, computerized systems, infrastructure, LANs and WANs including clinical, laboratory, manufacturing and distribution computer systems. It applies to all Sites, Locations and Departments within the company.

Equipment / Materials / Tools (SOP)

This section lists any required or recommended equipment, materials, tools, templates or guides. Provide the name and a brief description of the equipment, instruments, materials and/or tools to be used in the procedures or testing covered by the Unit Testing SOP. This is for reference purposes and requires the user to utilize the specific item or a designated equivalent. The purpose and use is explained elsewhere in the document. Substitution of comparable equipment may be made in accordance with procedures for Change Control.

If the information is included in a job aid or operating manual, a description of where the information may be found with a specific reference to the document number, title, date and revision number should be specified.

Warnings / Notes / General Information / Safety (SOP)

Warnings, notes, general or safety information applicable to the Unit Testing SOP are listed in this section.

Policies, Guidelines, References (SOP)

This section lists references to appropriate regulatory requirements, SOPs, policies, guidelines, or other relevant documents referred to in the Unit Testing SOP. The name and a brief description of applicable related SOPs, corporate policies, procedures and guidelines, Quality Control monographs, books, technical papers, publications, specific regulations, batch records, vendor or system manuals, and all other relevant documents that are required by the Unit Testing SOP should be included. This is for reference purposes and requires the user to use the specific item or a designated equivalent. The purpose and use are explained elsewhere in the document.

SOPs cite references and incorporate them in whole by reference rather than specifying details that duplicate these items. This is done so that changes in policy, guidelines or reference materials require no changes to SOPs themselves and the most current version of the referenced material is used. Examples are:

- Policy ABC-123, Administrative Policy for Electronic Records; Electronic Signatures
- Corporate, Plan for Compliance with 21 CFR Part 11, Electronic Records; Electronic Signatures
- SOP-456, Electronic Records; Electronic Signatures
- FDA, 21 CFR Part 11, Electronic Records; Electronic Signatures
- FDA, General Principals of Software Validation; Final Guidance for Industry and FDA Staff, January 11, 2002

Assumptions / Exclusions / Limitations (SOP)

Any and all assumptions, exclusions or limitations within the context of the SOP should be discussed. Specific issues should be directly addressed in this section of the Unit testing SOP. For example:

"Informal Unit Testing for proof of concept or preliminary testing is excluded from compliance with this SOP."

Glossary of Terms (SOP)

A Glossary of Terms should be included that defines all of the terms and acronyms that are used in this document or that are specific to the operation being described and may not be generally known to the reader. The reader may also be referred to the FDA document, "Glossary of Computerized System and Software Development Terminology."

Roles and Responsibilities (SOP)

The Roles and Responsibilities of the Unit Testing protocol should be divided into two specific areas:

Creation and Maintenance

The System Developer or System Development Team of the computer or computerized system software application code is responsible for the preparation and maintenance of the Unit Testing.

It is common for a Unit Testing to go through many revisions during the course of a project, as it is considered a living document, especially during the project's early stages. As a normal process, it helps to ensure that the validation activities are in compliance with cGxPs, FDA regulations, company policies, industry standards, and guidelines.

The roles and responsibilities for preparing the Unit Testing protocol and other project documentation are addressed in accordance with established Standard Operating Procedures.

Use

The members of the System Development Team must be familiar with the Unit Testing in order to fulfill the project requirements.

The Unit Testing protocol is one of the governing documents for the testing and qualification of the software application unit or module level code it covers.

Procedure (SOP

The Unit Testing SOP contains instructions and procedures that enable the creation of a Unit Testing protocol document. Unit Testing should be required in order to provide the qualification requirements for the required software at the unit or module level that can be used by a System Developer to qualify the code.

Unit Testing is prepared as a stand-alone document for all projects. It is not good practice to include it as part of other documents in an attempt to reduce the amount of workload or paperwork. This only causes more work over the length of the project resulting from misunderstandings and confusion. Clear, concise documents that address specific parts of any given project are desirable.

Unit Testing provides a means to test and record the proper operation of software components at the unit or module level.

The Unit Testing protocol provides an overall technical description of the qualification requirements of a software unit or module.

Chapter 16 – Unit Testing

Unit Testing verifies that the software application code meets the requirements and the design specifications for the specified unit or module code.

Purpose (protocol)

This section clearly states the software or firmware application with its Unit Testing qualification requirements and information as to why it is considered for use in a regulated system. This short introduction should be concise and present details so that an unfamiliar reader can understand the purpose of the document. For example:

"The purpose of the Unit Testing Protocol is to verify and document the proper operation of the specified software or firmware unit or module level code. These specifications are provided in the User Requirements, Functional Require ments and Software Design Specification documents and include:

- Application Software
- Firmware"

Scope (protocol)

The Scope section identifies the limitations of a specific Unit Testing protocol in relationship to validating a specific software application, computer or computerized system. For example, if part of the system consists of a Local Area Network (LAN) that has already been qualified, then the qualification of the LAN may be excluded by stating that it is being addressed elsewhere.

Equipment / Materials / Tools (protocol)

This section lists any required or recommended equipment, materials, tools, templates or guides. Provide the name and a brief description of the equipment, instruments, materials and/or tools to be used in the procedures or testing covered by the Unit Testing protocol. This is for reference purposes and requires the user to utilize the specific item or a designated equivalent. The purpose and use is explained elsewhere in the document. Substitution of comparable equipment may be made in accordance with procedures for Change Control.

If the information is included in a job aid or operating manual, a description of where the information may be found with a specific reference to the document number, title, date and revision number should be specified.

Warnings / Notes / General Information / Safety (protocol)

Warnings, notes, general or safety information applicable to the Unit Testing protocol are listed in this section.

Policies / Guidelines / References (protocol)

This section lists references to appropriate regulatory requirements, SOPs, policies, guidelines, or other relevant documents referred to in the Unit Testing protocol. The name and a brief description of applicable related SOPs, corporate policies, procedures and guidelines, Quality Control monographs, books, technical papers, publications, specific regulations, batch records, vendor or system manuals, and all other relevant documents that are required by the Unit Testing protocol should be included. This is for reference purposes and requires the user to use the specific item or a designated equivalent. The purpose and use are explained elsewhere in the document.

Assumptions / Exclusions / Limitations (protocol)

Discuss any assumptions, exclusions or limitations within the context of the Unit Testing. Specific issues should be directly addressed in this section of the Unit Testing protocol.

Glossary of Terms (protocol)

The Glossary of Terms should define all of the terms and acronyms that are used in the document or that are specific to the operation being described and may not be generally known to the reader. The reader may also be referred to the FDA document, "Glossary of Computerized System and Software Development Terminology," or other specific glossaries that may be available.

Titles, terms, and definitions that are used in the corporate policies, procedures and guidelines should be defined.

Roles and Responsibilities (protocol)

Validation activities are the result of a cooperative effort involving many people, including the System Owners, End Users, the Computer Validation Group, Subject Matter Experts (SME) and Quality Assurance (Unit).

This section clearly identifies the roles, responsibilities and individual(s) responsible for the creation and use of the Unit Testing protocol. It specifies the participants by name, title, department or company affiliation and responsibility in a table. Responsibilities are specific for the procedures and activities delineated in each specific document.

The System Developer or Development Team is responsible for the development and understanding of the Unit Testing. They acquire and coordinate support from in-house and/or external consultants, the Computer Validation Group, Quality Assurance (Unit) and system developers.

The System Development Team manager is responsible to verify that the details in the Unit Testing are compliant with applicable departmental, interdepartmental and corporate policies and procedures and that the approach is in accordance with cGxP regulations and good business practices, especially 21 CFR Part 11. The System Development Team manager reviews the executed protocol for completeness.

Only qualified Testers execute the protocol.

The System Developer or Development Team writes a Unit Testing Summary Report and ensures that it is reviewed and approved by the System Development Team manager.

The System Development Team manager ensures that changes required as a result of deviations during testing and subsequent remediation are captured in the Summary Report and incorporated in subsequent tests. If the subsequent protocol execution is begun prior to the approval of the Unit Testing summary report, then the System Development Team manager certifies that deviations and remediation have no effect on the subsequent testing. Otherwise, the affected Unit Testing protocol tests need to be reviewed, repeated as necessary, or re-written and regression testing performed.

Chapter 16 – Unit Testing

Procedure (protocol)

Background Information

This section of the Unit Testing protocol provides an overview of the business operations for the system including workflow, material flow and the control philosophy of what functions, calculations and processes are to be performed by the system, as well as interfaces that must be provided to the system or application. It outlines the activities required for commissioning the equipment.

The execution of the protocol verifies compliance with the Unit Testing.

Instructions on Testing Procedures

This section describes the procedures that the tester is to follow. It must include how the tests will be documented, how deviations will be handled and other items relevant to the execution of the protocol. Test procedures are in accordance with a Validation Plan.

Execution of each test procedure observes the following general instructions:

All signatures or initials are written in permanent ink and dated. Ditto marks or other methods to indicate repeated acceptance of a test must not be used (i.e., individual signatures/initials are required).

Each individual involved must be qualified to perform the assigned tasks. *Qualified* means that the person's training, education and experience are appropriate for the task and are suitably documented.

Each test must list the action to be taken, the expected results (also referred to as the acceptance criteria), the observed results, the disposition of the test (e.g. - pass or fail), and the signature/date of the person performing the test.

Results must be documented by one individual and reviewed by a second.

The test results are recorded in the spaces provided as each test is executed. When the test listed on a particular page is completed, the person performing the test signs the space labeled "Performed By" on the bottom of the page. If the test was not completed satisfactorily (i.e., a component fails to meet expected criteria), a "Discrepancy Report" is prepared.

Discrepancy Reports are numbered sequentially and noted on the appropriate protocol page. Each Discrepancy Report summarizes the observed problem and indicates the actions taken to resolve it. Once all items on a page are completely resolved and the re-tested procedure passes, the tester signs and dates the bottom of the page. When the overall test is completed, a person other than the tester reviews the work, accepts it as complete by signing on the "Verified By" line at the bottom of the page noting any exceptions.

All pages of the document must have a page number and the total number of pages of the document. This includes attachments, which are also referenced in the primary document. Attachments must be numbered, signed and dated and marked in such a manner that they can be associated with a specific test step.

A signature sheet (where the person's signature, printed name, initials, and affiliation are recorded) is included for each test document to identify all individuals involved in the execution of the test protocol.

Protocol test deviations or failures identified during the testing are addressed on a case-by-case basis. It is the responsibility of the site team to evaluate and resolve these deviations or failures.

Test deviations or failures are initially documented in the test script and then recorded in a Discrepancy / Deviation Report and a Log. This record includes a description of the problem, when it was detected, the circumstances of detection, and who detected it. The site validation team reviews the failures and decides on appropriate actions. The team's review is documented in the Discrepancy / Deviation Report and Log, which contains the identification of the failure, the date reviewed, assessment of the failure, assessment of the impact of the failure on the validation effort, and the team's recommendation for resolution. The log also indicates when and how the deviation or failure was resolved. For each failure, the team determines whether the failure impacts site or core software. Impact on core software must be communicated to all other site teams.

Deviations / Failures are categorized in accordance with a defined scheme. An example of such a scheme:

- Critical

An issue that is critical to cGxP compliance; affects compliance with company SOPs or regulatory requirements; in any way compromises the integrity of the data; or that could directly impact the identity, strength, quality, or purity of a drug product, the operation of a medical device, data that is subject to regulatory review, or the safety of individuals who operate the system.

- Major

An issue for which there is an acceptable workaround solution.

- Minor

Errors in text fields that do not fall into the Critical or Major category and are unlikely to be misleading.

- Cosmetic

Trivial errors for which no action is required.

Outcomes of the Review Process:

- Change in Software Configuration or Code

This is in accordance with a Change Control and/or Configuration Management SOP.

- Modify System Requirements or Specifications

For major documentation changes, the document is revised, the revision level incremented and the document sent for re-approval. For minor test script changes a deviation report is generated. If such a change is required, the team may elect to escalate the problem to the implementation team for further study or to management.

- Develop a Manual Procedure to Work Around the Problem

The workaround is written into one of the user SOPs.

- Modify the Test Script

If after evaluation it is determined that the system is responding to a test as required, but the test script contains either the wrong instructions or an incorrect expected response, the team may recommend a change in the test script. This requires the team to document the change in the test script and retest after the change is approved.

Following execution of the protocols, the completed documents are reviewed for completeness and compliance with the appropriate procedures. If the protocol document is complete and compliant, the reviewer signs and dates it. This includes the review of any attached test outputs.

Tester Identification

This section identifies each person participating in the testing. The following information is entered in a table for identification purposes:

- Full Name – printed
- Signature
- Initials
- Affiliation

Qualification Tasks

This section of the protocol requires a series of processes and tests that must be performed as part of the Unit Testing. Each procedure and/or test section should consist of the following sections:

- Test Section Number and Title
- Purpose of test
- Execution procedures, if appropriate
- Functional Requirement being tested
- Design Specification being tested
- Procedure or Test

Each Test Step should consist of the following sections:

- Test Step number identification
- Action to be performed
- Expected Results (acceptance criteria)
- Actual Results (observations)
- Pass / Fail Results
- Tester Signature / Date
- Comments
- Reviewer Signature / Date

The Procedure or Test section indicates the actions that the tester is expected to perform. These are specific instructions on how to accomplish each required action or task.

The Expected Results section indicates what the tester is expected to observe as a result of the action.

The Actual Results is an entry of what the tester observes, whether the test performs as expected (passes) or does not (fails).

The Pass / Fails Results section should clearly state the conclusion of the test step by the Tester entering the word "Pass" or "Fail." The use of the letter "P" for pass or the letter "F" for fail *is* acceptable.

The Tester Signature / Date section is where the tester initials and dates the results, indicating that the test was performed by that person.

The Comments section may be used to record comments or explanations as required.

The Reviewer Signature / Date section is where the Reviewer signs or initials, and dates each test or procedure indicating that the item has been reviewed and is acceptable as recorded.

If the procedure is a gathering of information, such asserial numbers, there is no need for Expected Results and Pass / Fail sections. The initials of the Tester and date will suffice.

The procedure for the handling of test deviations and failures is specified in the Unit Testing protocol's Instruction on Testing Procedures section, which is written in accordance with the section Instructions on Testing Procedures of this SOP.

Discrepancy / Deviation Report

This section provides for a form that is used as a Discrepancy / Deviation Report. Whenever a test or procedure results in a discrepancy or deviation, a report must be made detailing the results and explaining the resolution. The form must contain the following sections:

- Report Number
- Discrepancy / Deviation
- Test Section Number and Title
- Test Step Number
- Signature /Date
- Resolution
- Signature / Date
- QA(U) Signature / Date

The Report Number section is a sequential number derived from the Discrepancy / Deviation Log form where the report is to be logged and tracked.

The Discrepancy / Deviation section is used to enter the details of the test or procedure results.

The Test Section Number and Title is used to enter the test section number and title of the test or procedure being reported.

The Test Step Number is used to enter the test or procedure step number being reported.

The Signature / Date section is filled in by the person who issued the report.

The Resolution section is used to enter the details of the resolution or workaround employed to remediate the deviation or discrepancy.

The Signature / Date section is filled in by the person who issued the remediation report.

The QA(U) Signature / Date section is used by the QA(U) individual who reviewed and approved the report.

Discrepancy / Deviation Log

This section provides for a form for use as a Discrepancy / Deviation Log. It must contain a table with the following columns:

- Report Number
- Unit Test
- Test Section Number and Title
- Test Step Number
- Deviation / Discrepancy
- Date / Initials

The Report Number is a sequential number used to identify the reports associated with the specific qualification protocol.

The Unit Test section is used to identify the protocol that the report represents.

The Test Section Number and Title is used to enter the Test Section Number and Title of the test or procedure being reported.

The Test Step Number is used to enter the test or procedure step number being reported.

The Deviation / Discrepancy item is used to enter a brief description of the deviation or discrepancy being logged.

The Date / Initials section is filled in by the person issuing the remediation Report Number.

Document Management and Storage (protocol)

A Document Management and Storage section is included in each Unit Testing protocol. It defines the Document Management and Storage methodology that is followed for issue, revision, replacement, control and management of each Unit Testing protocol. This methodology is in accordance with an established Change Control SOP and a Document Management and Storage SOP.

Revision History (protocol)

A Revision History section is included in each Unit Testing protocol. The Unit Testing protocol Version Number and all Revisions made to the document are recorded in this section. These changes are in accordance with a Change Control Procedure SOP.

The Revision History section consists of a summary of the changes, a brief statement explaining the reasons the document was revised, and the name and affiliation of the person responsible for the entries. It includes the numbers of all items the Unit Testing protocol replaces and is written so that the original Unit Testing protocol is not needed to understand the revisions or the revised document being issued.

When the Unit Testing protocol is a new document, a "000" as the Revision Number is entered with a statement such as "Initial Issue" in the comments section.

Regulatory or audit commitments require preservation from being inadvertently lost or modified in subsequent revisions. All entries or modifications made as a result of regulatory or audit commitments require indication

as such in the Revision History and should indicate the organization, date and applicable references. A complete and accurate audit trail must be developed and maintained throughout the project lifecycle.

Attachments (protocol)

If needed, an Attachments section is included in each Unit Testing protocol in accordance with the SOP on SOPs. An Attachment Index and Attachment section follow the text of a Unit Testing protocol. The top of each attachment page indicates "Attachment # x," where 'x' is the sequential number of each attachment. The second line of each attachment page contains the attachment title. The number of pages and date of issue corresponding to each attachment are listed in the Attachment Index.

Attachments are placed directly after the Attachment Index and are sequentially numbered (e.g., Attachment # 1, Attachment # 2) with each page of an attachment containing the corresponding document number, page numbering, and total number of pages in the attachment (e.g., Page x of y). Each attachment is labeled consistent with the title of the attachment itself.

Attachments are treated as part of the Unit Testing protocol. If an attachment needs to be changed, the entire Unit Testing protocol requires a new revision number and is reissued. New attachments or revisions of existing attachments are in accordance with an established Change Control SOP and a Document Management and Storage SOP.

Document Management and Storage *(SOP)*

A Document Management and Storage section is included in the Unit Testing SOP. It defines the Document Management and Storage methodology that is followed for issue, revision, replacement, control and management of this SOP. This methodology is in accordance with an established Change Control SOP and a Document Management and Storage SOP.

Revision History *(SOP)*

A Revision History section is included in the Unit Testing SOP. The Unit Testing SOP Version Number and all Revisions made to the document are recorded in this section. These changes are in accordance with a Change Control Procedure SOP.

The Revision History section consists of a summary of the changes, a brief statement explaining the reasons the document was revised, and the name and affiliation of the person responsible for the entries. It includes the numbers of all items the Unit Testing SOP replaces and is written so that the original Unit Testing SOP is not needed to understand the revisions or the revised document being issued.

When the Unit Testing SOP is a new document, a "000" as the Revision Number is entered with a statement such as "Initial Issue" in the comments section.

Regulatory or audit commitments require preservation from being inadvertently lost or modified in subsequent revisions. All entries or modifications made as a result of regulatory or audit commitments require indication as such in the Revision History and should indicate the organization, date and applicable references. A complete and accurate audit trail must be developed and maintained throughout the project lifecycle.

Attachments *(SOP)*

If needed, an Attachments section is included in the Unit Testing SOP in accordance with the SOP on SOPs. An Attachment Index and Attachment section follow the text of the Unit Testing SOP. The top of each attachment page indicates "Attachment # x," where 'x' is the sequential number of each attachment. The second line of each attachment page contains the attachment title. The number of pages and date of issue corresponding to each attachment are listed in the Attachment Index.

Attachments are placed directly after the Attachment Index and are sequentially numbered (e.g., Attachment # 1, Attachment # 2) with each page of an attachment containing the corresponding document number, page numbering, and total number of pages in the attachment (e.g., Page x of y). Each attachment is labeled consistent with the title of the attachment itself.

Attachments are treated as part of the Unit Testing SOP. If an attachment needs to be changed, the entire Unit Testing SOP requires a new revision number and is reissued. New attachments or revisions of existing attachments are in accordance with an established Change Control SOP and a Document Management and Storage SOP.

Chapter 17 – Factory Acceptance Testing

In this chapter we will discuss the Standard Operating Procedure(SOP) for creating Factory Acceptance Testing (FAT) plans and protocols for a validation project where an outside vendor is supplying products. This SOP defines the method to be used for specifying the Factory Acceptance Testing plan and protocols that a System Owner will employ in testing software applications, computer and computerized systems provided by a vendor prior to delivery, installation and implementation. Typically, this process is a scaled down version of the System Development Life Cycle (SDLC). As previously determined by a Vendor Audit, the vendor will already be qualified and have a full SLDC methodology implemented.

See Chapter 11 – Vendor Audit.

Best industry practices and FDA Guidance documents indicate that in order to demonstrate that a computer or computerized system performs as it purports and is intended to perform, a methodical plan is needed that details how this may be proven. The Factory Acceptance Testing protocol defines the tests and methods employed in qualifying software applications, computer or computerized systems provided by an outside vendor.

This SOP is used for all externally acquired software, hardware, equipment, and systems that create, modify, maintain, retrieve, transmit or archive records in electronic format. It is used in conjunction with an overlying corporate policy to provide sustainable compliance with the regulation. The System Owner uses it to create and execute the Factory Acceptance Testing protocol that defines the tests and methods employed in qualifying software applications, computer or computerized systems provided by an outside vendor.

The System Owner uses this SOP to create and execute the Factory Acceptance Testing protocol that defines the tests and methods employed in qualifying software applications, computer or computerized systems provided by an outside vendor that controls equipment and/or produces electronic data or records in support of all regulated laboratory, manufacturing, clinical or distribution activities.

Title Block, Headers and Footers (SOP)

The Factory Acceptance Testing SOP should have a title block, title or cover page along with a header and/or footer on each page of the document that clearly identifies the document. The minimum information that should be displayed for the Factory Acceptance Testing SOP is as follows:

- Document Title
- Document Number
- Document Revision Number
- Project Name and Number
- Site/Location/Department Identification
- Date Document was issued, last revised or draft status
- Effective Date
- Pagination: page of total pages (e.g., Page x of y)

The Document Title should be in large type font that enables clear identification. It should clearly and concisely describe the contents and intent of the document. Since this document will be part of a set or package of documents within a project, it should be consistent with other documents used for the project and specify that it is the Factory Acceptance Testing SOP. This information should be on every page of the document.

The Document Number should be in large type font that enables clear identification. Since the Factory Acceptance Testing SOP will be part of a set or package of documents within a project, it should be consistent with other documents used for the project. This information should appear on every page of the document.

The Document Revision Number, which identifies the revision of the document, should appear on each page of the document. Many documents will go through one or more changes over time, particularly during the draft stage and the review and approval cycle. It is important to maintain accurate revision control because even a subtle change in a document can have an immense impact on the outcomes of activities. The SOP on SOPs should govern the numbering methodology.

See Chapter 6 – "The SOP on SOPs."

The Project Name and Number, if applicable, should be indicated on the cover or title page. This information identifies the overall project to which this document belongs.

The Site/Location/Department Identification should be indicated on the cover or title page. It identifies the Site/Location/Department that has responsibility for this document and its associated project.

The Date that the Document was issued, last revised or issued with draft status should be on each page of the document (e.g., 12-Dec-01). This will clearly identify the date on which the document was last modified.

The Effective Date of the Document should be on the title or cover page of the document (e.g., 12-Dec-01). This will clearly identify when to begin using the current version of the document.

Pagination, page of total pages (e.g., Page x of y), should be on each page of the document. This will help identify whether or not all of the pages of the document are present. Pages intentionally left blank should clearly be designated as such.

Approval Section *(SOP)*

The Approval Section should contain a table with the Author and Approvers printed names, titles and a place to sign and date the SOP. The SOP on SOPs should govern who the approvers of the document should be. The functional department (e.g., Production / Quality Assurance / Engineering) of the signatory should be placed below each printed name and title along with a statement associated with each approver indicating the person's role or qualification in the approval process. Indicate the significance of the associated signature (e.g. This document meets the requirements of Corporate Policy 9055, Electronic Records; Electronic Signatures, dated 22-Mar-2000).

The completion (signing) of this section indicates that the contents have been reviewed and approved by the listed individuals.

Table of Contents *(SOP)*

The SOP should contain a Table of Contents, which appears towards the beginning of the document. This Table of Contents provides a way to easily locate various topics contained in the document.

Purpose *(SOP))*

This introduction to the SOP concisely describes the document's purpose in sufficient detail so that a non-technical reader can understand its purpose.

The Purpose section should clearly state that the intention of the document is to provide a Standard Operating Procedure to be used to specify the requirements for the preparation and execution of a Factory Acceptance Testing protocol for a new or upgraded computer or computerized system acquired from an outside vendor. These requirements are necessary for the Factory Acceptance Testing protocol to be effective and compliant with company polices and government regulation. As one of the foundations of the software application, computer or computerized system, the Factory Acceptance Testing protocol needs to define the scope of work, responsibilities and deliverables in order for a software application, computer and/or computerized system to demonstrate that it meets its intended use in a reliable, consistent and compliant manner prior to installation, qualification and use. The Factory Acceptance Testing protocol document derived from this SOP will become the standard that is used to qualify specific software, hardware, or system. For example:

"The purpose of Factory Acceptance Testing is to provide and perform a series of tests at the vendor's facility to demonstrate that the application or system described in the User and/or Functional Requirements Specification for the item being acquired performs in accordance with those specifications and within the limits possible under factory conditions."

Scope *(SOP)*

The Scope section identifies the limitations of the Factory Acceptance Testing SOP. It limits the scope to all GxP systems and is applicable to all computers, computerized systems, infrastructure, LANs and WANs including clinical, laboratory, manufacturing and distribution computer systems. It applies to all Sites, Locations and Departments within the company.

All FAT is performed in accordance with an established project plan and is fully documented.

Equipment / Materials / Tools *(SOP)*

This section lists any required or recommended equipment, materials, tools, templates or guides. Provide the name and a brief description of the equipment, instruments, materials and/or tools to be used in the procedures or testing covered by the Factory Acceptance Testing SOP. This is for reference purposes and requires the user to utilize the specific item or a designated equivalent. The purpose and use is explained elsewhere in the document. Substitution of comparable equipment may be made in accordance with procedures for Change Control.

If the information is included in a job aid or operating manual, a description of where the information may be found with a specific reference to the document number, title, date and revision number should be specified.

Warnings / Notes / General Information / Safety *(SOP)*

Warnings, notes, general or safety information applicable to the Factory Acceptance Testing SOP are listed in this section.

Policies, Guidelines, References *(SOP)*

This section lists references to appropriate regulatory requirements, SOPs, policies, guidelines, or other relevant documents referred to in the Factory Acceptance Testing SOP. The name and a brief description of applicable related SOPs, corporate policies, procedures and guidelines, Quality Control monographs, books, technical papers, publications, specific regulations, batch records, vendor or system manuals, and all other relevant documents that are required by the Factory Acceptance Testing SOP should be included. This is for reference purposes and requires the user to use the specific item or a designated equivalent. The purpose and use are explained elsewhere in the document.

SOPs cite references and incorporate them in whole by reference rather than specifying details that duplicate these items. This is done so that changes in policy, guidelines or reference materials require no changes to SOPs themselves and the most current version of the referenced material is used. Examples are:

- Policy ABC-123, Administrative Policy for Electronic Records; Electronic Signatures
- Corporate, Plan for Compliance with 21 CFR Part 11, Electronic Records; Electronic Signatures
- SOP-456, Electronic Records; Electronic Signatures
- FDA, 21 CFR Part 11, Electronic Records; Electronic Signatures
- FDA, General Principals of Software Validation; Final Guidance for Industry and FDA Staff, January 11, 2002

Assumptions / Exclusions / Limitations *(SOP)*

Any and all assumptions, exclusions or limitations within the context of the SOP should be discussed. Specific issues should be directly addressed in this section of the Factory Acceptance Testing SOP. For example:

> "LANs, WANs and other infrastructure specific to the Example Pharmaceuticals Corporation are excluded from testing."

Glossary of Terms *(SOP*

A Glossary of Terms should be included that defines all of the terms and acronyms that are used in this document or that are specific to the operation being described and may not be generally known to the reader. The reader may also be referred to the FDA document, "Glossary of Computerized System and Software Development Terminology."

Roles and Responsibilities *(SOP)*

The Roles and Responsibilities of the Factory Acceptance Testing protocol should be divided into two specific areas:

Creation and Maintenance

The System Owner of the computer or computerized system is responsible for the preparation, maintenance and execution of the Factory Acceptance Testing protocol.

It is common for a Factory Acceptance Testing protocol to go through some revisions during the course of a project in its early stages. As a normal process, the Factory Acceptance Testing protocol helps to ensure that the validation activities are in compliance with cGxPs, FDA regulations, company policies and industry standards and guidelines.

The roles and responsibilities for preparing the Factory Acceptance Testing protocol and other project documentation are addressed in accordance with established Standard Operating Procedures.

Use

The members of the Project Team must be familiar with the Factory Acceptance Testing in order to fulfill the project requirements.

Chapter 17 – Factory Acceptance Testing

The Factory Acceptance Testing protocol is one of the governing documents employed in the testing and qualification of the software applications, computer and computerized systems provided by an outside vendor prior to delivery, installation and operation.

Procedure *(SOP)*

The SOP should contain instructions and procedures that enable the creation and use of a Factory Acceptance Testing protocol document. The Factory Acceptance Testing protocol is required in order to provide the qualification requirements needed to qualify a software, computer or computerized system provided by a vendor prior to delivery, installation and operation.

The Factory Acceptance Testing is prepared as a stand-alone document for all projects. It is not good practice to include it as part of other documents in an attempt to reduce the amount of workload or paperwork. This only causes more work over the length of the project resulting from misunderstandings and confusion. Clear, concise documents that address specific parts of any given project are desirable.

Factory Acceptance Testing provides a means to test and record the proper operation of a software, computer or computerized system supplied by a vendor prior to delivery, installation and operation.

The Factory Acceptance Testing protocol provides an overall technical description of the qualification requirements.

Purpose (protocol)

This section should clearly state the software or firmware application with its Factory Acceptance Testing qualification requirements and information as to why it is considered for use in a regulated system. This short introduction should be concise and present details so that an unfamiliar reader can understand the purpose of the document. For example:

> "The purpose of the Factory Acceptance Testing Protocol is to verify and document the proper operation of the Model ABC-1234 High Throughput Liquid Chromatography Computerized System supplied by the XYZ Manufacturing Company. The requirements are specified in the User Requirements and Functional Requirements Specification documents and include:
>
> - Hardware
> - Application Software
> - Firmware"

Scope (protocol)

The Scope section identifies the limitations of a specific Factory Acceptance Testing protocol in relationship to validating a specific software application, computer or computerized system. For example, if part of the system consists of a Local Area Network (LAN) that is not available for FAT, then the qualification of the LAN interface functionality may be excluded by stating that it is being addressed elsewhere. If only a part of the system will be validated, then this section would be used to explain the reason for this decision.

Equipment / Materials / Tools (protocol)

This section lists any required or recommended equipment, materials, tools, templates or guides. Provide the name and a brief description of the equipment, instruments, materials and/or tools to be used in the procedures or testing covered by the Factory Acceptance Testing protocol. This is for reference purposes and requires the user to utilize the specific item or a designated equivalent. The purpose and use is explained elsewhere in the document. Substitution of comparable equipment may be made in accordance with procedures for Change Control.

If the information is included in a job aid or operating manual, a description of where the information may be found with a specific reference to the document number, title, date and revision number should be specified.

Warnings / Notes / General Information / Safety (protocol)

Warnings, notes, general or safety information applicable to the Factory Acceptance Testing protocol are listed in this section.

Policies / Guidelines / References (protocol)

This section lists references to appropriate regulatory requirements, SOPs, policies, guidelines, or other relevant documents referred to in the Factory Acceptance Testing protocol. The name and a brief description of applicable related SOPs, corporate policies, procedures and guidelines, Quality Control monographs, books, technical papers, publications, specific regulations, batch records, vendor or system manuals, and all other relevant documents that are required by the Factory Acceptance Testing protocol should be included. This is for reference purposes and requires the user to use the specific item or a designated equivalent. The purpose and use are explained elsewhere in the document.

Assumptions / Exclusions / Limitations (protocol)

Discuss any assumptions, exclusions or limitations within the context of the Factory Acceptance Testing. Specific issues should be directly addressed in this section of the Factory Acceptance Testing protocol.

Glossary of Terms (protocol)

The Glossary of Terms should define all of the terms and acronyms that are used in the document or that are specific to the operation being described and may not be generally known to the reader. The reader may also be referred to the FDA document, "Glossary of Computerized System and Software Development Terminology," or other specific glossaries that may be available.

Titles, terms, and definitions that are used in the corporate policies, procedures and guidelines should be defined.

Roles and Responsibilities (protocol)

Factory Acceptance Testing activities are the result of a cooperative effort involving many people, including the System Owner, End-Users, the Computer Validation Group, Subject Matter Experts (SME) and Quality Assurance (Unit).

Clearly identify the roles, responsibilities, and individual(s) responsible for the creation and use of the Factory Acceptance Test protocol. Specify the participants by name, title, department or company affiliation and responsibility in a table. Responsibilities should be specific for the procedures and activities delineated in each specific document.

The System Owner is responsible for the development and understanding of the FAT. He or she acquires and coordinates support from in-house and/or external consultants, the Computer Validation Group, Quality Assurance (Unit) and the product's vendor.

Chapter 17 – Factory Acceptance Testing

The System Owner, with the assistance of SME's, CVG and QA(U), is responsible to verify that the details in the Factory Acceptance Testing protocol are compliant with applicable departmental, interdepartmental and corporate policies and procedures and that the approach is in accordance with cGxP regulations and good business practices, especially 21 CFR Part 11. The System Owner reviews the executed protocol for completeness.

Only qualified Testers execute the protocol.

The System Owner, with the assistance of SME's, CVG and QA(U), writes a Factory Acceptance Testing Summary Report and ensures that it is reviewed and approved by the System Owner, CVG and QA(U).

The System Owner ensures that changes required as a result of deviations during testing and subsequent remediation are captured in the Summary Report and incorporated in subsequent tests. If subsequent protocol execution is begun prior to approval, then the System Owner or designee certifies that deviations and remediations have no effect on the subsequent testing. Otherwise, the affected Factory Acceptance Testing protocol tests need to be reviewed, repeated as necessary, or re-written and regression testing performed.

Procedure (protocol)

Introduction

This section of the Factory Acceptance Testing plan and protocol provides an overview of the business operations for the system including workflow, material flow and the control philosophy of what functions, calculations and processes are to be performed by the system, as well as interfaces that the system or application must be provided. It outlines the activities required for performing Factory Acceptance Testing of the software, hardware and/or equipment.

The execution of the protocol verifies compliance with the requirements specifications and procurement documents.

Validation Program Overview

Present an overview of the validation program that is being employed. Explain how this program establishes documented evidence that what is being acquired does what it is intended and purports to do.

Facility Description

This section describes the facilities and specific attributes of the site where the system is being observed and tested. Specify the name, address at which the testing is to take place, telephone number, FAX number and principal contact of the company. Include a description of the general layout of the area of the facility where the testing is to be performed with a floor plan, if available.

Critical Utilities and Equipment

Provide an inventory and description of utilities and equipment that are critical to the operation and performance of the system under test. Be specific about items such as power, environment, test and system instrumentation. Describe equipment that is part of the testing but not necessarily part of the system being delivered.

Validation Test Methods, Validation Cycle and Document Flow Diagrams

Similar to a Validation Plan, provide a test plan in both narrative and diagrammatic form with the workflow or processes that are to be followed in performing the acceptance testing. The diagram or chart portion contains the functions and flows of the process in the format of a Network Diagram – PERT (Program Evaluation and Review Technique) or CPM (CriticalPath Method).

See Chapter 8 – Validation Plan.

Design Review

Review the design specifications for the software application, computer and/or computerized system, as produced by the vendor, to assure conformance with the requirements and specifications for the project. This comparison is performed prior to actual testing so that design deviations may be acknowledged and addressed accordingly.

See Chapter 14 – Design Qualification.

Installation Qualification

Using an Installation Qualification (IQ) protocol that consists of a requirement to test and document the installation of the hardware and/or software components of what is being acquired, qualify the installation. The IQ must meet the requirements and specifications of the project. In cases where the vendor is providing the protocols, care must be taken to ensure that the project requirements are met.

See Chapter 18 – Hardware Installation Qualification and Chapter 20 – Software Installation Qualification.

Operational Qualification

Using an Operational Qualification (OQ) protocol that consists of a requirement to test and document the integrated system against the project requirements and specifications of what is being acquired, qualify the operation of the system. The purpose of this testing is to verify that the system operates as purported and intended. In cases where the vendor is providing the protocols, care must be taken to ensure that the project requirements are met.

See Chapter 19 – Hardware Operational Qualification and Chapter 21 – Software Operational Qualification.

Performance Qualification

Using a Performance Qualification (PQ) protocol that consists of verifying the correct operation of the integrated system in the production environment including human use and interaction with the system of what is being acquired, qualify the performance of the system. The Performance Qualification procedure includes manuals and SOPs, procedures surrounding the use of the system and verification that users and administrators are capable of being trained on the use and administration of the system prior to the system being placed in production. Training must be capable of being documented in accordance with company policies and procedures. The PQ must include Process Qualification if a turnkey system is being acquired. In cases where the vendor is providing the protocols, care must be taken to ensure that the project requirements are met.

See Chapter 22 – Performance Qualification.

Timelines

Specify the timelines for each activity. Provide a Gantt type chart containing work breakdown structures with deliverables and milestones that are required to be met. In this way, everyone involved in this phase of the project will know ahead of time what is expected of him or her and can plan accordingly.

See Chapter 7 – Project Plan.

Manuals and SOPs

Review the manuals and SOPs that govern the vendor's operation along with those specific to the software application, computer and/or computerized system.

See Chapter 31 – Manuals and SOPs.

Validation Maintenance, Periodic Review and Re-Validation

Confirm that the vendor has demonstrated the implementation of policies, practices and procedures that are used to provide sustainable compliance with GxPs, 21 CFR Part 11 and other applicable regulations for the items being acquired.

Once the computer system is validated and approved for acquisition, the computer system must be capable of being maintained in a sustainable compliant validated state. This is in accordance with appropriate cGxPs, which include, but are not limited to, Change Control, Document Management and Storage, Configuration Management and Periodic Review.

See Chapter 27 – Change Control, Chapter 28 – Document Management and Storage, Chapter 29 – Configuration Management, and Chapter 30 – Periodic Review.

If this work was performed during the Vendor Audit specifically for what is being acquired, it may be excluded from Factory Acceptance Testing if the results were satisfactory. A statement to that effect is placed in the Assumptions / Exclusions / Limitations section of the protocol.

Summary Reports

The System Owner prepares Summary Reports for each stage completed. A report is prepared for the Design Review, IQ, OQ, PQ and conclusion of FAT. Ensure that it is reviewed and approved by at least the System Owner, CVG and QA(U).

See Chapter 25 – Summary Reports.

Instructions on Testing Procedures

This section describes the procedures that the tester is to follow. It must include how the tests will be documented, how deviations will be handled and other items relevant to the execution of the protocol. Test procedures are in accordance with this FAT plan and qualification protocols.

Execution of each test procedure should observe the following general instructions:

All signatures or initials are written in permanent ink and dated. Ditto marks or other methods to indicate repeated acceptance of a test must not be used (i.e., individual signatures/initials are required).

Each individual involved must be qualified to perform the assigned tasks. *Qualified* means that the person's training, education and experience are appropriate for the task and are suitably documented.

Each test must list the action to be taken, the expected results (also referred to as the acceptance criteria), the observed results, the disposition of the test (e.g. - pass or fail), and the signature/date of the person performing the test.

Results must be documented by one individual and reviewed by a second.

The test results are recorded in the spaces provided as each test is executed. When the test listed on a particular page is completed, the person performing the test signs the space labeled "Performed By" on the bottom of the page. If the test was not completed satisfactorily (i.e., a component fails to meet expected criteria), a "Discrepancy Report" is prepared.

Discrepancy Reports are numbered sequentially and noted on the appropriate protocol page. Each Discrepancy Report summarizes the observed problem and indicates the actions taken to resolve it. Once all items on a page are completely resolved and the re-tested procedure passes, the tester signs and dates the bottom of the page. When the overall test is completed, a person other than the tester reviews the work, accepts it as complete by signing on the "Verified By" line at the bottom of the page noting any exceptions.

All pages of the document must have a page number and the total number of pages of the document. This includes attachments, which are also be referenced in the primary document. Attachments must be numbered, signed, dated, and marked in such a manner that they can be associated with a specific test step.

A signature sheet (where the person's signature, printed name, initials and affiliation are recorded) is included for each test document to identify all individuals involved in the execution of the test protocol.

Protocol test deviations or failures identified during the testing are addressed on a case-by-case basis. It is the responsibility of the site team to evaluate and resolve these deviations and/or failures.

Test deviations or failures are initially documented in the test script and then recorded in a Discrepancy / Deviation Report and a Log. This record includes a description of the problem, when it was detected, the circumstances of detection, and who detected it. The site validation team reviews the deviations or failures and decides on appropriate actions. The team's review is documented in the Discrepancy / Deviation Report and Log, which contains the identification of the failure, the date reviewed, assessment of the failure, assessment of the impact of the failure on the validation effort, and the team's recommendation for resolution. The log also indicates when and how the deviation or failure was resolved. For each failure, the team determines whether the failure impacts site or core software. Failures that impact core software are communicated to all other site teams.

Deviations / Failures are categorized in accordance with a defined scheme. An example of such a scheme:

- Critical

An issue that is critical to cGxP compliance; affects compliance with company SOPs or regulatory requirements; in any way compromises the integrity of the data; or that could directly impact the identity, strength, quality, or purity of a drug product, the operation of a medical device, data that is subject to regulatory review, or the safety of individuals who operate the system.

- Major

An issue for which there is an acceptable workaround solution.

- Minor

Errors in text fields that do not fall into the Critical or Major category and are unlikely to be misleading.

- Cosmetic

Trivial errors for which no action is required.

Outcomes of the Review Process:

- Change in Software Configuration or Code

This is in accordance with a Change Control and/or Configuration Management SOP.

- Modify System Requirements or Specifications

For major documentation changes, the document is revised, the revision level incremented and the document sent for re-approval. For minor test script changes a deviation report is generated. If such a change is required, the team may elect to escalate the problem to the implementation team for further study or to management.

- Develop a Manual Procedure to Work Around the Problem

The workaround is written into one of the user SOPs.

- Modify the Test Script

If after evaluation it is determined that the system is responding to a test as required, but the test script contains either the wrong instructions or an incorrect expected response, the team may recommend a change in the test script. This requires the team to document the change in the test script and retest after the change is approved.

Following execution of the protocols, the completed documents are reviewed for completeness and compliance with the appropriate procedures. If the protocol document is complete and compliant, the reviewer signs and dates it. This includes the review of any attached test outputs.

Tester Identification

This section identifies each person participating in the testing. The following information is entered in a table for identification purposes:

- Full Name – printed
- Signature
- Initials
- Affiliation

Qualification Tasks

This section of the protocol requires a series of processes and tests that must be performed as part of the Factory Acceptance Testing. Each procedure and/or test section should consist of the following sections:

- Test Section Number and Title
- Purpose of test
- Execution procedures, if appropriate
- Functional Requirement being tested
- Design Specification being tested
- Procedure or Test

Each Test Step should consist of the following sections:

- Test Step number identification
- Action to be performed
- Expected Results (acceptance criteria)
- Actual Results (observations)
- Pass / Fail Results
- Tester Signature / Date
- Comments
- Reviewer Signature / Date

The Procedure or Test section indicates the actions that the tester is expected to perform. These are specific instructions on how to accomplish each required action or task.

The Expected Results section indicates what the tester is expected to observe as a result of the action.

The Actual Results is an entry of what the tester observes, whether the test performs as expected (passes) or does not (fails).

The Pass / Fails Results section should clearly state the conclusion of the test step by the Tester entering the word "Pass" or "Fail." The use of the letter "P" for pass or the letter "F" for fail *is* acceptable.

The Tester Signature / Date section is where the tester initials and dates the results, indicating that the test was performed by that person.

The Comments section may be used to record comments or explanations as required.

The Reviewer Signature / Date section is where the Reviewer signs or initials, and dates each test or procedure indicating that the item has been reviewed and is acceptable as recorded.

If the procedure is a gathering of information, such as serial numbers, there is no need for Expected Results and Pass / Fail sections. The initials of the Tester and date will suffice.

The procedure for the handling of test deviations and failures is specified in the Factory Acceptance Testing protocol's Instruction on Testing Procedures section, which is written in accordance with the section Instructions on Testing Procedures of this SOP.

Discrepancy / Deviation Report

This section provides for a form that is used as a Discrepancy / Deviation Report. Whenever a test or procedure results in a discrepancy or deviation, a report must be made detailing the results and explaining the resolution. The form must contain the following sections:

- Report Number
- Discrepancy / Deviation
- Test Section Number and Title
- Test Step Number
- Signature /Date
- Resolution
- Signature / Date
- QA(U) Signature / Date

The Report Number section is a sequential number derived from the Discrepancy / Deviation Log form where the report is to be logged and tracked.

The Discrepancy / Deviation section is used to enter the details of the test or procedure results.

The Test Section Number and Title is used to enter the test section number and title of the test or procedure being reported.

The Test Step Number is used to enter the test or procedure step number being reported.

The Signature / Date section is filled in by the person who issued the report.

The Resolution section is used to enter the details of the resolution or work around employed to remediate the deviation or discrepancy.

The Signature / Date section is filled in by the person who issued the remediation report.

The QA(U) Signature / Date section is used by the QA(U) individual who reviewed and approved the report.

Discrepancy / Deviation Log

This section provides for a form for use as a Discrepancy / Deviation Log. It must contain a table with the following columns:

- Report Number
- Factory Acceptance Test
- Test Section Number and Title
- Test Step number
- Deviation / Discrepancy
- Date / Initials

The Report Number is a sequential number used to identify the reports associated with the specific qualification protocol.

The Factory Acceptance Test section is used to identify the protocol that the report represents.

The Test Section Number and Title is used to enter the Test Section Number and Title of the test or procedure being reported.

The Test Step Number is used to enter the test or procedure step number being reported.

The Deviation / Discrepancy item is used to enter a brief description of the deviation or discrepancy being logged.

The Date / Initials section is filled in by the person issuing the remediation Report Number.

Document Management and Storage (protocol)

A Document Management and Storage section is included in each Factory Acceptance Testing protocol. It defines the Document Management and Storage methodology that is followed for issue, revision, replacement, control and management of each Factory Acceptance Testing protocol. This methodology is in accordance with an established Change Control SOP and a Document Management and Storage SOP.

Revision History (protocol)

A Revision History section is included in each Factory Acceptance Testing protocol. The Factory Acceptance Testing protocol Version Number and all Revisions made to the document are recorded in this section. These changes are in accordance with a Change Control Procedure SOP.

The Revision History section consists of a summary of the changes, a brief statement explaining the reasons the document was revised, and the name and affiliation of the person responsible for the entries. It includes the numbers of all items the Factory Acceptance Testing protocol replaces and is written so that the original Factory Acceptance Testing protocol is not needed to understand the revisions or the revised document being issued.

When the Factory Acceptance Testing protocol is a new document, a "000" as the Revision Number is entered with a statement such as "Initial Issue" in the comments section.

Regulatory or audit commitments require preservation from being inadvertently lost or modified in subsequent revisions. All entries or modifications made as a result of regulatory or audit commitments require indication as such in the Revision History and should indicate the organization, date and applicable references. A complete and accurate audit trail must be developed and maintained throughout the project lifecycle.

Attachments (protocol)

If needed, an Attachments section is included in each Factory Acceptance Testing protocol in accordance with the SOP on SOPs. An Attachment Index and Attachment section follow the text of a Factory Acceptance Testing protocol. The top of each attachment page indicates "Attachment # x," where 'x' is the sequential number of each attachment. The second line of each attachment page contains the attachment title. The number of pages and date of issue corresponding to each attachment are listed in the Attachment Index.

Attachments are placed directly after the Attachment Index and are sequentially numbered (e.g., Attachment # 1, Attachment # 2) with each page of an attachment containing the corresponding document number, page numbering, and total number of pages in the attachment (e.g., Page x of y). Each attachment is labeled consistent with the title of the attachment itself.

Attachments are treated as part of the Factory Acceptance Testing protocol. If an attachment needs to be changed, the entire Factory Acceptance Testing protocol requires a new revision number and is reissued. New attachments or revisions of existing attachments are in accordance with an established Change Control SOP and a Document Management and Storage SOP.

Document Management and Storage *(SOP)*

A Document Management and Storage section is included in the Factory Acceptance Testing SOP. It defines the Document Management and Storage methodology that is followed for issue, revision, replacement, control and management of this SOP. This methodology is in accordance with an established Change Control SOP and a Document Management and Storage SOP.

Revision History *(SOP)*

A Revision History section is included in the Factory Acceptance Testing SOP. The Factory Acceptance Testing SOP Version Number and all Revisions made to the document are recorded in this section. These changes are in accordance with a Change Control Procedure SOP.

The Revision History section consists of a summary of the changes, a brief statement explaining the reasons the document was revised, and the name and affiliation of the person responsible for the entries. It includes the numbers of all items the Factory Acceptance Testing SOP replaces and is written so that the original Factory Acceptance Testing SOP is not needed to understand the revisions or the revised document being issued.

When the Factory Acceptance Testing SOP is a new document, a "000" as the Revision Number is entered with a statement such as "Initial Issue" in the comments section.

Regulatory or audit commitments require preservation from being inadvertently lost or modified in subsequent revisions. All entries or modifications made as a result of regulatory or audit commitments require indication as such in the Revision History and should indicate the organization, date and applicable references. A complete and accurate audit trail must be developed and maintained throughout the project lifecycle.

Attachments *(SOP)*

If needed, an Attachments section is included in the Factory Acceptance Testing SOP in accordance with the SOP on SOPs. An Attachment Index and Attachment section follow the text of the Factory Acceptance Testing SOP. The top of each attachment page indicates "Attachment # x," where 'x' is the sequential number of each attachment. The second line of each attachment page contains the attachment title. The number of pages and date of issue corresponding to each attachment are listed in the Attachment Index.

Attachments are placed directly after the Attachment Index and are sequentially numbered (e.g., Attachment # 1, Attachment # 2) with each page of an attachment containing the corresponding document number, page numbering, and total number of pages in the attachment (e.g., Page x of y). Each attachment is labeled consistent with the title of the attachment itself.

Attachments are treated as part of the Factory Acceptance Testing SOP. If an attachment needs to be changed, the entire Factory Acceptance Testing SOP requires a new revision number and is reissued. New attachments or revisions of existing attachments are in accordance with an established Change Control SOP and a Document Management and Storage SOP.

Chapter 18 – Hardware Installation Qualification

In this chapter we will discuss the Standard Operating Procedure (SOP) for creating a Hardware Installation Qualification protocol for a validation project. This SOP defines the method to be used for specifying the Hardware Installation Qualification that a computer operations group will use in configuring, qualifying, and implementing computer and computer related hardware.

Best industry practices and FDA Guidance documents indicate that in order to demonstrate that a computer or computerized system performs as it purports and is intended to perform, a methodical plan is needed that details how this may be proven. The Hardware Installation Qualification protocol defines the specifications that are to be used in configuring, qualifying, and implementing computer and computer related hardware in fulfillment of the requirements of the User and Functional Requirements protocols.

This SOP is used for all systems that create, modify, maintain, retrieve, transmit or archive records in electronic format. It is used in conjunction with an overlying corporate policy to provide sustainable compliance with the regulations. The project team uses it to create the Hardware Installation Qualification protocol from which computer operations configures, qualifies and implements hardware into a production environment.

The System Owners, End-Users, and Subject Matter Experts (SME) use this SOP to create the Hardware Installation Qualification protocol for computer or computer related hardware that controls equipment and/or produces electronic data or records in support of all regulated laboratory, manufacturing, clinical or distribution activities. The protocols produced in accordance with this SOP are used by Computer Operations to configure, implement and qualify them.

Title Block, Headers and Footers (SOP)

The Hardware Installation Qualification SOP should have a title block, title or cover page along with a header and/or footer on each page of the document that clearly identifies the document. The minimum information that should be displayed for the Hardware Installation Qualification SOP is as follows:

- Document Title
- Document Number
- Document Revision Number
- Project Name and Number
- Site/Location/Department Identification
- Date Document was issued, last revised or draft status
- Effective Date
- Pagination: page of total pages (e.g., Page x of y)

The Document Title should be in large type font that enables clear identification. It should clearly and concisely describe the contents and intent of the document. Since this document will be part of a set or package of documents within a project, it should be consistent with other documents used for the project and specify that it is the Hardware Installation Qualification SOP. This information should be on every page of the document.

The Document Number should be in large type font that enables clear identification. Since the Hardware Installation Qualification SOP will be part of a set or package of documents within a project, it should be consistent with other documents used for the project. This information should appear on every page of the document.

The Document Revision Number, which identifies the revision of the document, should appear on each page of the document. Many documents will go through one or more changes over time, particularly during the draft stage and the review and approval cycle. It is important to maintain accurate revision control because even a subtle change in a document can have an immense impact on the outcomes of activities. The SOP on SOPs should govern the numbering methodology.

See Chapter 6 – The SOP on SOPs.

The Project Name and Number, if applicable, should be indicated on the cover or title page. This information identifies the overall project to which this document belongs.

The Site/Location/Department Identification should be indicated on the cover or title page. It identifies the Site/Location/Department that has responsibility for this document and its associated project.

The Date that the Document was issued, last revised or issued with draft status should be on each page of the document (e.g., 12-Dec-01). This will clearly identify the date on which the document was last modified.

The Effective Date of the Document should be on the title or cover page of the document (e.g., 12-Dec-01). This will clearly identify when to begin using the current version of the document.

Pagination, page of total pages (e.g., Page x of y), should be on each page of the document. This will help identify whether or not all of the pages of the document are present. Pages intentionally left blank should clearly be designated as such.

Approval Section *(SOP)*

The Approval Section should contain a table with the Author and Approvers printed names, titles and a place to sign and date the SOP. The SOP on SOPs should govern who the approvers of the document should be. The functional department (e.g., Production / Quality Assurance / Engineering) of the signatory should be placed below each printed name and title along with a statement associated with each approver indicating the person's role or qualification in the approval process. Indicate the significance of the associated signature (e.g. This document meets the requirements of Corporate Policy 9055, Electronic Records; Electronic Signatures, dated 22-Mar-2000).

The completion (signing) of this section indicates that the contents have been reviewed and approved by the listed individuals.

Table of Contents *(SOP)*

The SOP should contain a Table of Contents, which appears towards the beginning of the document. This Table of Contents provides a way to easily locate various topics contained in the document.

Purpose *(SOP)****)***

This introduction to the SOP concisely describes the document's purpose in sufficient detail so that a non-technical reader can understand its purpose.

The Purpose section should clearly state that the intention of the document is to provide a Standard Operating Procedure to be used to specify the requirements for the preparation of Hardware Installation Qualification protocol for a new or upgraded computer or computerized system (how the hardware is supposed to be installed) along with the required activities and deliverables. These requirements are necessary for the

Hardware Installation Qualification protocol to be effective and compliant with company polices and government regulation. As one of the foundations of the software application, computer or computerized system, the Hardware Installation Qualification protocol needs to define the scope of work, responsibilities and deliverables in order to create the software application, computer and/or computerized system and demonstrate that it meets its intended use in a reliable, consistent and compliant manner. The Hardware Installation Qualification protocol document derived from this SOP will become the standard that is used to configure and deploy a specific hardware item.

Scope *(SOP)*

The Scope section identifies the limitations of the Hardware Installation Qualification SOP. It limits the scope to all GxP systems and is applicable to all computers, computerized systems, infrastructure, LANs and WANs including clinical, laboratory, manufacturing and distribution computer systems. It applies to all Sites, Locations and Departments within the company.

Equipment / Materials / Tools *(SOP*

This section lists any required or recommended equipment, materials, tools, templates or guides. Provide the name and a brief description of the equipment, instruments, materials and/or tools to be used in the procedures or testing covered by the Hardware Installation Qualification SOP. This is for reference purposes and requires the user to utilize the specific item or a designated equivalent. The purpose and use is explained elsewhere in the document. Substitution of comparable equipment may be made in accordance with procedures for Change Control

If the information is included in a job aid or operating manual, a description of where the information may be found with a specific reference to the document number, title, date and revision number should be specified.

Warnings / Notes / General Information / Safety *(SOP)*

Warnings, notes, general or safety information applicable to the Hardware Installation Qualification SOP are listed in this section.

Policies, Guidelines, References *(SOP)*

This section lists references to appropriate regulatory requirements, SOPs, policies, guidelines, or other relevant documents referred to in the Hardware Installation Qualification SOP. The name and a brief description of applicable related SOPs, corporate policies, procedures and guidelines, Quality Control monographs, books, technical papers, publications, specific regulations, batch records, vendor or system manuals, and all other relevant documents that are required by the Hardware Installation Qualification SOP should be included. This is for reference purposes and requires the user to use the specific item or a designated equivalent. The purpose and use are explained elsewhere in the document.

SOPs cite references and incorporate them in whole by reference rather than specifying details that duplicate these items. This is done so that changes in policy, guidelines or reference materials require no changes to SOPs themselves and the most current version of the referenced material is used. Examples are:

- Policy ABC-123, Administrative Policy for Electronic Records; Electronic Signatures
- Corporate, Plan for Compliance with 21 CFR Part 11, Electronic Records; Electronic Signatures
- SOP-456, Electronic Records; Electronic Signatures
- FDA, 21 CFR Part 11, Electronic Records; Electronic Signatures

- FDA, General Principals of Software Validation; Final Guidance for Industry and FDA Staff, January 11, 2002

Assumptions / Exclusions / Limitations *(SOP)*

Any and all assumptions, exclusions or limitations within the context of the SOP should be discussed. Specific issues should be directly addressed in this section of the Hardware Installation Qualification SOP. For example:

"Only operating systems and firmware should normally be addressed."

Glossary of Terms *(SOP)*

A Glossary of Terms should be included that defines all of the terms and acronyms that are used in this document or that are specific to the operation being described and may not be generally known to the reader. The reader may also be referred to the FDA document, "Glossary of Computerized System and Software Development Terminology."

Roles and Responsibilities *(SOP)*

The Roles and Responsibilities of the Hardware Installation Qualification protocol should be divided into two specific areas:

Creation and Maintenance

The System Owner of the computer or computerized system and Computer Operations is responsible for the preparation and maintenance of the Hardware Installation Qualification.

It is common for a Hardware Installation Qualification to go through several revisions during the course of a project, as it is considered a living document, especially during the project's early stages. As a normal process, it helps to ensure that the validation activities are in compliance with cGxPs, FDA regulations, company policies, industry standards and guidelines.

The roles and responsibilities for preparing the Hardware Installation Qualification protocol and other project documentation are addressed in accordance with established Standard Operating Procedures.

Use

The members of the Project Team must be familiar with the Hardware Installation Qualification in order to fulfill the project requirements. The Quality Assurance (Unit), Project Team and Computer Operations use the Hardware Installation Qualification in concert with other project documentation as a contract that details the project and quality expectations.

The Hardware Installation Qualification is one of the governing documents for the qualification and testing of the hardware it covers.

Procedure *(SOP)*

The SOP should contain instructions and procedures that enable the creation of a Hardware Installation Qualification protocol document. The Hardware Installation Qualification is required in order to provide the installation qualification tests for the specified hardware that can be used by a Computer Operations Group to install, qualify, and implement the hardware.

The Hardware Installation Qualification is prepared as a stand-alone document for all projects. It is not good practice to include it as part of other documents in an attempt to reduce the amount of workload or paperwork. This only causes more work over the length of the project resulting from misunderstandings and confusion. Clear, concise documents that address specific parts of any given project are desirable.

IQ provides a means to test and record the proper installation of the hardware and software components of a system. This will also provide a baseline record of the system at startup or at its last upgrade that can be used for determining what changes may be required in the future. It ensures that the manufacturer's requirements, regulations and codes have been met.

The Hardware Installation Qualification protocol provides an overall technical description of the installation qualification requirements of a piece of hardware. It provides the final information and documentation requirements that are used by Computer Operations to install, configure, and test a piece of hardware with its associated infrastructure software or firmware.

IQ verifies that the computer system meets the hardware installation requirements and the manufacturer's design specifications for items such as electrical requirements, environmental conditions, and local electrical and fire codes. It must document that the equipment is calibrated and ma intained in a calibrated state during the validation process.

IQ requires the identification and inventorying of documentation for all hardware components, operating and related systems, installation procedures, configuration data and diagrams (e.g., blueprints, piping & instrumentation diagrams (P&ID), wiring diagrams) used to build the equipment. The information must be recorded and verified as correct reflecting the actual "as built" configuration of the equipment.

If the system relies on existing components or infrastructure outside of End-User control, such as a server or network managed by Computer Operations, the appropriate group in accordance with approved procedures qualifies these. The End-User is not required to duplicate these efforts. However, it is required that the End-User reference and include copies of these documents in the validation package.

Post Execution Approval (protocol)

This section contains a place for the QA(U) approval of the executed version of the protocol. The QA(U) signature in this section indicates that the IQ protocol has been properly executed from a compliance standpoint. All sections were completed, all tests and re-tests properly executed, no blank spaces exist and all deviations were properly remediated. It is also construed to mean that each page of the protocol was reviewed. Should QA(U) determine that even though the procedure was properly followed but the IQ failed, it should be so noted upon signing. The phrase "Disapproved for Use" along with a brief explanation is clearly entered in association with the signature.

Purpose (protocol)

This section should clearly state the equipment with its hardware installation qualification requirements and information as to why it is considered for use in a regulated system. This short introduction should be concise and present details so that an unfamiliar reader can understand the purpose of the document. For example:

> "The purpose of the Hardware Installation Qualification Protocol is to verify and document the purchase and/or proper installation of system hardware. The requirements are provided in the Hardware System Design Specification document and include:

- System Hardware
- Interface Hardware
- Peripheral Hardware
- Operating System Software
- Firmware
- Environmental Requirements"

Scope (protocol)

The Scope section identifies the limitations of a specific Hardware Installation Qualification protocol in relationship to validating a specific software application, computer or computerized system. For example, if part of the system consists of a Local area network (LAN) that has already been qualified, then the qualification of the LAN may be excluded by stating that it is being addressed elsewhere.

Equipment / Materials / Tools (protocol)

This section lists any required or recommended equipment, materials, tools, templates or guides. Provide the name and a brief description of the equipment, instruments, materials and/or tools to be used in the procedures or testing covered by the Hardware Installation Qualification protocol. This is for reference purposes and requires the user to utilize the specific item or a designated equivalent. The purpose and use is explained elsewhere in the document. Substitution of comparable equipment may be made in accordance with procedures for Change Control.

If the information is included in a job aid or operating manual, a description of where the information may be found with a specific reference to the document number, title, date and revision number should be specified.

Warnings / Notes / General Information / Safety (protocol)

Warnings, notes, general or safety information applicable to the Hardware Installation Qualification protocol are listed in this section.

Policies / Guidelines / References (protocol)

This section lists references to appropriate regulatory requirements, SOPs, policies, guidelines, or other relevant documents referred to in the Hardware Installation Qualification protocol. The name and a brief description of applicable related SOPs, corporate policies, procedures and guidelines, Quality Control monographs, books, technical papers, publications, specific regulations, batch records, vendor or system manuals, and all other relevant documents that are required by the SOP should be included. This is for reference purposes and requires the user to use the specific item or a designated equivalent. The purpose and use are explained elsewhere in the document.

Assumptions / Exclusions / Limitations (protocol)

Discuss any assumptions, exc lusions or limitations within the context of the Hardware Installation Qualification. Specific issues should be directly addressed in this section of the Hardware Installation Qualification protocol.

Glossary of Terms (protocol)

The Glossary of Terms should define all of the terms and acronyms that are used in the document or that are specific to the operation being described and may not be generally known to the reader. The reader may also be referred to the FDA document, "Glossary of Computerized System and Software Development Terminology," or other specific glossaries that may be available.

Titles, terms, and definitions that are used in the corporate policies, procedures and guidelines should be defined.

Roles and Responsibilities (protocol)

Validation activities are the result of a cooperative effort involving many people, including the System Owners, End Users, the Computer Validation Group, Subject Matter Experts (SME) and Quality Assurance (Unit).

This section requires clear identification of the roles, responsibilities and individual(s) responsible for the creation and use of the Hardware Installation Qualification. It should specify the participants by name, title, department or company affiliation and responsibility in a table. Responsibilities should be specific for the procedures and activities delineated in each specific document.

The System Owners or End-Users are responsible for the development and understanding of the Hardware Installation Qualification. They should acquire and coordinate support from in-house and/or external consultants, the Computer Validation Group, Quality Assurance (Unit) and system developers.

The QA(U) is responsible to verify that the details in the Hardware Installation Qualification are compliant with applicable departmental, interdepartmental and corporate policies and procedures and that the approach is in accordance with cGxP regulations and good business practices, especially 21 CFR Part 11. The QA(U) should review the executed protocol for completeness.

Computer Operations is responsible for understanding the Computer System Design Specifications document and working with the System Owner/End-Users to develop a Hardware Installation Qualification protocol for the hardware that will satisfy the business needs as indicated in the User and Functional Requirements.

Only qualified Testers execute the protocol.

The System Owner / End-User should write an IQ Summary Report and ensure that it is reviewed and approved by the System Owner, CVG and QA(U).

The Subject Matter Expert (SME) should ensure that changes required as a result of deviations during testing and subsequent remediation are captured in the summary report and incorporated into the Hardware Operational Qualification protocol. If the OQ protocol execution is begun prior to the approval of the IQ summary report, then the SME should certify that deviations and remediations have no effect on the OQ. Otherwise, the affected OQ protocol tests need to be reviewed, repeated as necessary, or re-written and regression testing performed.

Procedure (protocol)

Background Information

This section of the Hardware Installation Qualification protocol provides an overview of the business operations for the system including workflow, material flow and the control philosophy of what functions, calculations and processes are to be performed by the system, as well as interfaces that must be provided to the system or application. It outlines the activities required for commissioning the equipment.

The execution of the protocol verifies compliance with the Hardware Installation Qualification.

Instructions on Testing Procedures

This section describes the procedures that the tester is to follow. It must include how the tests will be documented, how deviations will be handled and other items relevant to the execution of the protocol. Test procedures are in accordance with a Validation Plan.

All IQ testing is done in a qualified pre-production environment unless it is the only system currently running on the equipment during testing. It is not good practice to perform qualification testing of new or revised software, computer or computerized systems in the production environment. Should a major failure occur, such as a system crash, then the previously qualified items running in the production environment must be re-qualified.

Execution of each test procedure should observe the following general instructions:

All signatures or initials are written in permanent ink and dated. Ditto marks or other methods to indicate repeated acceptance of a test must not be used (i.e., individual signatures/initials are required).

Each individual involved must be qualified to perform the assigned tasks. *Qualified* means that the person's training, education and experience are appropriate for the task and are suitably documented.

Each test must list the action to be taken, the expected results (also referred to as the acceptance criteria), the observed results, the disposition of the test (e.g. - pass or fail), and the signature/date of the person performing the test.

Results must be documented by one individual and reviewed by a second.

The test results are recorded in the spaces provided as each test is executed. When the test listed on a particular page is completed, the person performing the test signs the space labeled "Performed By" on the bottom of the page. If the test was not completed satisfactorily (i.e., a component fails to meet expected criteria), a "Discrepancy Report" is prepared.

Discrepancy Reports are numbered sequentially and noted on the appropriate protocol page. Each Discrepancy Report summarizes the observed problem and indicates the actions taken to resolve it. Once all items on a page are completely resolved and the re-tested procedure passes, the tester signs and dates the bottom of the page. When the overall test is completed, a person other than the tester reviews the work, accepts it as complete by signing on the "Verified By" line at the bottom of the page noting any exceptions.

Screen prints should be used wherever possible as proof of execution, especially when expected and unexpected errors occur.

All pages of the document must have a page number and the total number of pages of the document. This includes attachments, which are also referenced in the primary document. Attachments must be numbered, signed and dated and marked in such a manner that they can be associated with a specific test step.

A signature sheet (where the person's signature, printed name, initials, and affiliation are recorded) is included for each test document to identify all individuals involved in the execution of the test protocol.

Protocol test deviations or failures identified during the testing are addressed on a case-by-case basis. It is the responsibility of the site team to evaluate and resolve these deviations or failures.

Test deviations or failures are initially documented in the test script and then recorded in a Discrepancy / Deviation Report and a Log. This record includes a description of the problem, when it was detected, the circumstances of detection, and who detected it. The site validation team reviews the failures and decides on appropriate actions. The team's review is documented in the Discrepancy / Deviation Report and Log, which contains the identification of the failure, the date reviewed, assessment of the failure, assessment of the impact of the failure on the validation effort, and the team's recommendation for resolution. The log also indicates when and how the deviation or failure was resolved. For each failure, the team determines whether the failure impacts site or core software. Impact on core software must be communicated to all other site teams.

Deviations / Failures are categorized in accordance with a defined scheme. An example of such a scheme:

- Critical

An issue that is critical to cGxP compliance; affects compliance with company SOPs or regulatory requirements; in any way compromises the integrity of the data; or that could directly impact the identity, strength, quality, or purity of a drug product, the operation of a medical device, data that is subject to regulatory review, or the safety of individuals who operate the system.

- Major

An issue for which there is an acceptable workaround solution.

- Minor

Errors in text fields that do not fall into the Critical or Major category and are unlikely to be misleading.

- Cosmetic

Trivial errors for which no action is required.

Outcomes of the Review Process:

- Change in Software Configuration or Code

This is in accordance with a Change Control and/or Configuration Management SOP.

- Modify System Requirements or Specifications

For major documentation changes, the document is revised, the revision level incremented and the document sent for re-approval. For minor test script changes a deviation report is generated. If such a change is required, the team may elect to escalate the problem to the implementation team for further study or to management.

- Develop a Manual Procedure to Work Around the Problem

The workaround is written into one of the user SOPs

- Modify the Test Script

If after evaluation it is determined that the system is responding to a test as required, but the test script contains either the wrong instructions or an incorrect expected response, the team may recommend a change in the test script. This requires the team to document the change in the test script and retest after the change is approved.

Following execution of the protocols, the completed documents are reviewed for completeness and compliance with the appropriate procedures. If the protocol document is complete and compliant, the reviewer signs and dates it. This includes the review of any attached test outputs.

Tester Identification

This section identifies each person participating in the testing. The following information is entered in a table for identification purposes:

- Full Name – printed
- Signature
- Initials
- Affiliation

Measuring Devices

This section provides information about the equipment used for testing (e.g., volt meters, ammeters, ohmmeters, temperature and humidity recorders, air flow gages, weights, or measures). All devices must be calibrated and certified traceable to the National Institute of Standards and Technology or equivalent. The following information is recorded: manufacturer, model number, Preventive Maintenance Operation (PMO) number, last PMO date, and calibration expiration date.

Equipment Inventory

This section lists items of equipment to be inventoried and installed, and provides a table to record their verification.

Software Inventory

This section lists s oftware or firmware to be inventoried and installed, and provides a table to record their verification.

Document Inventory

This section lists documentation to be inventoried and installed, and provides a table to record their verification.

Site Design and Environmental Conditions Verification

This section specifies the required site design and environmental conditions that are to be verified and provides a table to record their verification.

Qualification Tasks

This section of the protocol requires a series of processes and tests that must be performed as part of the Hardware Installation Qualification. Each procedure and/or test section should consist of the following sections:

- Test Section Number and Title
- Purpose of test
- Execution procedures, if appropriate
- Functional Requirement being tested
- Design Specification being tested
- Procedure or Test

Each Test Step should consist of the following sections:

- Test Step number identification
- Action to be performed
- Expected Results (acceptance criteria)
- Actual Results (observations)
- Pass / Fail Results
- Tester Signature / Date
- Comments
- Reviewer Signature / Date

The Procedure or Test section indicates the actions that the tester is expected to perform. These are specific instructions on how to accomplish each required action or task.

The Expected Results section indicates what the tester is expected to observe as a result of the action.

The Actual Results is an entry of what the tester observes, whether the test performs as expected (passes) or does not (fails).

Screen prints should be required wherever possible as proof of execution, especially when expected and unexpected errors occur.

The Pass / Fail Results section should clearly state the conclusion of the test step by the Tester entering the word "Pass" or "Fail." The use of the letter "P" for pass or the letter "F" for fail *is* acceptable.

The Tester Signature / Date section is where the tester initials and dates the results, indicating that the test was performed by that person.

The Comments section may be used to record comments or explanations as required.

The Reviewer Signature / Date section is where the Reviewer signs or initials, and dates each test or procedure indicating that the item has been reviewed and is acceptable as recorded.

If the procedure is a gathering of information, such as serial numbers, there is no need for Expected Results and Pass / Fail sections. The initials of the Tester and date will suffice.

The procedure for the handling of test deviations and failures is specified in the Hardware IQ protocol's Instruction on Testing Procedures section, which is written in accordance with the section Instructions on Testing Procedures of this SOP.

Discrepancy / Deviation Report

This section provides for a form that is used as a Discrepancy / Deviation Report. Whenever a test or procedure results in a discrepancy or deviation, a report must be made detailing the results and explaining the resolution. The form must contain the following sections:

- Report Number
- Discrepancy / Deviation
- Test Section Number and Title
- Test Step Number
- Signature /Date
- Resolution
- Signature / Date
- QA(U) Signature / Date

The Report Number section is a sequential number derived from the Discrepancy / Deviation Log form where the report is to be logged and tracked.

The Discrepancy / Deviation section is used to enter the details of the test or procedure results.

The Test Section Number and Title is used to enter the test section number and title of the test or procedure being reported.

The Test Step Number is used to enter the test or procedure step number being reported.

The Signature / Date section is filled in by the person who issued the report.

The Resolution section is used to enter the details of the resolution or work around employed to remediate the deviation or discrepancy.

The Signature / Date section is filled in by the person who issued the remediation report.

The QA(U) Signature / Date section is used by the QA(U) individual who reviewed and approved the report.

Discrepancy / Deviation Log

This section provides for a form for use as a Discrepancy / Deviation Log. It must contain a table with the following columns:

- Report Number
- IQ
- Test Section Number and Title
- Test Step number
- Deviation / Discrepancy
- Date / Initials

The Report Number is a sequential number used to identify the reports associated with the specific qualification protocol.

The IQ section is used to identify the protocol that the report represents.

The Test Section Number and Title is used to enter the Test Section Number and Title of the test or procedure being reported.

The Test Step Number is used to enter the test or procedure step number being reported.

The Deviation / Discrepancy item is used to enter a brief description of the deviation or discrepancy being logged.

The Date / Initials section is filled in by the person issuing the remediation Report Number.

Document Management and Storage (protocol)

A Document Management and Storage section is included in each Hardware Installation Qualification protocol. It defines the Document Management and Storage methodology that is followed for issue, revision, replacement, control and management of each Hardware Installation Qualification protocol. This methodology is in accordance with an established Change Control SOP and a Document Management and Storage SOP.

Revision History (protocol)

A Revision History section is included in each Hardware Installation Qualification protocol. The Hardware Installation Qualification protocol Version Number and all Revisions made to the document are recorded in this section. These changes are in accordance with a Change Control Procedure SOP.

The Revision History section consists of a summary of the changes, a brief statement explaining the reasons the document was revised, and the name and affiliation of the person responsible for the entries. It includes the numbers of all items the Hardware Installation Qualification protocol replaces and is written so that the original Hardware Installation Qualification protocol is not needed to understand the revisions or the revised document being issued.

When the Hardware Installation Qualification protocol is a new document, a "000" as the Revision Number is entered with a statement such as "Initial Issue" in the comments section.

Regulatory or audit commitments require preservation from being inadvertently lost or modified in subsequent revisions. All entries or modifications made as a result of regulatory or audit commitments require indication as such in the Revision History and should indicate the organization, date and applicable references. A complete and accurate audit trail must be developed and maintained throughout the project lifecycle.

Attachments (protocol)

If needed, an Attachments section is included in each Hardware Installation Qualification protocol in accordance with the SOP on SOPs. An Attachment Index and Attachment section follow the text of a Hardware Installation Qualification protocol. The top of each attachment page indicates "Attachment # x," where 'x' is the sequential number of each attachment. The second line of each attachment page contains the attachment title. The number of pages and date of issue corresponding to each attachment are listed in the Attachment Index.

Attachments are placed directly after the Attachment Index and are sequentially numbered (e.g., Attachment # 1, Attachment # 2) with each page of an attachment containing the corresponding document number, page numbering, and total number of pages in the attachment (e.g., Page x of y). Each attachment is labeled consistent with the title of the attachment itself.

Attachments are treated as part of the Hardware Installation Qualification protocol. If an attachment needs to be changed, the entire Hardware Installation Qualification protocol requires a new revision number and is reissued. New attachments or revisions of existing attachments are in accordance with an established Change Control SOP and a Document Management and Storage SOP.

Document Management and Storage *(SOP)*

A Document Management and Storage section is included in the Hardware Installation Qualification SOP. It defines the Document Management and Storage methodology that is followed for issue, revision, replacement, control and management of this SOP. This methodology is in accordance with an established Change Control SOP and a Document Management and Storage SOP.

Revision History *(SOP)*

A Revision History section is included in the Hardware Installation Qualification SOP. The Hardware Installation Qualification SOP Version Number and all Revisions made to the document are recorded in this section. These changes are in accordance with a Change Control Procedure SOP.

The Revision History section consists of a summary of the changes, a brief statement explaining the reasons the document was revised, and the name and affiliation of the person responsible for the entries. It includes the numbers of all items the Hardware Installation Qualification SOP replaces and is written so that the original Hardware Installation Qualification SOP is not needed to understand the revisions or the revised document being issued.

When the Hardware Installation Qualification SOP is a new document, a "000" as the Revision Number is entered with a statement such as "Initial Issue" in the comments section.

Regulatory or audit commitments require preservation from being inadvertently lost or modified in subsequent revisions. All entries or modifications made as a result of regulatory or audit commitments require indication as such in the Revision History and should indicate the organization, date and applicable references. A complete and accurate audit trail must be developed and maintained throughout the project lifecycle.

Attachments *(SOP)*

If needed, an Attachments section is included in the Hardware Installation Qualification SOP in accordance with the SOP on SOPs. An Attachment Index and Attachment section follow the text of the Hardware Installation Qualification SOP. The top of each attachment page indicates "Attachment # x," where 'x' is the sequential number of each attachment. The second line of each attachment page contains the attachment title. The number of pages and date of issue corresponding to each attachment are listed in the Attachment Index.

Attachments are placed directly after the Attachment Index and are sequentially numbered (e.g., Attachment # 1, Attachment # 2) with each page of an attachment containing the corresponding document number, page numbering, and total number of pages in the attachment (e.g., Page x of y). Each attachment is labeled consistent with the title of the attachment itself.

Attachments are treated as part of the Hardware Installation Qualification SOP. If an attachment needs to be changed, the entire Hardware Installation Qualification SOP requires a new revision number and is reissued. New attachments or revisions of existing attachments are in accordance with an established Change Control SOP and a Document Management and Storage SOP.

Chapter 19 – Hardware Operational Qualification

In this chapter we will discuss the Standard Operating Procedure (SOP) for creating a Hardware Operational Qualification protocol for a validation project. This SOP defines the method to be used for specifying the Hardware Operational Qualification that a computer operations group will use in qualifying computer and computer related hardware.

Best industry practices and FDA Guidance documents indicate that in order to demonstrate that a computer or computerized system performs as it purports and is intended to perform, a methodical plan is needed that details how this may be proven. The Hardware Operational Qualification protocol defines the specifications that are to be used in qualifying computer and computer related hardware in fulfillment of the requirements of the User and Functional Requirements protocols and the Hardware Design Specification.

This SOP is used for all systems that create, modify, maintain, retrieve, transmit or archive records in electronic format. It is used in conjunction with an overlying corporate policy to provide sustainable compliance with the regulation. The project team uses it to create the Hardware Operational Qualification protocol from which Computer Operations qualifies the hardware.

The System Owners, End-Users, and Subject Matter Experts (SME) use this SOP to create the Hardware Operational Qualification protocol for computer or computer related hardware that controls equipment and/or produces electronic data or records in support of all regulated laboratory, manufacturing, clinical or distribution activities. The protocols produced in accordance with this SOP are used by Computer Operations to qualify them.

Title Block, Headers and Footers (SOP)

The Hardware Operational Qualification SOP should have a title block, title or cover page along with a header and/or footer on each page of the document that clearly identifies the document. The minimum information that should be displayed for the Hardware Operational Qualification SOP is as follows:

- Document Title
- Document Number
- Document Revision Number
- Project Name and Number
- Site/Location/Department Identification
- Date Document was issued, last revised or draft status
- Effective Date
- Pagination: page of total pages (e.g., Page x of y)

The Document Title should be in large type font that enables clear identification. It should clearly and concisely describe the contents and intent of the document. Since this document will be part of a set or package of documents within a project, it should be consistent with other documents used for the project and specify that it is the Hardware Operational Qualification SOP. This information should be on every page of the document.

The Document Number should be in large type font that enables clear identification. Since the Hardware Operational Qualification SOP will be part of a set or package of documents within a project, it should be consistent with other documents used for the project. This information should appear on every page of the document.

The Document Revision Number, which identifies the revision of the document, should appear on each page of the document. Many documents will go through one or more changes over time, particularly during the draft stage and the review and approval cycle. It is important to maintain accurate revision control because even a subtle change in a document can have an immense impact on the outcomes of activities. The SOP on SOPs should govern the numbering methodology.

See Chapter 6 – "The SOP on SOPs."

The Project Name and Number, if applicable, should be indicated on the cover or title page. This information identifies the overall project to which this document belongs.

The Site/Location/Department Identification should be indicated on the cover or title page. It identifies the Site/Location/Department that has responsibility for this document and its associated project.

The Date that the Document was issued, last revised or issued with draft status should be on each page of the document (e.g., 12-Dec-01). This will clearly identify the date on which the document was last modified.

The Effective Date of the Document should be on the title or cover page of the document (e.g., 12-Dec-01). This will clearly identify when to begin using the current version of the document.

Pagination, page of total pages (e.g., Page x of y), should be on each page of the document. This will help identify whether or not all of the pages of the document are present. Pages intentionally left blank should clearly be designated as such.

Approval Section *(SOP)*

The Approval Section should contain a table with the Author and Approvers printed names, titles and a place to sign and date the SOP. The SOP on SOPs should govern who the approvers of the document should be. The functional department (e.g., Production / Quality Assurance / Engineering) of the signatory should be placed below each printed name and title along with a statement associated with each approver indicating the person's role or qualification in the approval process. Indicate the significance of the associated signature (e.g., This document meets the requirements of Corporate Policy 9055, Electronic Records; Electronic Signatures, dated 22-Mar-2000).

The completion (signing) of this section indicates that the contents have been reviewed and approved by the listed individuals.

Table of Contents *(SOP)*

The SOP should contain a Table of Contents, which appears towards the beginning of the document. This Table of Contents provides a way to easily locate various topics contained in the document.

Purpose *(SOP))*

This introduction to the SOP concisely describes the document's purpose in sufficient detail so that a non-technical reader can understand its purpose.

The Purpose section should clearly state that the intention of the document is to provide a Standard Operating Procedure to be used to specify the requirements for the preparation of Hardware Operational Qualification protocol for a new or upgraded computer or computerized system (how the hardware is supposed to be installed) along with the required activities and deliverables. These requirements are necessary for the

Chapter 19 – Hardware Operational Qualification

Hardware Operational Qualification protocol to be effective and compliant with company polices and government regulation. As one of the foundations of the software application, computer or computerized system, the Hardware Operational Qualification protocol needs to define the scope of work, responsibilities and deliverables in order to create the software application, computer and/or computerized system and demonstrate that it meets its intended use in a reliable, consistent and compliant manner. The Hardware Operational Qualification protocol document derived from this SOP will become the standard that is used to configure and deploy a specific hardware item.

Scope *(SOP)*

The Scope section identifies the limitations of the Hardware Operational Qualification SOP. It limits the scope to all GxP systems and is applicable to all computers, computerized systems, infrastructure, LANs and WANs including clinical, laboratory, manufacturing and distribution computer systems. It applies to all Sites, Locations and Departments within the company.

Equipment / Materials / Tools *(SOP)*

This section lists any required or recommended equipment, materials, tools, templates or guides. Provide the name and a brief description of the equipment, instruments, materials and/or tools to be used in the procedures or testing covered by the Hardware Operational Qualification SOP. This is for reference purposes and requires the user to utilize the specific item or a designated equivalent. The purpose and use is explained elsewhere in the document. Substitution of comparable equipment may be made in accordance with procedures for Change Control.

If the information is included in a job aid or operating manual, a description of where the information may be found with a specific reference to the document number, title, date and revision number should be specified.

Warnings / Notes / General Information / Safety *(SOP)*

Warnings, notes, general or safety information applicable to the Hardware Operational Qualification SOP are listed in this section.

Policies, Guidelines, References *(SOP)*

This section lists references to appropriate regulatory requirements, SOPs, policies, guidelines, or other relevant documents referred to in the Hardware Operational Qualification SOP. The name and a brief description of applicable related SOPs, corporate policies, procedures and guidelines, Quality Control monographs, books, technical papers, publications, specific regulations, batch records, vendor or system manuals, and all other relevant documents that are required by the Hardware Operational Qualification SOP should be included. This is for reference purposes and requires the user to use the specific item or a designated equivalent. The purpose and use are explained elsewhere in the document.

SOPs cite references and incorporate them in whole by reference rather than specifying details that duplicate these items. This is done so that changes in policy, guidelines or reference materials require no changes to SOPs themselves and the most current version of the referenced material is used. Examples are:

- Policy ABC-123, Administrative Policy for Electronic Records; Electronic Signatures
- Corporate, Plan for Compliance with 21 CFR Part 11, Electronic Records; Electronic Signatures
- SOP-456, Electronic Records; Electronic Signatures
- FDA, 21 CFR Part 11, Electronic Records; Electronic Signatures

- FDA, General Principals of Software Validation; Final Guidance for Industry and FDA Staff, January 11, 2002

Assumptions / Exclusions / Limitations *(SOP)*

Any and all assumptions, exclusions or limitations within the context of the SOP should be discussed. Specific issues should be directly addressed in this section of the Hardware Operational Qualification SOP. For example:

"Only operating systems and firmware should normally be addressed."

Glossary of Terms *(SOP)*

A Glossary of Terms should be included that defines all of the terms and acronyms that are used in this document or that are specific to the operation being described and may not be generally known to the reader. The reader may also be referred to the FDA document, "Glossary of Computerized System and Software Development Terminology."

Roles and Responsibilities *(SOP)*

The Roles and Responsibilities of the Hardware Operational Qualification protocol should be divided into two specific areas:

Creation and Maintenance

The System Owner of the computer or computerized system and Computer Operations should be responsible for the preparation and maintenance of the Hardware Operational Qualification.

It is common for a Hardware Operational Qualification to go through several revisions during the course of a project, as it is considered a living document, especially during the project's early stages. As a normal process, it helps to ensure that the validation activities are in compliance with cGxPs, FDA regulations, company policies and industry standards and guidelines.

The roles and responsibilities for preparing the Hardware Operational Qualification protocol and other project documentation should be addressed in accordance with established Standard Operating Procedures.

Use

The members of the Project Team must be familiar with the Hardware Operational Qualification in order to fulfill the project requirements. The Quality Assurance (Unit), Project Team and Computer Operations use the Hardware Operational Qualification in concert with other project documentation as a contract that details the project and quality expectations.

The Hardwa re Operational Qualification is one of the governing documents for the qualification and testing of the hardware it covers.

Procedure *(SOP)*

The SOP should contain instructions and procedures that enable the creation of a Hardware Operational Qualification protocol document. The Hardware Operational Qualification should be required in order to provide the operational qualification requirements for the required hardware that can be used by a Computer Operations Group to qualify the hardware for use.

The Hardware Operational Qualification should be prepared as a stand-alone document for all projects. It is not good practice to include it as part of other documents in an attempt to reduce the amount of workload or paperwork. This only causes more work over the length of the project resulting from misunderstandings and confusion. Clear, concise documents that address specific parts of any given project are desirable.

OQ provides a means to test and record the proper operation of the hardware and software components of a system. This will also provide a baseline record of the system at startup or at its last upgrade that can be used for determining what changes may be required in the future. It ensures that the hardware functions as purported and intended.

The Hardware Operational Qualification protocol should provide an overall technical description of the operational qualification requirements of a piece of hardware. It provides the final information and documentation requirements that are used by Computer Operations to test a piece of hardware with its associated infrastructure software or firmware for operation.

If the system relies on existing components or infrastructure outside of End-User control, such as a server or network managed by Computer Operations, these should be qualified by the appropriate group in accordance with approved procedures. The End-User is not required to duplicate these efforts. However, it is required that the End-User reference and include copies of these documents in the validation package.

Post Execution Approval (protocol)

This section contains a place for the QA(U) approval of the executed version of the protocol. The QA(U) signature in this section indicates that the OQ protocol has been properly executed from a compliance standpoint. All sections were completed, all tests and re-tests properly executed, no blank spaces exist and all deviations were properly remediated. It should also be construed to mean that each page of the protocol was reviewed. Should QA(U) determine that the even though the procedure was properly followed but the OQ failed, it should be so noted upon signing. The phrase "Disapproved for Use" along with a brief explanation should be clearly entered in association with the signature.

Purpose (protocol)

This section should clearly state the equipment with its hardware operational qualification requirements and information as to why it is considered for use in a regulated system. This short introduction should be concise and present details so that an unfamiliar reader can understand the purpose of the document. For example:

> "The purpose of the Hardware Operational Qualification Protocol is to verify and document the proper operational of system hardware. The requirements are provided in the Hardware System Design Specification document and include:
>
> - System Hardware
> - Interface Hardware
> - Peripheral Hardware
> - Operating System Software
> - Firmware"

Scope (protocol)

The Scope section identifies the limitations of a specific Hardware Operational Qualification protocol in relationship to validating a specific software application, computer or computerized system. For example, if part of the system consists of a Local area network (LAN) that has already been qualified, then the qualification of the LAN may be excluded by stating that it is being addressed elsewhere.

Equipment / Materials / Tools (protocol)

This section lists any required or recommended equipment, materials, tools, templates or guides. Provide the name and a brief description of the equipment, instruments, materials and/or tools to be used in the procedures or testing covered by the Hardware Operational Qualification protocol. This is for reference purposes and requires the user to utilize the specific item or a designated equivalent. The purpose and use is explained elsewhere in the document. Substitution of comparable equipment may be made in accordance with procedures for Change Control.

If the information is included in a job aid or operating manual, a description of where the information may be found with a specific reference to the document number, title, date and revision number should be specified.

Warnings / Notes / General Information / Safety (protocol)

Warnings, notes, general or safety information applicable to the Hardware Operational Qualification protocol are listed in this section.

Policies / Guidelines / References (protocol)

This section lists references to appropriate regulatory requirements, SOPs, policies, guidelines, or other relevant documents referred to in the Hardware Operational Qualification protocol. The name and a brief description of applicable related SOPs, corporate policies, procedures and guidelines, Quality Control monographs, books, technical papers, publications, specific regulations, batch records, vendor or system manuals, and all other relevant documents that are required by the Hardware Operational Qualification protocol should be included. This is for reference purposes and requires the user to use the specific item or a designated equivalent. The purpose and use are explained elsewhere in the document.

Assumptions / Exclusions / Limitations (protocol)

Discuss any assumptions, exclusions or limitations within the context of the Hardware Operational Qualification. Specific issues should be directly addressed in this section of the Hardware Operational Qualification protocol.

Glossary of Terms (protocol)

The Glossary of Terms should define all of the terms and acronyms that are used in the document or that are specific to the operation being described and may not be generally known to the reader. The reader may also be referred to the FDA document, "Glossary of Computerized System and Software Development Terminology," or other specific glossaries that may be available.

Titles, terms, and definitions that are used in the corporate policies, procedures and guidelines should be defined.

Chapter 19 – Hardware Operational Qualification

Roles and Responsibilities (protocol)

Validation activities are the result of a cooperative effort involving many people, including the System Owners, End Users, the Computer Validation Group, Subject Matter Experts (SME) and Quality Assurance (Unit).

This section clearly identifies the roles, responsibilities and individual(s) responsible for the creation and use of the Hardware Operational Qualification protocol. It specifies the participants by name, title, department or company affiliation and responsibility in a table. Responsibilities are specific for the procedures and activities delineated in each specific document.

The System Owners or End-Users are responsible for the development and understanding of the Hardware Operational Qualification. They acquire and coordinate support from in-house and/or external consultants, the Computer Validation Group, Quality Assurance (Unit) and system developers.

The QA(U) is responsible for verifying that the details in the Hardware Operational Qualification are compliant with applicable departmental, interdepartmental and corporate policies and procedures and that the approach is in accordance with cGxP regulations and good business practices, especially 21 CFR Part 11. The QA(U) should review the executed protocol for completeness.

Computer Operations is responsible for understanding the Computer System Design Specifications document and working with the System Owner/End-Users to develop a Hardware Operational Qualification protocol for the hardware that will satisfy the business needs and requirements as indicated in the User and Functional Requirements.

Only qualified Testers execute the protocol.

The System Owner / End-User writes an OQ Summary Report and ensures that it is reviewed and approved by the System Owner, CVG and QA(U).

The Subject Matter Expert (SME) ensures that changes required as a result of deviations during testing and subsequent remediation are captured in the summary report and incorporated into the Hardware Operational Qualification protocol. If the OQ protocol execution is begun prior to the approval of the OQ summary report, then the SME certifies that deviations and remediations have no effect on the PQ. Otherwise, the affected OQ protocol tests need to be reviewed, repeated as necessary, or re-written and regression testing performed.

Procedure (protocol)

Background Information

This section of the Hardware Operational Qualification protocol provides an overview of the business operations for the system including workflow, material flow and the control philosophy of what functions, calculations and processes are to be performed by the system, as well as interfaces that must be provided to the system or application. It outlines the activities required for commissioning the equipment.

The execution of the protocol verifies compliance with the Hardware Operational Qualification.

Instructions on Testing Procedures

This section describes the procedures that the tester is to follow. It must include how the tests will be documented, how deviations will be handled and other items relevant to the execution of the protocol. Test procedures are in accordance with a Validation Plan.

All OQ testing is done in a qualified pre-production environment unless it is the only system currently running on the equipment during testing. It is not good practice to perform qualification testing of new or revised software, computer or computerized systems in the production environment. Should a major failure occur, such as a system crash, then the previously qualified items running in the production environment must be re-qualified.

Execution of each test procedure should observe the following general instructions:

All signatures or initials are written in permanent ink and dated. Ditto marks or other methods to indicate repeated acceptance of a test must not be used (i.e., individual signatures/initials are required).

Each individual involved must be qualified to perform the assigned tasks. *Qualified* means that the person's training, education and experience are appropriate for the task and are suitably documented.

Each test must list the action to be taken, the expected results (also referred to as the acceptance criteria), the observed results, the disposition of the test (e.g., pass or fail), and the signature/date of the person performing the test.

Results must be documented by one individual and reviewed by a second.

The test results are recorded in the spaces provided as each test is executed. When the test listed on a particular page is completed, the person performing the test signs the space labeled "Performed By" on the bottom of the page. If the test was not completed satisfactorily (i.e., a component fails to meet expected criteria), a "Discrepancy Report" is prepared.

Discrepancy Reports are numbered sequentially and noted on the appropriate protocol page. Each Discrepancy Report summarizes the observed problem and indicates the actions taken to resolve it. Once all items on a page are completely resolved and the re-tested procedure passes, the tester signs and dates the bottom of the page. When the overall test is completed, a person other than the tester reviews the work, accepts it as complete by signing on the "Verified By" line at the bottom of the page noting any exceptions.

Screen prints should be used wherever possible as proof of execution, especially when expected and unexpected errors occur.

All pages of the document must have a page number and the total number of pages of the document. This includes attachments, which are also referenced in the primary document. Attachments must be numbered, signed and dated and marked in such a manner that they can be associated with a specific test step.

A signature sheet (where the person's signature, printed name, initials, and affiliation are recorded) is included for each test document to identify all individuals involved in the execution of the test protocol.

Protocol test deviations or failures identified during the testing are addressed on a case-by-case basis. It is the responsibility of the site team to evaluate and resolve these deviations or failures.

Test deviations or failures are initially documented in the test script and then recorded in a Discrepancy / Deviation Report and a Log. This record includes a description of the problem, when it was detected, the circumstances of detection, and who detected it. The site validation team reviews the failures and decides on appropriate actions. The team's review is documented in the Discrepancy / Deviation Report and Log, which contains the identification of the failure, the date reviewed, assessment of the failure, assessment of the impact of the failure on the validation effort, and the team's recommendation for resolution. The log also indicates when and how the deviation or failure was resolved. For each failure, the team determines whether the failure impacts site or core software. Impact on core software must be communicated to all other site teams.

Deviations / Failures are categorized in accordance with a defined scheme. An example of such a scheme:

- Critical

An issue that is critical to cGxP compliance; affects compliance with company SOPs or regulatory requirements; in any way compromises the integrity of the data; or that could directly impact the identity, strength, quality, or purity of a drug product, the operation of a medical device, data that is subject to regulatory review, or the safety of individuals who operate the system.

- Major

An issue for which there is an acceptable workaround solution.

- Minor

Errors in text fields that do not fall into the Criticalor Major category and are unlikely to be misleading.

- Cosmetic

Trivial errors for which no action is required.

Outcomes of the Review Process:

- Change in Software Configuration or Code

This is in accordance with a Change Control and/or Configuration Management SOP.

- Modify System Requirements or Specifications

For major documentation changes, the document is revised, the revision level incremented and the document sent for re-approval. For minor test script changes a deviation report is generated. If such a change is required, the team may elect to escalate the problem to the implementation team for further study or to management.

- Develop a Manual Procedure to Work Around the Problem

The workaround is written into one of the user SOPs

- Modify the Test Script

If after evaluation it is determined that the system is responding to a test as required, but the test script contains either the wrong instructions or an incorrect expected response, the team may recommend a change in the test script. This requires the team to document the change in the test script and retest after the change is approved.

Following execution of the protocols, the completed documents are reviewed for completeness and compliance with the appropriate procedures. If the protocol document is complete and compliant, the reviewer signs and dates it. This includes the review of any attached test outputs.

Tester Identification

This section identifies each person participating in the testing. The following information is entered in a table for identification purposes:

- Full Name – printed
- Signature
- Initials
- Affiliation

Qualification Tasks

This section of the protocol requires a series of processes and tests that must be performed as part of the Hardware Operational Qualification. Each procedure and/or test section should consist of the following sections:

- Test Section Number and Title
- Purpose of test
- Execution procedures, if appropriate
- Functional Requirement being tested
- Design Specification being tested
- Procedure or Test

Each Test Step should consist of the following sections:

- Test Step number identification
- Action to be performed
- Expected Results (acceptance criteria)
- Actual Results (observations)
- Pass / Fail Results
- Tester Signature / Date
- Comments
- Reviewer Signature / Date

The Procedure or Test section indicates the actions that the tester is expected to perform. These are specific instructions on how to accomplish each required action or task.

The Expected Results section indicates what the tester is expected to observe as a result of the action.

The Actual Results is an entry of what the tester observes, whether the test performs as expected (passes) or does not (fails).

Screen prints should be required wherever possible as proof of execution, especially when expected and unexpected errors occur.

The Pass / Fails Results section should clearly state the conclusion of the test step by the Tester entering the word "Pass" or "Fail." The use of the letter "P" for pass or the letter "F" for fail *is* acceptable.

The Tester Signature / Date section is where the tester initials and dates the results, indicating that the test was performed by that person.

The Comments section may be used to record comments or explanations as required.

The Reviewer Signature / Date section is where the Reviewer signs or initials, and dates each test or procedure indicating that the item has been reviewed and is acceptable as recorded.

If the procedure is a gathering of information, such as serial numbers, there is no need for Expected Results and Pass / Fail sections. The initials of the Tester and date will suffice.

The procedure for the handling of test deviations and failures is specified in the Hardware OQ protocol's Instruction on Testing Procedures section, which is written in accordance with the section Instructions on Testing Procedures of this SOP.

Discrepancy / Deviation Report

This section provides for a form that is used as a Discrepancy / Deviation Report. Whenever a test or procedure results in a discrepancy or deviation, a report must be made detailing the results and explaining the resolution. The form must contain the following sections:

- Report Number
- Discrepancy / Deviation
- Test Section Number and Title
- Test Step Number
- Signature /Date
- Resolution
- Signature / Date
- QA(U) Signature / Date

The Report Number section is a sequential number derived from the Discrepancy / Deviation Log form where the report is to be logged and tracked.

The Discrepancy / Deviation section is used to enter the details of the test or procedure results.

The Test Section Number and Title is used to enter the test section number and title of the test or procedure being reported.

The Test Step Number is used to enter the test or procedure step number being reported.

The Signature / Date section is filled in by the person who issued the report.

The Resolution section is used to enter the details of the resolution or work around employed to remediate the deviation or discrepancy.

The Signature / Date section is filled in by the person who issued the remediation report.

The QA(U) Signature / Date section is used by the QA(U) individual who reviewed and approved the report.

Discrepancy / Deviation Log

This section provides for a form for use as a Discrepancy / Deviation Log. It must contain a table with the following columns:

- Report Number
- OQ
- Test Section Number and Title
- Test Step number
- Deviation / Discrepancy
- Date / Initials

The Report Number is a sequential number used to identify the reports associated with the specific qualification protocol.

The OQ section is used to identify the protocol that the report represents.

The Test Section Number and Title is used to enter the Test Section Number and Title of the test or procedure being reported.

The Test Step Number is used to enter the test or procedure step number being reported.

The Deviation / Discrepancy item is used to enter a brief description of the deviation or discrepancy being logged.

The Date / Initials section is filled in by the person issuing the remediation Report Number.

Document Management and Storage (protocol)

A Document Management and Storage section is included in each Hardware Operational Qualification protocol. It defines the Document Management and Storage methodology that is followed for issue, revision, replacement, control and management of each Hardware Operational Qualification protocol. This methodology is in accordance with an established Change Control SOP and a Document Management and Storage SOP.

Revision History (protocol)

A Revision History section is included in each Hardware Operational Qualification protocol. The Hardware Operational Qualification protocol Version Number and all Revisions made to the document are recorded in this section. These changes are in accordance with a Change Control Procedure SOP.

The Revision History section consists of a summary of the changes, a brief statement explaining the reasons the document was revised, and the name and affiliation of the person responsible for the entries. It includes the numbers of all items the Hardware Operational Qualification protocol replaces and is written so that the original Hardware Operational Qualification protocol is not needed to understand the revisions or the revised document being issued.

When the Hardware Operational Qualification protocol is a new document, a "000" as the Revision Number is entered with a statement such as "Initial Issue" in the comments section.

Regulatory or audit commitments require preservation from being inadvertently lost or modified in subsequent revisions. All entries or modifications made as a result of regulatory or audit commitments require indication as such in the Revision History and should indicate the organization, date and applicable references. A complete and accurate audit trail must be developed and maintained throughout the project lifecycle.

Attachments (protocol)

If needed, an Attachments section is included in each Hardware Operational Qualification protocol in accordance with the SOP on SOPs. An Attachment Index and Attachment section follow the text of a Hardware Operational Qualification protocol. The top of each attachment page indicates "Attachment # x," where 'x' is the sequential number of each attachment. The second line of each attachment page contains the attachment title. The number of pages and date of issue corresponding to each attachment are listed in the Attachment Index.

Attachments are placed directly after the Attachment Index and are sequentially numbered (e.g., Attachment # 1, Attachment # 2) with each page of an attachment containing the corresponding document number, page numbering, and total number of pages in the attachment (e.g., Page x of y). Each attachment is labeled consistent with the title of the attachment itself.

Attachments are treated as part of the Hardware Operational Qualification protocol. If an attachment needs to be changed, the entire Hardware Operational Qualification protocol requires a new revision number and is reissued. New attachments or revisions of existing attachments are in accordance with an established Change Control SOP and a Document Management and Storage SOP.

Document Management and Storage *(SOP)*

A Document Management and Storage section is included in the Hardware Operational Qualification SOP. It defines the Document Management and Storage methodology that is followed for issue, revision, replacement, control and management of this SOP. This methodology is in accordance with an established Change Control SOP and a Document Management and Storage SOP.

Revision History *(SOP)*

A Revision History section is included in the Hardware Operational Qualification SOP. The Hardware Operational Qualification SOP Version Number and all Revisions made to the document are recorded in this section. These changes are in accordance with a Change Control Procedure SOP.

The Revision History section consists of a summary of the changes, a brief statement explaining the reasons the document was revised, and the name and affiliation of the person responsible for the entries. It includes the numbers of all items the Hardware Operational Qualification SOP replaces and is written so that the original Hardware Operational Qualification SOP is not needed to understand the revisions or the revised document being issued.

When the Hardware Operational Qualification SOP is a new document, a "000" as the Revision Number is entered with a statement such as "Initial Issue" in the comments section.

Regulatory or audit commitments require preservation from being inadvertently lost or modified in subsequent revisions. All entries or modifications made as a result of regulatory or audit commitments require indication as such in the Revision History and should indicate the organization, date and applicable references. A complete and accurate audit trail must be developed and maintained throughout the project lifecycle.

Attachments *(SOP)*

If needed, an Attachments section is included in the Hardware Operational Qualification SOP in accordance with the SOP on SOPs. An Attachment Index and Attachment section follow the text of the Hardware Operational Qualification SOP. The top of each attachment page indicates "Attachment # x," where 'x' is the sequential number of each attachment. The second line of each attachment page contains the attachment title. The number of pages and date of issue corresponding to each attachment are listed in the Attachment Index.

Attachments are placed directly after the Attachment Index and are sequentially numbered (e.g., Attachment # 1, Attachment # 2) with each page of an attachment containing the corresponding document number, page numbering, and total number of pages in the attachment (e.g., Page x of y). Each attachment is labeled consistent with the title of the attachment itself.

Attachments are treated as part of the Hardware Operational Qualification SOP. If an attachment needs to be changed, the entire Hardware Operational Qualification SOP requires a new revision number and is reissued. New attachments or revisions of existing attachments are in accordance with an established Change Control SOP and a Document Management and Storage SOP.

Chapter 20 – Software Installation Qualification

In this chapter we will discuss the Standard Operating Procedure (SOP) for creating a Software Installation Qualification protocol for a validation project. This SOP defines the method to be used for specifying the Software Installation Qualification that a computer operations group will use in configuring, qualifying, and implementing application software, firmware, utilities and drivers.

Best industry practices and FDA Guidance documents indicate that in order to demonstrate that a computer or computerized system performs as it purports and is intended to perform, a methodical plan is needed that details how this may be proven. The Software Installation Qualification protocol defines the processes, procedures and tests that are to be used in configuring, implementing and qualifying computer and computer related software, firmware, utility or driver in fulfillment of the requirements of the User and Functional Requirements protocols and Software Design Specifications.

This SOP is used for all systems that create, modify, maintain, retrieve, transmit or archive records in electronic format. It is used in conjunction with an overlying corporate policy to provide sustainable compliance with the regulation. The project team uses it to create the Software Installation Qualification protocol from which computer operations configures, qualifies, and implements the software, firmware, utility or driver.

System Owners, End-Users, and Subject Matter Experts (SME) use this SOPto create the Software Installation Qualification protocol for computer or computer related software that controls equipment and/or produces electronic data or records in support of all regulated laboratory, manufacturing, clinical or distribution activities. The protocols produced in accordance with this SOP are used by Computer Operations to configure, qualify, and implement software, firmware, utility or driver.

Title Block, Headers and Footers (SOP)

The Software Installation Qualification SOP should have a title block, title or cover page along with a header and/or footer on each page of the document that clearly identifies the document. The minimum information that should be displayed for the Software Installation Qualification SOP is as follows:

- Document Title
- Document Number
- Document Revision Number
- Project Name and Number
- Site/Location/Department Identification
- Date Document was issued, last revised or draft status
- Effective Date
- Pagination: page of total pages (e.g., Page x of y)

The Document Title should be in large type font that enables clear identification. It should clearly and concisely describe the contents and intent of the document. Since this document will be part of a set or package of documents within a project, it should be consistent with other documents used for the project and specify that it is the Software Installation Qualification SOP. This information should be on every page of the document.

The Document Number should be in large type font that enables clear identification. Since the Software Installation Qualification SOP will be part of a set or package of documents within a project, it should be consistent with other documents used for the project. This information should appear on every page of the document.

The Document Revision Number, which identifies the revision of the document, should appear on each page of the document. Many documents will go through one or more changes over time, particularly during the draft stage and the review and approval cycle. It is important to maintain accurate revision control because even a subtle change in a document can have an immense impact on the outcomes of activities. The SOP on SOPs should govern the numbering methodology.

See Chapter 6 – The SOP on SOPs.

The Project Name and Number, if applicable, should be indicated on the cover or title page. This information identifies the overall project to which this document belongs.

The Site/Location/Department Identification should be indicated on the cover or title page. It identifies the Site/Location/Department that has responsibility for this document and its associated project.

The Date that the Document was issued, last revised or issued with draft status should be on each page of the document (e.g., 12-Dec-01). This will clearly identify the date on which the document was last modified.

The Effective Date of the Document should be on the title or cover page of the document (e.g., 12-Dec-01). This will clearly identify when to begin using the current version of the document.

Pagination, page of total pages (e.g., Page x of y), should be on each page of the document. This will help identify whether or not all of the pages of the document are present. Pages intentionally left blank should clearly be designated as such.

Approval Section *(SOP)*

The Approval Section should contain a table with the Author and Approvers printed names, titles and a place to sign and date the SOP. The SOP on SOPs should govern who the approvers of the document should be. The functional department (e.g., Production / Quality Assurance / Engineering) of the signatory should be placed below each printed name and title along with a statement associated with each approver indicating the person's role or qualification in the approval process. Indicate the significance of the associated signature (e.g., This document meets the requirements of Corporate Policy 9055, Electronic Records; Electronic Signatures, dated 22-Mar-2000).

The completion (signing) of this section indicates that the contents have been reviewed and approved by the listed individuals.

Table of Contents *(SOP)*

The SOP should contain a Table of Contents, which appears towards the beginning of the document. This Table of Contents provides a way to easily locate various topics contained in the document.

Purpose *(SOP)*

This introduction to the SOP concisely describes the document's purpose in sufficient detail so that a non-technical reader can understand its purpose.

The Purpose section should clearly state that the intention of the document is to provide a Standard Operating Procedure to be used to specify the requirements for the preparation of Software Installation Qualification protocol for a new or upgraded computer or computerized system (how the software, firmware, utility or driver is supposed to be installed) along with the required activities and deliverables. These requirements are necessary for the Software Installation Qualification protocol to be effective and compliant with company polices and government regulation. As one of the foundations of the software application, computer or computerized system, the Software Installation Qualification protocol needs to define the scope of work, responsibilities and deliverables in order to create the software application, computer and/or computerized system and demonstrate that it meets its intended use in a reliable, consistent and compliant manner. The Software Installation Qualification protocol document derived from this SOP will become the standard that is used to configure and deploy a specific software, firmware, utility, or driver.

Scope *(SOP)*

The Scope section identifies the limitations of the Software Installation Qualification SOP. It limits the scope to all GxP systems and is applicable to all computers, computerized systems, infrastructure, LANs and WANs including clinical, laboratory, manufacturing and distribution computer systems. It applies to all Sites, Locations and Departments within the company.

Equipment / Materials / Tools *(SOP)*

This section lists any required or recommended equipment, materials, tools, templates or guides. Provide the name and a brief description of the equipment, instruments, materials and/or tools to be used in the procedures or testing covered by the Software Installation Qualification SOP. This is for reference purposes and requires the user to utilize the specific item or a designated equivalent. The purpose and use is explained elsewhere in the document. Substitution of comparable equipment may be made in accordance with procedures for Change Control.

If the information is included in a job aid or operating manual, a description of where the information may be found with a specific reference to the document number, title, date and revision number should be specified.

Warnings / Notes / General Information / Safety *(SOP)*

Warnings, notes, general or safety information applicable to the Software Installation Qualification SOP are listed in this section.

Policies, Guidelines, References *(SOP)*

This section lists references to appropriate regulatory requirements, SOPs, policies, guidelines, or other relevant documents referred to in the Software Installation Qualification SOP. The name and a brief description of applicable related SOPs, corporate policies, procedures and guidelines, Quality Control monographs, books, technical papers, publications, specific regulations, batch records, vendor or system manuals, and all other relevant documents that are required by the Software Installation Qualification SOP should be included. This is for reference purposes and requires the user to use the specific item or a designated equivalent. The purpose and use are explained elsewhere in the document.

SOPs cite references and incorporate them in whole by reference rather than specifying details that duplicate these items. This is done so that changes in policy, guidelines or reference materials require no changes to SOPs themselves and the most current version of the referenced material is used. Examples are:

- Policy ABC-123, Administrative Policy for Electronic Records; Electronic Signatures
- Corporate, Plan for Compliance with 21 CFR Part 11, Electronic Records; Electronic Signatures
- SOP-456, Electronic Records; Electronic Signatures
- FDA, 21 CFR Part 11, Electronic Records; Electronic Signatures

- FDA, General Principals of Software Validation; Final Guidance for Industry and FDA Staff, January 11, 2002

***Assumptions / Exclusions / Limitations** (SOP)*

Any and all assumptions, exclusions or limitations within the context of the SOP should be discussed. Specific issues should be directly addressed in this section of the Software Installation Qualification SOP. For example:

"Only application software, firmware, utilities, and drivers should normally be addressed."

***Glossary of Terms** (SOP)*

A Glossary of Terms should be included that defines all of the terms and acronyms that are used in this document or that are specific to the operation being described and may not be generally known to the reader. The reader may also be referred to the FDA document, "Glossary of Computerized System and Software Development Terminology."

***Roles and Responsibilities** (SOP)*

The Roles and Responsibilities of the Software Installation Qualification protocol should be divided into two specific areas:

Creation and Maintenance

The System Owner and End-Users of the computer or computerized system are responsible for the preparation and maintenance of the Software Installation Qualification.

It is common for a Software Installation Qualification to go through several revis ions during the course of a project, as it is considered a living document, especially during the project's early stages. As a normal process, it helps to ensure that the validation activities are in compliance with cGxPs, FDA regulations, company policies, industry standards, and guidelines.

The roles and responsibilities for preparing the Software Installation Qualification protocol and other project documentation are addressed in accordance with established Standard Operating Procedures.

Use

The members of the Project Team must be familiar with the Software Installation Qualification in order to fulfill the project requirements. The Quality Assurance (Unit), Project Team and Computer Operations use the Software Installation Qualification in concert with other project documentation as a contract that details the project and quality expectations.

The Software Installation Qualification is one of the governing documents for the qualification and testing of the software, firmware, utility, or driver it covers.

***Procedure** (SOP)*

The SOP should contain instructions and procedures that enable the creation of a Software Installation Qualification protocol document. The Software Installation Qualification is required in order to provide the installation qualification requirements for the required software that can be used by a Computer Operations Group to install, qualify, and implement a software application, firmware, utility or driver.

The Software Installation Qualification is prepared as a stand-alone document for all projects. It is not good practice to include it as part of other documents in an attempt to reduce the amount of workload or paperwork. This only causes more work over the length of the project resulting from misunderstandings and confusion. Clear, concise documents that address specific parts of any given project are desirable.

IQ provides a means to test and record the proper installation of the software and software components of a system. This will also provide a baseline record of the system at startup or at its last upgrade that can be used for determining what changes may be required in the future. It ensures that the requirements, regulations and codes have been met.

The Software Installation Qualification protocol provides an overall technical description of the installation qualification requirements of a piece of software. It provides the final information and documentation requirements that are used by Computer Operations to configure, test, and install a software application, firmware, utility or driver.

Installation Qualification requires the identification and inventorying of documentation for all software components, utilities, drivers and related systems, installation procedures, configuration data and diagrams (e.g., data maps, entity relation diagrams) used to build the application. The information is recorded, verified correct reflecting the actual "as built" configuration of the system.

If the system relies on existing components or infrastructure outside of End-User control, such as a server or network managed by Computer Operations, these are qualified by the appropriate group in accordance with approved procedures. The End-User is not required to duplicate these efforts. However, it is required that the End-User reference and include copies of these documents in the validation package.

Post Execution Approval (protocol)

This section contains a place for the QA(U) approval of the executed version of the protocol. The QA(U) signature in this section indicates that the IQ protocol has been properly executed from a compliance standpoint. All sections were completed, all tests and re-tests properly executed, no blank spaces exist and all deviations were properly remediated. It also means that each page of the protocol was reviewed. Should QA(U) determine that the even though the procedure was properly followed but the IQ failed, it is so noted upon signing. The phrase "Disapproved for Use" along with a brief explanation is clearly entered in association with the signature.

Purpose (protocol)

Clearly state the software, firmware, utility or drivers with their installation qualification requirements and information as to why it is considered for use in a regulated system. This short introduction should be concise and present details so that an unfamiliar reader can understand the purpose of the document. For example:

> "The purpose of the Software Installation Qualification Protocol is to verify and document the purchase and/or proper installation of XYZ Software Application with its associated drivers. The requirements are provided in the Software System Design Specification document and include:
>
> - Application Software
> - Interface Software
> - Utility Software
> - Device Drivers"

Scope (protocol)

The Scope section identifies the limitations of a specific Software Installation Qualification protocol in relationship to validating a specific software application, computer or computerized system. For example, if part of the system consists of a Local area network (LAN) that has already been qualified, then the qualification of the LAN may be exc luded by stating that it is being addressed elsewhere.

Equipment / Materials / Tools (protocol)

This section lists any required or recommended equipment, materials, tools, templates or guides. Provide the name and a brief description of the equipment, instruments, materials and/or tools to be used in the procedures or testing covered by the Software Installation Qualification protocol. This is for reference purposes and requires the user to utilize the specific item or a designated equivalent. The purpose and use is explained elsewhere in the document. Substitution of comparable equipment may be made in accordance with procedures for Change Control.

If the information is included in a job aid or operating manual, a description of where the information may be found with a specific reference to the document number, title, date and revision number should be specified.

Warnings / Notes / General Information / Safety (protocol)

Warnings, notes, general or safety information applicable to the Software Installation Qualification protocol are listed in this section.

Policies / Guidelines / References (protocol)

This section lists references to appropriate regulatory requirements, SOPs, policies, guidelines, or other relevant documents referred to in the Software Installation Qualification protocol . The name and a brief description of applicable related SOPs, corporate policies, procedures and guidelines, Quality Control monographs, books, technical papers, publications, specific regulations, batch records, vendor or system manuals, and all other relevant documents that are required by the Software Installation Qualification protocol should be included. This is for reference purposes and requires the user to use the specific item or a designated equivalent. The purpose and use are explained elsewhere in the document.

Assumptions / Exclusions / Limitations (protocol)

Discuss any assumptions, exclusions or limitations within the context of the Software Installation Qualification. Specific issues should be directly addressed in this section of the Software Installation Qualification protocol.

Glossary of Terms (protocol)

The Glossary of Terms should define all of the terms and acronyms that are used in the document or that are specific to the operation being described and may not be generally known to the reader. The reader may also be referred to the FDA document, "Glossary of Computerized System and Software Development Terminology," or other specific glossaries that may be available.

Titles, terms, and definitions that are used in the corporate policies, procedures and guidelines should be defined.

Chapter 20 – Software Installation Qualification

Roles and Responsibilities (protocol)

Validation activities are the result of a cooperative effort involving many people, including the System Owners, End Users, the Computer Validation Group, Subject Matter Experts (SME) and Quality Assurance (Unit.

This section requires clear identification of the roles, responsibilities and individual(s) responsible for the creation and use of the Software Installation Qualification. It specifies the participants by name, title, department or company affiliation and responsibility in a table. Responsibilities are specific for the procedures and activities delineated in each specific document.

The System Owners or End-Users are responsible for the development and understanding of the Software Installation Qualification. They acquire and coordinate support from in-house and/or external consultants, the Computer Validation Group, Quality Assurance (Unit) and System Developers.

The QA(U) is responsible for verifying that the details in the Software Installation Qualification are compliant with applicable departmental, interdepartmental and corporate policies and procedures, and that the approach is in accordance with cGxP regulations and good business practices, especially 21 CFR Part 11. The QA(U) reviews the executed protocol for completeness.

Computer Operations is responsible for understanding the Computer System Design Specifications document and working with the System Owner/End-Users to develop a Software Installation qualification protocol for the software, firmware, utility, or driver that will satisfy the business needs and requirements as indicated in the User and Functional Requirements.

Only qualified Testers execute the protocol.

The System Owner / End-User writes an IQ Summary Report and ensures that it is reviewed and approved by the System Owner, CVG and QA(U).

The Subject Matter Expert (SME) ensures that changes required as a result of deviations during testing and subsequent remediation are captured in the summary report and incorporated into the Software Operational Qualification protocol. If the OQ protocol execution is begun prior to the approval of the IQ summary report, then the SME certifies that deviations and remediations have no effect on the OQ. Otherwise, the affected OQ protocol tests need to be reviewed, repeated as necessary, or re-written and regression testing performed.

Procedure (protocol)

Background Information

This section of the Software Installation Qualification Testing protocol provides an overview of the business operations for the system including workflow, material flow and the control philosophy of what functions, calculations and processes are to be performed by the system, as well as interfaces that must be provided to the system or application. It outlines the activities required for commissioning the equipment.

The execution of the protocol verifies compliance with the Software Installation Qualification.

Instructions on Testing Procedures

This section describes the procedures that the tester is to follow. It must include how the tests will be documented, how deviations will be handled and other items relevant to the execution of the protocol. Test procedures should be in accordance with a Validation Plan.

All IQ testing is done in a qualified pre-production environment unless it is the only system currently running on the equipment during testing. It is not good practice to perform qualification testing of new or revised software, computer or computerized systems in the production environment. Should a major failure occur, such as a system crash, then the previously qualified items running in the production environment must be re-qualified.

Execution of each test procedure should observe the following general instructions:

All signatures or initials are written in permanent ink and dated. Ditto marks or other methods to indicate repeated acceptance of a test must not be used (i.e., individual signatures/initials are required).

Each individual involved must be qualified to perform the assigned tasks. *Qualified* means that the person's training, education and experience are appropriate for the task and are suitably documented.

Each test must list the action to be taken, the expected results (also referred to as the acceptance criteria), the observed results, the disposition of the test (e.g., pass or fail), and the signature/date of the person performing the test.

Results must be documented by one individual and reviewed by a second.

The test results are recorded in the spaces provided as each test is executed. When the test listed on a particular page is completed, the person performing the test signs the space labeled "Performed By" on the bottom of the page. If the test was not completed satisfactorily (i.e., a component fails to meet expected criteria), a "Discrepancy Report" is prepared.

Discrepancy Reports are numbered sequentially and noted on the appropriate protocol page. Each Discrepancy Report summarizes the observed problem and indicates the actions taken to resolve it. Once all items on a page are completely resolved and the re-tested procedure passes, the tester signs and dates the bottom of the page. When the overall test is completed, a person other than the tester reviews the work, accepts it as complete by signing on the "Verified By" line at the bottom of the page noting any exceptions.

Screen prints should be used wherever possible as proof of execution, especially when expected and unexpected errors occur.

All pages of the document must have a page number and the total number of pages of the document. This includes attachments, which are also referenced in the primary document. Attachments must be numbered, signed and dated and marked in such a manner that they can be associated with a specific test step.

A signature sheet (where the person's signature, printed name, initials, and affiliation are recorded) is included for each test document to identify all individuals involved in the execution of the test protocol.

Protocol test deviations or failures identified during the testing are addressed on a case-by-case basis. It is the responsibility of the site team to evaluate and resolve these deviations or failures.

Test deviations or failures are initially documented in the test script and then recorded in a Discrepancy / Deviation Report and a Log. This record includes a description of the problem, when it was detected, the circumstances of detection, and who detected it. The site validation team reviews the failures and decides on appropriate actions. The team's review is documented in the Discrepancy / Deviation Report and Log, which contains the identification of the failure, the date reviewed, assessment of the failure, assessment of the impact of the failure on the validation effort, and the team's recommendation for resolution. The log also indicates when and how the deviation or failure was resolved. For each failure, the team determines whether the failure impacts site or core software. Impact on core software must be communicated to all other site teams.

Deviations / Failures are categorized in accordance with a defined scheme. An example of such a scheme:

- Critical

An issue that is critical to cGxP compliance; affects compliance with company SOPs or regulatory requirements; in any way compromises the integrity of the data; or that could directly impact the identity, strength, quality, or purity of a drug product, the operation of a medical device, data that is subject to regulatory review, or the safety of individuals who operate the system.

- Major

An issue for which there is an acceptable workaround solution.

- Minor

Errors in text fields that do not fall into the Critical or Major category and are unlikely to be misleading.

- Cosmetic

Trivial errors for which no action is required.

Outcomes of the Review Process:

- Change in Software Configuration or Code

This is in accordance with a Change Control and/or Configuration Management SOP.

- Modify System Requirements or Specifications

For major documentation changes, the document is revised, the revision level incremented and the document sent for re-approval. For minor test script changes a deviation report is generated. If such a change is required, the team may elect to escalate the problem to the implementation team for further study or to management.

- Develop a Manual Procedure to Work Around the Problem

The workaround is written into one of the user SOPs

- Modify the Test Script

If after evaluation it is determined that the system is responding to a test as required, but the test script contains either the wrong instructions or an incorrect expected response, the team may recommend a change in the test script. This requires the team to document the change in the test script and retest after the change is approved.

Following execution of the protocols, the completed documents are reviewed for completeness and compliance with the appropriate procedures. If the protocol document is complete and compliant, the reviewer signs and dates it. This includes the review of any attached test outputs.

Tester Identification

This section identifies each person participating in the testing. The following information is entered in a table for identification purposes:

- Full Name – printed
- Signature
- Initials
- Affiliation

Software Inventory

This section specifies the required software, firmware, utility, or driver to be inventoried and installed, and provides a table to record their verification.

Document Inventory

This section specifies the required documentation to be inventoried and provides a table to record their verification.

Qualification Tasks

This section of the protocol requires a series of processes and tests that must be performed as part of the Software Installation Qualification. Each procedure and/or test section should consist of the following sections:

- Test Section Number and Title
- Purpose of test
- Execution procedures, if appropriate
- Functional Requirement being tested
- Design Specification being tested
- Procedure or Test

Each Test Step should consist of the following sections:

- Test Step number identification
- Action to be performed
- Expected Results (acceptance criteria)
- Actual Results (observations)
- Pass / Fail Results
- Tester Signature / Date
- Comments
- Reviewer Signature / Date

The Procedure or Test section indicates the actions that the tester is expected to perform. These are specific instructions on how to accomplish each required action or task.

The Expected Results section indicates what the tester is expected to observe as a result of the action.

The Actual Results is an entry of what the tester observes, whether the test performs as expected (passes) or does not (fails).

Screen prints should be required wherever possible as proof of execution, especially when expected and unexpected errors occur.

The Pass / Fails Results section should clearly state the conclusion of the test step by the Tester entering the word "Pass" or "Fail." The use of the letter "P" for pass or the letter "F" for fail *is* acceptable.

The Tester Signature / Date section is where the tester initials and dates the results, indicating that the test was performed by that person.

The Comments section may be used to record comments or explanations as required.

The Reviewer Signature / Date section is where the Reviewer signs or initials, and dates each test or procedure indicating that the item has been reviewed and is acceptable as recorded.

If the procedure is a gathering of information, such as serial numbers, there is no need for Expected Results and Pass / Fail sections. The initials of the Tester and date will suffice.

The procedure for the handling of test deviations and failures is specified in the Software IQ protocol's Instruction on Testing Procedures section, which is written in accordance with the section Instructions on Testing Procedures of this SOP.

Discrepancy / Deviation Report

This section provides for a form that is used as a Discrepancy / Deviation Report. Whenever a test or procedure results in a discrepancy or deviation, a report must be made detailing the results and explaining the resolution. The form must contain the following sections:

- Report Number
- Discrepancy / Deviation
- Test Section Number and Title
- Test Step Number
- Signature /Date
- Resolution
- Signature / Date
- QA(U) Signature / Date

The Report Number section is a sequential number derived from the Discrepancy / Deviation Log form where the report is to be logged and tracked.

The Discrepancy / Deviation section is used to enter the details of the test or procedure results.

The Test Section Number and Title is used to enter the test section number and title of the test or procedure being reported.

The Test Step Number is used to enter the test or procedure step number being reported.

The Signature / Date section is filled in by the person who issued the report.

The Resolution section is used to enter the details of the resolution or work around employed to remediate the deviation or discrepancy.

The Signature / Date section is filled in by the person who issued the remediation report.

The QA(U) Signature / Date section is used by the QA(U) individual who reviewed and approved the report.

Discrepancy / Deviation Log

This section provides for a form for use as a Discrepancy / Deviation Log. It must contain a table with the following columns:

- Report Number
- IQ
- Test Section Number and Title
- Test Step number
- Deviation / Discrepancy
- Date / Initials

The Report Number is a sequential number used to identify the reports associated with the specific qualification protocol.

The IQ section is used to identify the protocol that the report represents.

The Test Section Number and Title is used to enter the Test Section Number and Title of the test or procedure being reported.

The Test Step Number is used to enter the test or procedure step number being reported.

The Deviation / Discrepancy item is used to enter a brief description of the deviation or discrepancy being logged.

The Date / Initials section is filled in by the person issuing the remediation Report Number.

Document Management and Storage (protocol)

A Document Management and Storage section is included in each Software Installation Qualification protocol. It defines the Document Management and Storage methodology that is followed for issue, revision, replacement, control and management of each Software Installation Qualification protocol. This methodology is in accordance with an established Change Control SOP and a Document Management and Storage SOP.

Revision History (protocol)

A Revision History section is included in each Software Installation Qualification protocol. The Software Installation Qualification protocol Version Number and all Revisions made to the document are recorded in this section. These changes are in accordance with a Change Control Procedure SOP.

The Revision History section consists of a summary of the changes, a brief statement explaining the reasons the document was revised, and the name and affiliation of the person responsible for the entries. It includes the numbers of all items the Software Installation Qualification protocol replaces and is written so that the original

Software Installation Qualification protocol is not needed to understand the revisions or the revised document being issued.

When the Software Installation Qualification protocol is a new document, a "000" as the Revision Number is entered with a statement such as "Initial Issue" in the comments section.

Regulatory or audit commitments require preservation from being inadvertently lost or modified in subsequent revisions. All entries or modifications made as a result of regulatory or audit commitments require indication as such in the Revision History and should indicate the organization, date and applicable references. A complete and accurate audit trail must be developed and maintained throughout the project lifecycle.

Attachments (protocol)

If needed, an Attachments section is included in each Software Installation Qualification protocol in accordance with the SOP on SOPs. An Attachment Index and Attachment section follow the text of a Software Installation Qualification protocol. The top of each attachment page indicates "Attachment # x," where 'x' is the sequential number of each attachment. The second line of each attachment page contains the attachment title. The number of pages and date of issue corresponding to each attachment are listed in the Attachment Index.

Attachments are placed directly after the Attachment Index and are sequentially numbered (e.g., Attachment # 1, Attachment # 2) with each page of an attachment containing the corresponding document number, page numbering, and total number of pages in the attachment (e.g., Page x of y). Each attachment is labeled consistent with the title of the attachment itself.

Attachments are treated as part of the Software Installation Qualification protocol. If an attachment needs to be changed, the entire Software Installation Qualification protocol requires a new revision number and is reissued. New attachments or revisions of existing attachments are in accordance with an established Change Control SOP and a Document Management and Storage SOP.

Document Management and Storage *(SOP)*

A Document Management and Storage section is included in the Software Installation Qualification SOP. It defines the Document Management and Storage methodology that is followed for issue, revision, replacement, control and management of this SOP. This methodology is in accordance with an established Change Control SOP and a Document Management and Storage SOP.

Revision History *(SOP)*

A Revision History section is included in the Software Installation Qualification SOP. The Software Installation Qualification SOP Version Number and all Revisions made to the document are recorded in this section. These changes are in accordance with a Change Control Procedure SOP.

The Revision History section consists of a summary of the changes, a brief statement explaining the reasons the document was revised, and the name and affiliation of the person responsible for the entries. It includes the numbers of all items the Software Installation Qualification SOP replaces and is written so that the original Software Installation Qualification SOP is not needed to understand the revisions or the revised document being issued.

When the Software Installation Qualification SOP is a new document, a "000" as the Revision Number is entered with a statement such as "Initial Issue" in the comments section.

Regulatory or audit commitments require preservation from being inadvertently lost or modified in subsequent revisions. All entries or modifications made as a result of regulatory or audit commitments require indication as such in the Revision History and should indicate the organization, date and applicable references. A complete and accurate audit trail must be developed and maintained throughout the project lifecycle.

***Attachments** (SOP)*

If needed, an Attachments section is included in the Software Installation Qualification SOP in accordance with the SOP on SOPs. An Attachment Index and Attachment section follow the text of the Software Installation Qualification SOP. The top of each attachment page indicates "Attachment # x," where 'x' is the sequential number of each attachment. The second line of each attachment page contains the attachment title. The number of pages and date of issue corresponding to each attachment are listed in the Attachment Index.

Attachments are placed directly after the Attachment Index and are sequentially numbered (e.g., Attachment # 1, Attachment # 2) with each page of an attachment containing the corresponding document number, page numbering, and total number of pages in the attachment (e.g., Page x of y). Each attachment is labeled consistent with the title of the attachment itself.

Attachments are treated as part of the Software Installation Qualification SOP. If an attachment needs to be changed, the entire Software Installation Qualification SOP requires a new revision number and is reissued. New attachments or revisions of existing attachments are in accordance with an established Change Control SOP and a Document Management and Storage SOP.

Chapter 21 – Software Operational Qualification

In this chapter we will discuss the Standard Operating Procedure (SOP) for creating a Software Operational Qualification (OQ) protocol for a validation project. This SOP defines the method to be used for specifying the Software Operational Qualification that a computer operations group will use in qualifying computer software, firmware, utilities or drivers.

Best industry practices and FDA Guidance documents indicate that in order to demonstrate that a computer or computerized system performs as it purports and is intended to perform, a methodical plan is needed that details how this may be proven. The Software Operational Qualification protocol defines the specifications that are to be used in qualifying computer software, firmware, utilities, or drivers in fulfillment of the requirements of the User and Functional Requirements protocols and the Software Design Specification.

This SOP is used for all systems that create, modify, maintain, retrieve, transmit or archive records in electronic format. It is used in conjunction with an overlying corporate policy to provide sustainable compliance with the regulation. The project team uses it to create the Software Operational Qualification protocol from which computer operations qualifies the specified software, firmware, utility, or driver.

System Owners, End-Users, and Subject Matter Experts (SME) use this SOP to create the Software Operational Qualification protocol for computer software and firmware that controls equipment and/or produces of electronic data or records in support of all regulated laboratory, manufacturing, clinical or distribution activities. The protocols produced in accordance with this SOP are used by Computer Operations to qualify the specified software, firmware, utilities, or drivers.

Title Block, Headers and Footers (SOP)

The Software Operational Qualification SOP should have a title block, title or cover page along with a header and/or footer on each page of the document that clearly identifies the document. The minimum information that should be displayed for the Software Operational Qualification SOP is as follows:

- Document Title
- Document Number
- Document Revision Number
- Project Name and Number
- Site/Location/Department Identification
- Date Document was issued, last revised or draft status
- Effective Date
- Pagination: page of total pages (e.g., Page x of y)

The Document Title should be in large type font that enables clear identification. It should clearly and concisely describe the contents and intent of the document. Since this document will be part of a set or package of documents within a project, it should be consistent with other documents used for the project and specify that it is the Software Operational Qualification SOP. This information should be on every page of the document.

The Document Number should be in large type font that enables clear identification. Since the Software Operational Qualification SOP will be part of a set or package of documents within a project, it should be consistent with other documents used for the project. This information should appear on every page of the document.

The Document Revision Number, which identifies the revision of the document, should appear on each page of the document. Many documents will go through one or more changes over time, particularly during the draft stage and the review and approval cycle. It is important to maintain accurate revision control because even a subtle change in a document can have an immense impact on the outcomes of activities. The SOP on SOPs should govern the numbering methodology.

See Chapter 6 – "The SOP on SOPs."

The Project Name and Number, if applicable, should be indicated on the cover or title page. This information identifies the overall project to which this document belongs.

The Site/Location/Department Identification should be indicated on the cover or title page. It identifies the Site/Location/Department that has responsibility for this document and its associated project.

The Date that the Document was issued, last revised or issued with draft status should be on each page of the document (e.g., 12-Dec-01). This will clearly identify the date on which the document was last modified.

The Effective Date of the Document should be on the title or cover page of the document (e.g., 12-Dec-01). This will clearly identify when to begin using the current version of the document.

Pagination, page of total pages (e.g., Page x of y), should be on each page of the document. This will help identify whether or not all of the pages of the document are present. Pages intentionally left blank should clearly be designated as such.

Approval Section *(SOP)*

The Approval Section should contain a table with the Author and Approvers printed names, titles and a place to sign and date the SOP. The SOP on SOPs should govern who the approvers of the document should be. The functional department (e.g., Production / Quality Assurance / Engineering) of the signatory should be placed below each printed name and title along with a statement associated with each approver indicating the person's role or qualification in the approval process. Indicate the significance of the associated signature (e.g., This document meets the requirements of Corporate Policy 9055, Electronic Records; Electronic Signatures, dated 22-Mar-2000).

The completion (signing) of this section indicates that the contents have been reviewed and approved by the listed individuals.

Table of Contents *(SOP)*

The SOP should contain a Table of Contents, which appears towards the beginning of the document. This Table of Contents provides a way to easily locate various topics contained in the document.

Purpose *(SOP)*

This introduction to the SOP concisely describes the document's purpose in sufficient detail so that a non-technical reader can understand its purpose.

The Purpose section should clearly state that the intention of the document is to provide a Standard Operating Procedure to be used to specify the requirements for the preparation of Software Operational Qualification protocol for a new or upgraded computer or computerized system (how the software is supposed to be installed) along with the required activities and deliverables. These requirements are necessary for the

Software Operational Qualification protocol to be effective and compliant with company polices and government regulation. As one of the foundations of the software application, computer or computerized system, the Software Operational Qualification protocol needs to define the scope of work, responsibilities and deliverables in order to create the software application, computer and/or computerized system and demonstrate that it meets its intended use in a reliable, consistent and compliant manner. The Software Operational Qualification protocol document derived from this SOP will become the standard that is used to functionally operate a specific software or firmware application.

Scope *(SOP)*

The Scope section identifies the limitations of the Software Operational Qualification SOP. It limits the scope to all GxP systems and is applicable to all computers, computerized systems, infrastructure, LANs and WANs including clinical, laboratory, manufacturing and distribution computer systems. It applies to all Sites, Locations and Departments within the company.

Equipment / Materials / Tools *(SOP)*

This section lists any required or recommended equipment, materials, tools, templates or guides. Provide the name and a brief description of the equipment, instruments, materials and/or tools to be used in the procedures or testing covered by the Software Operational Qualification SOP. This is for reference purposes and requires the user to utilize the specific item or a designated equivalent. The purpose and use is explained elsewhere in the document. Substitution of comparable equipment may be made in accordance with procedures for Change Control.

If the information is included in a job aid or operating manual, a description of where the information may be found with a specific reference to the document number, title, date and revision number should be specified.

Warnings / Notes / General Information / Safety *(SOP)*

Warnings, notes, general or safety information applicable to the Software Operational Qualification SOP are listed in this section.

Policies, Guidelines, References *(SOP)*

This section lists references to appropriate regulatory requirements, SOPs, policies, guidelines, or other relevant documents referred to in the Software Operational Qualification SOP. The name and a brief description of applicable related SOPs, corporate policies, procedures and guidelines, Quality Control monographs, books, technical papers, publications, specific regulations, batch records, vendor or system manuals, and all other relevant documents that are required by the Software Operational Qualification SOP should be included. This is for reference purposes and requires the user to use the specific item or a designated equivalent. The purpose and use are explained elsewhere in the document.

SOPs cite references and incorporate them in whole by reference rather than specifying details that duplicate these items. This is done so that changes in policy, guidelines or reference materials require no changes to SOPs themselves and the most current version of the referenced material is used. Examples are:

- Policy ABC-123, Administrative Policy for Electronic Records; Electronic Signatures
- Corporate, Plan for Compliance with 21 CFR Part 11, Electronic Records; Electronic Signatures
- SOP-456, Electronic Records; Electronic Signatures
- FDA, 21 CFR Part 11, Electronic Records; Electronic Signatures

- FDA, General Principals of Software Validation; Final Guidance for Industry and FDA Staff, January 11, 2002

Assumptions / Exclusions / Limitations *(SOP)*

Any and all assumptions, exclusions or limitations within the context of the SOP should be discussed. Specific issues should be directly addressed in this section of the Software Operational Qualification SOP. For example:

"Only application software, firmware, utilities, and drivers should normally be addressed."

Glossary of Terms *(SOP)*

A Glossary of Terms should be included that defines all of the terms and acronyms that are used in this document or that are specific to the operation being described and may not be generally known to the reader. The reader may also be referred to the FDA document, "Glossary of Computerized System and Software Development Terminology."

Roles and Responsibilities *(SOP)*

The Roles and Responsibilities of the Software Operational Qualification protocol should be divided into two specific areas:

Creation and Maintenance

The System Owner of the computer or computerized system and Computer Operations is responsible for the preparation and maintenance of the Software Operational Qualification.

It is common for a Software Operational Qualification to go through several revisions during the course of a project, as it is considered a living document, especially during the project's early stages. As a normal process, it helps to ensure that the validation activities are in compliance with cGxPs, FDA regulations, company policies, industry standards, and guidelines.

The roles and responsibilities for preparing the Software Operational Qualification protocol and other project documentation are addressed in accordance with established Standard Operating Procedures.

Use

The members of the Project Team must be familiar with the Software Operational Qualification in order to fulfill the project requirements. The Quality Assurance (Unit), Project Team and Computer Operations use the Software Operational Qualification in concert with other project documentation as a contract that details the project and quality expectations.

The Software Operational Qualification is one of the governing documents for the qualification and testing of the software it covers.

Procedure *(SOP)*

The Software Operation Qualification SOP should contain instructions and procedures that enable the creation of a Software Operational Qualification protocol document. The Software Operational Qualification is required in order to provide the operational qualification requirements for the required software that can be used by a Computer Operations Group to qualify the specified software, firmware, utility, or driver for use.

The Software Operational Qualification is prepared as a stand-alone document for all projects. It is not good practice to include it as part of other documents in an attempt to reduce the amount of workload or paperwork. This only causes more work over the length of the project resulting from misunderstandings and confusion. Clear, concise documents that address specific parts of any given project are desirable.

OQ provides a means to test and record the proper operation of the software and software components of a system. This will also provide a baseline record of the system at startup or at its last upgrade that can be used for determining what changes may be required in the future. It ensures that the software functions as purported and intended.

The Software Operational Qualification protocol provides an overall technical description of the operational qualification requirements of software, firmware, utilities, or drivers. It provides the final information and documentation requirements that are used by Computer Operations to test a software application, firmware, utilities or drivers for proper operation.

If the system relies on existing components or infrastructure outside of End-User control, such as a server or network managed by Computer Operations, these are qualified by the appropriate group in accordance with approved procedures. The End-User is not required to duplicate these efforts. However, it is required that the End-User reference and include copies of these documents in the validation package.

Data used in qualification scripts must be identified, justified for its use and simulate actual use.

Boundary testing is performed. That is, data that is outside, on, and inside the boundaries of a given function to verify that these conditions are properly handled

Post Execution Approval (protocol)

This section contains a place for the QA(U) approval of the executed version of the protocol. The QA(U) signature in this section indicates that the OQ protocol has been properly executed from a compliance standpoint. All sections were completed, all tests and re-tests properly executed, no blank spaces exist and all deviations were properly remediated. It should also means that each page of the protocol was reviewed. Should QA(U) determine that the even though the procedure was properly followed but the OQ failed, it is so noted upon signing. The phrase "Disapproved for Use" along with a brief explanation is clearly entered in association with the signature.

Purpose (protocol)

This section should clearly state the software or firmware application with its operational qualification requirements and information as to why it is considered for use in a regulated system. This short introduction should be concise and present details so that an unfamiliar reader can understand the purpose of the document. For example:

> "The purpose of the Operational Qualification Protocol is to verify and document the proper operation of the specified software. These are provided in the User Requirements, Functional Requirements and Software Design Specification documents and include:
>
> - Application Software
> - Firmware"

Scope (protocol)

The Scope section identifies the limitations of a specific Software Operational Qualification protocol in relationship to validating a specific software application, computer or computerized system. For example, if part of the system consists of a Local area network (LAN) that has already been qualified, then the qualification of the LAN may be excluded by stating that it is being addressed elsewhere.

Equipment / Materials / Tools (protocol)

This section lists any required or recommended equipment, materials, tools, templates or guides. Provide the name and a brief description of the equipment, instruments, materials and/or tools to be used in the procedures or testing covered by the Software Operational Qualification protocol. This is for reference purposes and requires the user to utilize the specific item or a designated equivalent. The purpose and use is explained elsewhere in the document. Substitution of comparable equipment may be made in accordance with procedures for Change Control.

If the information is included in a job aid or operating manual, a description of where the information may be found with a specific reference to the document number, title, date and revision number should be specified.

Warnings / Notes / General Information / Safety (protocol)

Warnings, notes, general or safety information applicable to the Software Operational Qualification protocol are listed in this section.

Policies / Guidelines / References (protocol)

This section lists references to appropriate regulatory requirements, SOPs, policies, guidelines, or other relevant documents referred to in the Software Operational Qualification protocol. The name and a brief description of applicable related SOPs, corporate policies, procedures and guidelines, Quality Control monographs, books, technical papers, publications, specific regulations, batch records, vendor or system manuals, and all other relevant documents that are required by the Software Operational Qualification protocol should be included. This is for reference purposes and requires the user to use the specific item or a designated equivalent. The purpose and use are explained elsewhere in the document.

Assumptions / Exclusions / Limitations (protocol)

Discuss any assumptions, exclusions or limitations within the context of the Software Operational Qualification. Specific issues should be directly addressed in this section of the Software Operational Qualification protocol.

Glossary of Terms (protocol)

The Glossary of Terms should define all of the terms and acronyms that are used in the document or that are specific to the operation being described and may not be generally known to the reader. The reader may also be referred to the FDA document, "Glossary of Computerized System and Software Development Terminology," or other specific glossaries that may be available.

Titles, terms, and definitions that are used in the corporate policies, procedures and guidelines should be defined.

Chapter 21 – Software Operational Qualification

Roles and Responsibilities (protocol)

Validation activities are the result of a cooperative effort involving many people, including the System Owners, End Users, the Computer Validation Group, Subject Matter Experts (SME) and Quality Assurance (Unit).

This section requires clear identification of the roles, responsibilities, and individual(s) responsible for the creation and use of the Software Operational Qualification. It specifies the participants by name, title, department, or company affiliation and responsibility in a table. Responsibilities must be specific for the procedures and activities delineated in each specific document.

The System Owners or End-Users are responsible for the development and understanding of the Software Operational Qualification. They acquire and coordinate support from in-house and/or external consultants, the Computer Validation Group, Quality Assurance (Unit), and System Developers.

The QA(U) is responsible for verifying that the details in the Software Operational Qualification are compliant with applicable departmental, interdepartmental and corporate policies and procedures and that the approach is in accordance with cGxP regulations and good business practices, especially 21 CFR Part 11. The QA(U) reviews the executed protocol for completeness.

Computer Operations is responsible for understanding the Computer System Design Specifications document and working with the System Owner/End-Users to develop a Software Operational qualification protocol for the software that will satisfy the business needs and requirements as indicated in the User and Functional Requirements.

Only qualified Testers execute the protocol.

The System Owner / End-User writes an OQ Summary Report and ensures that it is reviewed and approved by the System Owner, CVG and QA(U).

The Subject Matter Expert (SME) ensures that changes required as a result of deviations during testing and subsequent remediation are captured in the summary report and incorporated into the Software Operational Qualification protocol. If the OQ protocol execution is begun prior to the approval of the OQ summary report, then the SME certifies that deviations and remediations have no effect on the PQ. Otherwise, the affected OQ protocol tests need to be reviewed, repeated as necessary, or re-written, and regression testing performed.

Procedure (protocol)

Background Information

This section of the Software Operational Qualification protocol provides an overview of the business operations for the system including workflow, material flow and the control philosophy of what functions, calculations and processes are to be performed by the system, as well as interfaces that must be provided to the system or application. It outlines the activities required for commissioning the equipment.

The execution of the protocol verifies compliance with the Software Operational Qualification.

Instructions on Testing Procedures

This section describes the procedures that the tester is to follow. It must include how the tests will be documented, how deviations will be handled and other items relevant to the execution of the protocol. Test procedures should be in accordance with a Validation Plan.

All OQ testing is done in a qualified pre-production environment unless it is the only system currently running on the equipment during testing. It is not good practice to perform qualification testing of new or revised software, computer or computerized systems in the production environment. Should a major failure occur, such as a system crash, then the previously qualified items running in the production environment must be re-qualified.

Execution of each test procedure should observe the following general instructions:

All signatures or initials are written in permanent ink and dated. Ditto marks or other methods to indicate repeated acceptance of a test must not be used (i.e., individual signatures/initials are required).

Each individual involved must be qualified to perform the assigned tasks. *Qualified* means that the person's training, education and experience are appropriate for the task and are suitably documented.

Each test must list the action to be taken, the expected results (also referred to as the acceptance criteria), the observed results, the disposition of the test (e.g., pass or fail), and the signature/date of the person performing the test.

Results must be documented by one individual and reviewed by a second.

The test results are recorded in the spaces provided as each test is executed. When the test listed on a particular page is completed, the person performing the test signs the space labeled "Performed By" on the bottom of the page. If the test was not completed satisfactorily (i.e., a component fails to meet expected criteria), a "Discrepancy Report" is prepared.

Discrepancy Reports are numbered sequentially and noted on the appropriate protocol page. Each Discrepancy Report summarizes the observed problem and indicates the actions taken to resolve it. Once all items on a page are completely resolved and the re-tested procedure passes, the tester signs and dates the bottom of the page. When the overall test is completed, a person other than the tester reviews the work, accepts it as complete by signing on the "Verified By" line at the bottom of the page noting any exceptions.

Screen prints should be used wherever possible as proof of execution, especially when expected and unexpected errors occur.

All pages of the document must have a page number and the total number of pages of the document. This includes attachments, which are also referenced in the primary document. Attachments must be numbered, signed and dated and marked in such a manner that they can be associated with a specific test step.

A signature sheet (where the person's signature, printed name, initials, and affiliation are recorded) is included for each test document to identify all individuals involved in the execution of the test protocol.

Protocol test deviations or failures identified during the testing are addressed on a case-by-case basis. It is the responsibility of the site team to evaluate and resolve these deviations or failures.

Test deviations or failures are initially documented in the test script and then recorded in a Discrepancy / Deviation Report and a Log. This record includes a description of the problem, when it was detected, the circumstances of detection, and who detected it. The site validation team reviews the failures and decides on appropriate actions. The team's review is documented in the Discrepancy / Deviation Report and Log, which contains the identification of the failure, the date reviewed, assessment of the failure, assessment of the impact of the failure on the validation effort, and the team's recommendation for resolution. The log also indicates when and how the deviation or failure was resolved. For each failure, the team determines whether the failure impacts site or core software. Impact on core software must be communicated to all other site teams.

Deviations / Failures are categorized in accordance with a defined scheme. An example of such a scheme:

- Critical

An issue that is critical to cGxP compliance; affects compliance with company SOPs or regulatory requirements; in any way compromises the integrity of the data; or that could directly impact the identity, strength, quality, or purity of a drug product, the operation of a medical device, data that is subject to regulatory review, or the safety of individuals who operate the system.

- Major

An issue for which there is an acceptable workaround solution.

- Minor

Errors in text fields that do not fall into the Criticalor Major category and are unlikely to be misleading.

- Cosmetic

Trivial errors for which no action is required.

Outcomes of the Review Process:

- Change in Software Configuration or Code

This is in accordance with a Change Control and/or Configuration Management SOP.

- Modify System Requirements or Specifications

For major documentation changes, the document is revised, the revision level incremented and the document sent for re-approval. For minor test script changes a deviation report is generated. If such a change is required, the team may elect to escalate the problem to the implementation team for further study or to management.

- Develop a Manual Procedure to Work Around the Problem

The workaround is written into one of the user SOPs

- Modify the Test Script

If after evaluation it is determined that the system is responding to a test as required, but the test script contains either the wrong instructions or an incorrect expected response, the team may recommend a change in the test script. This requires the team to document the change in the test script and retest after the change is approved.

Following execution of the protocols, the completed documents are reviewed for completeness and compliance with the appropriate procedures. If the protocol document is complete and compliant, the reviewer signs and dates it. This includes the review of any attached test outputs.

Tester Identification

This section identifies each person participating in the testing. The following information is entered in a table for identification purposes:

- Full Name – printed
- Signature
- Initials
- Affiliation

Qualification Tasks

This section of the protocol requires a series of processes and tests that must be performed as part of the Software Operational Qualification. Each procedure and/or test section should consist of the following sections:

- Test Section Number and Title
- Purpose of test
- Execution procedures, if appropriate
- Functional Requirement being tested
- Design Specification being tested
- Procedure or Test

Each Test Step should consist of the following sections:

- Test Step number identification
- Action to be performed
- Expected Results (acceptance criteria)
- Actual Results (observations)
- Pass / Fail Results
- Tester Signature / Date
- Comments
- Reviewer Signature / Date

The Procedure or Test section indicates the actions that the tester is expected to perform. These are specific instructions on how to accomplish each required action or task.

The Expected Results section indicates what the tester is expected to observe as a result of the action.

The Actual Results is an entry of what the tester observes, whether the test performs as expected (passes) or does not (fails).

Screen prints should be required wherever possible as proof of execution, especially when expected and unexpected errors occur.

The Pass / Fails Results section should clearly state the conclusion of the test step by the Tester entering the word "Pass" or "Fail." The use of the letter "P" for pass or the letter "F" for fail *is* acceptable.

The Tester Signature / Date section is where the tester initials and dates the results, indicating that the test was performed by that person.

The Comments section may be used to record comments or explanations as required.

The Reviewer Signature / Date section is where the Reviewer signs or initials, and dates each test or procedure indicating that the item has been reviewed and is acceptable as recorded.

If the procedure is a gathering of information, such as serial numbers, there is no need for Expected Results and Pass / Fail sections. The initials of the Tester and date will suffice.

The procedure for the handling of test deviations and failures is specified in the Software OQ protocol's Instruction on Testing Procedures section, which is written in accordance with the section Instructions on Testing Procedures of this SOP.

Discrepancy / Deviation Report

This section provides for a form that is used as a Discrepancy / Deviation Report. Whenever a test or procedure results in a discrepancy or deviation, a report must be made detailing the results and explaining the resolution. The form must contain the following sections:

- Report Number
- Discrepancy / Deviation
- Test Section Number and Title
- Test Step Number
- Signature /Date
- Resolution
- Signature / Date
- QA(U) Signature / Date

The Report Number section is a sequential number derived from the Discrepancy / Deviation Log form where the report is to be logged and tracked.

The Discrepancy / Deviation section is used to enter the details of the test or procedure results.

The Test Section Number and Title is used to enter the test section number and title of the test or procedure being reported.

The Test Step Number is used to enter the test or procedure step number being reported.

The Signature / Date section is filled in by the person who issued the report.

The Resolution section is used to enter the details of the resolution or work around employed to remediate the deviation or discrepancy.

The Signature / Date section is filled in by the person who issued the remediation report.

The QA(U) Signature / Date section is used by the QA(U) individual who reviewed and approved the report.

Discrepancy / Deviation Log

This section provides for a form for use as a Discrepancy / Deviation Log. It must contain a table with the following columns:

- Report Number
- OQ
- Test Section Number and Title
- Test Step number
- Deviation / Discrepancy
- Date / Initials

The Report Number is a sequential number used to identify the reports associated with the specific qualification protocol.

The OQ section is used to identify the protocol that the report represents.

The Test Section Number and Title is used to enter the Test Section Number and Title of the test or procedure being reported.

The Test Step Number is used to enter the test or procedure step number being reported.

The Deviation / Discrepancy item is used to enter a brief description of the deviation or discrepancy being logged.

The Date / Initials section is filled in by the person issuing the remediation Report Number.

Document Management and Storage (protocol)

A Document Management and Storage section is included in each Software Operational Qualification protocol. It defines the Document Management and Storage methodology that is followed for issue, revision, replacement, control and management of each Software Operational Qualification protocol. This methodology is in accordance with an established Change Control SOP and a Document Management and Storage SOP.

Revision History (protocol)

A Revision History section is included in each Software Operational Qualification protocol. The Software Operational Qualification protocol Version Number and all Revisions made to the document are recorded in this section. These changes are in accordance with a Change Control Procedure SOP.

The Revision History section consists of a summary of the changes, a brief statement explaining the reasons the document was revised, and the name and affiliation of the person responsible for the entries. It includes the numbers of all items the Software Operational Qualification protocol replaces and is written so that the original Software Operational Qualification protocol is not needed to understand the revisions or the revised document being issued.

When the Software Operational Qualification protocol is a new document, a "000" as the Revision Number is entered with a statement such as "Initial Issue" in the comments section.

Regulatory or audit commitments require preservation from being inadvertently lost or modified in subsequent revisions. All entries or modifications made as a result of regulatory or audit commitments require indication as such in the Revision History and should indicate the organization, date and applicable references. A complete and accurate audit trail must be developed and maintained throughout the project lifecycle.

Attachments (protocol)

If needed, an Attachments section is included in each Software Operational Qualification protocol in accordance with the SOP on SOPs. An Attachment Index and Attachment section follow the text of a Software Operational Qualification protocol. The top of each attachment page indicates "Attachment # x," where 'x' is the sequential number of each attachment. The second line of each attachment page contains the attachment title. The number of pages and date of issue corresponding to each attachment are listed in the Attachment Index.

Attachments are placed directly after the Attachment Index and are sequentially numbered (e.g., Attachment # 1, Attachment # 2) with each page of an attachment containing the corresponding document number, page numbering, and total number of pages in the attachment (e.g., Page x of y). Each attachment is labeled consistent with the title of the attachment itself.

Attachments are treated as part of the Software Operational Qualification protocol. If an attachment needs to be changed, the entire Software Operational Qualification protocol requires a new revision number and is reissued. New attachments or revisions of existing attachments are in accordance with an established Change Control SOP and a Document Management and Storage SOP.

Document Management and Storage *(SOP)*

A Document Management and Storage section is included in the Software Operational Qualification SOP. It defines the Document Management and Storage methodology that is followed for issue, revision, replacement, control and management of this SOP. This methodology is in accordance with an established Change Control SOP and a Document Management and Storage SOP.

Revision History (SOP)

A Revision History section is included in the Software Operational Qualification SOP. The Software Operational Qualification SOP Version Number and all Revisions made to the document are recorded in this section. These changes are in accordance with a Change Control Procedure SOP.

The Revision History section consists of a summary of the changes, a brief statement explaining the reasons the document was revised, and the name and affiliation of the person responsible for the entries. It includes the numbers of all items the Software Operational Qualification SOP replaces and is written so that the original Software Operational Qualification SOP is not needed to understand the revisions or the revised document being issued.

When the Software Operational Qualification SOP is a new document, a "000" as the Revision Number is entered with a statement such as "Initial Issue" in the comments section.

Regulatory or audit commitments require preservation from being inadvertently lost or modified in subsequent revisions. All entries or modifications made as a result of regulatory or audit commitments require indication as such in the Revision History and should indicate the organization, date and applicable references. A complete and accurate audit trail must be developed and maintained throughout the project lifecycle.

Attachments *(SOP)*

If needed, an Attachments section is included in the Software Operational Qualification SOP in accordance with the SOP on SOPs. An Attachment Index and Attachment section follow the text of the Software Operational Qualification SOP. The top of each attachment page indicates "Attachment # x," where 'x' is the sequential number of each attachment. The second line of each attachment page contains the attachment title. The number of pages and date of issue corresponding to each attachment are listed in the Attachment Index.

Attachments are placed directly after the Attachment Index and are sequentially numbered (e.g., Attachment # 1, Attachment # 2) with each page of an attachment containing the corresponding document number, page numbering, and total number of pages in the attachment (e.g., Page x of y). Each attachment is labeled consistent with the title of the attachment itself.

Attachments are treated as part of the Software Operational Qualification SOP. If an attachment needs to be changed, the entire Software Operational Qualification SOP requires a new revision number and is reissued. New attachments or revisions of existing attachments are in accordance with an established Change Control SOP and a Document Management and Storage SOP.

Chapter 22 – Performance Qualification

In this chapter we will discuss the Standard Operating Procedure (SOP) for creating a Performance Qualification protocol for a validation project. This SOP defines the method to be used for specifying the Performance Qualification that a computer operations group will use in qualifying a software application, computer or computerized system.

Best industry practices and FDA Guidance documents indicate that in order to demonstrate that a computer or computerized system performs as it purports and is intended to perform, a methodical plan is needed that details how this may be proven. The Performance Qualification protocol defines the specifications that are to be used in qualifying a software application, computer or computerized system in fulfillment of the requirements of the User and Functional Requirements protocols and the Design Specifications.

This SOP is used for all systems that create, modify, maintain, retrieve, transmit or archive records in electronic format. It is used in conjunction with an overlying corporate policy to provide sustainable compliance with the regulation. The project team uses it to create the Performance Qualification protocol from which the System Owner has it qualified.

System Owners, End-Users and Subject Matter Experts (SME) use this SOP to create the Performance Qualification protocol for a software application, computer or computerized system that controls equipment and/or produces electronic data or records in support of all regulated laboratory, manufacturing, clinical or distribution activities. The protocols produced in accordance with this SOP are used by the System Owner to qualify the system.

Title Block, Headers and Footers (SOP)

The Performance Qualification SOP should have a title block, title or cover page along with a header and/or footer on each page of the document that clearly identifies the document. The minimum information that should be displayed for the Performance Qualification SOP is as follows:

- Document Title
- Document Number
- Document Revision Number
- Project Name and Number
- Site/Location/Department Identification
- Date Document was issued, last revised or draft status
- Effective Date
- Pagination: page of total pages (e.g., Page x of y)

The Document Title should be in large type font that enables clear identification. It should clearly and concisely describe the contents and intent of the document. Since this document will be part of a set or package of documents within a project, it should be consistent with other documents used for the project and specify that it is the Performance Qualification SOP. This information should be on every page of the document.

The Document Number should be in large type font that enables clear identification. Since the Performance Qualification SOP will be part of a set or package of documents within a project, it should be consistent with other documents used for the project. This information should appear on every page of the document.

The Document Revision Number, which identifies the revision of the document, should appear on each page of the document. Many documents will go through one or more changes over time, particularly during the draft stage and the review and approval cycle. It is important to maintain accurate revision control because even a subtle change in a document can have an immense impact on the outcomes of activities. The SOP on SOPs should govern the numbering methodology.

See Chapter 6 – "The SOP on SOPs."

The Project Name and Number, if applicable, should be indicated on the cover or title page. This information identifies the overall project to which this document belongs.

The Site/Location/Department Identification should be indicated on the cover or title page. It identifies the Site/Location/Department that has responsibility for this document and its associated project.

The Date that the Document was issued, last revised or issued with draft status should be on each page of the document (e.g., 12-Dec-01). This will clearly identify the date on which the document was last modified.

The Effective Date of the Document should be on the title or cover page of the document (e.g., 12-Dec-01). This will clearly identify when to begin using the current version of the document.

Pagination, page of total pages (e.g., Page x of y), should be on each page of the document. This will help identify whether or not all of the pages of the document are present. Pages intentionally left blank should clearly be designated as such.

Approval Section *(SOP)*

The Approval Section should contain a table with the Author and Approvers printed names, titles and a place to sign and date the SOP. The SOP on SOPs should govern who the approvers of the document should be. The functional department (e.g., Production / Quality Assurance / Engineering) of the signatory should be placed below each printed name and title along with a statement associated with each approver indicating the person's role or qualification in the approval process. Indicate the significance of the associated signature (e.g., This document meets the requirements of Corporate Policy 9055, Electronic Records; Electronic Signatures, dated 22-Mar-2000).

The completion (signing) of this section indicates that the contents have been reviewed and approved by the listed individuals.

Table of Contents *(SOP)*

The SOP should contain a Table of Contents, which appears towards the beginning of the document. This Table of Contents provides a way to easily locate various topics contained in the document.

Purpose *(SOP))*

This introduction to the SOP concisely describes the document's purpose in sufficient detail so that a non-technical reader can understand its purpose.

The Purpose section should clearly state that the intention of the document is to provide a Standard Operating Procedure to be used to specify the requirements for the preparation of Performance Qualification protocol for a new or upgraded computer or computerized system along with the required activities and deliverables. These requirements are necessary for the Performance Qualification protocol to be effective and compliant with company polices and government regulation. As one of the foundations of the software application, computer or computerized system, the Performance Qualification protocol needs to define the scope of work, responsibilities and deliverables in order to create the software application, computer and/or computerized system and demonstrate that it meets its intended use in a reliable, consistent and compliant manner. It should provide sufficient technical information on how a computer or computerized system is to perform in actual use as purported and intended within a specified range including all users, software, firmware, hardware, peripherals, associated equipment and operating procedures. It must be based on the User Requirements and Functional Requirements against which the system will be tested. The Performance Qualification protocol document derived from this SOP will become the standard that is used to functionally operate a specific software or firmware application.

Scope (SOP)

The Scope section identifies the limitations of the Performance Qualification SOP. It limits the scope to all GxP systems and is applicable to all computers, computerized systems, infrastructure, LANs and WANs including clinical, laboratory, manufacturing and distribution computer systems. It applies to all Sites, Locations and Departments within the company.

Equipment / Materials / Tools (SOP)

This section lists any required or recommended equipment, materials, tools, templates or guides. Provide the name and a brief description of the equipment, instruments, materials and/or tools to be used in the procedures or testing covered by the Performance Qualification SOP. This is for reference purposes and requires the user to utilize the specific item or a designated equivalent. The purpose and use is explained elsewhere in the document. Substitution of comparable equipment may be made in accordance with procedures for Change Control.

If the information is included in a job aid or operating manual, a description of where the information may be found with a specific reference to the document number, title, date and revision number should be specified.

Warnings / Notes / General Information / Safety (SOP)

Warnings, notes, general or safety information applicable to the Performance Qualification SOP are listed in this section.

Policies, Guidelines, References (SOP)

This section lists references to appropriate regulatory requirements, SOPs, policies, guidelines, or other relevant documents referred to in the Performance Qualification SOP. The name and a brief description of applicable related SOPs, corporate policies, procedures and guidelines, Quality Control monographs, books, technical papers, publications, specific regulations, batch records, vendor or system manuals, and all other relevant documents that are required by the Performance Qualification SOP should be included. This is for reference purposes and requires the user to use the specific item or a designated equivalent. The purpose and use are explained elsewhere in the document.

SOPs cite references and incorporate them in whole by reference rather than specifying details that duplicate these items. This is done so that changes in policy, guidelines or reference materials require no changes to SOPs themselves and the most current version of the referenced material is used. Examples are:

- Policy ABC-123, Administrative Policy for Electronic Records; Electronic Signatures
- Corporate, Plan for Compliance with 21 CFR Part 11, Electronic Records; Electronic Signatures
- SOP-456, Electronic Records; Electronic Signatures
- FDA, 21 CFR Part 11, Electronic Records; Electronic Signatures

- FDA, General Principals of Software Validation; Final Guidance for Industry and FDA Staff, January 11, 2002

Assumptions / Exclusions / Limitations *(SOP)*

Any and all assumptions, exclusions or limitations within the context of the SOP should be discussed. Specific issues should be directly addressed in this section of the Performance Qualification SOP. For example:

"Only application software, firmware, utilities, and drivers should normally be addressed."

Glossary of Terms *(SOP)*

A Glossary of Terms should be included that defines all of the terms and acronyms that are used in this document or that are specific to the operation being described and may not be generally known to the reader. The reader may also be referred to the FDA document, "Glossary of Computerized System and Software Development Terminology."

Roles and Responsibilities *(SOP)*

The Roles and Responsibilities of the Performance Qualification protocol should be divided into two specific areas:

Creation and Maintenance

The System Owner of the computer or computerized system and Computer Operations is responsible for the preparation and maintenance of the Performance Qualification.

It is common for a Performance Qualification to go through several revisions during the course of a project, as it is considered a living document, especially during the project's early stages. As a normal process, it helps to ensure that the validation activities are in compliance with cGxPs, FDA regulations, company policies and industry standards and guidelines.

The roles and responsibilities for preparing the Performance Qualification protocol and other project documentation are addressed in accordance with established Standard Operating Procedures.

Use

The members of the Project Team must be familiar with the Performance Qualification in order to fulfill the project requirements. The Quality Assurance (Unit), Project Team and Computer Operations use the Performance Qualification in concert with other project documentation as a contract that details the project and quality expectations.

The Performance Qualification is one of the governing documents for the qualification and testing of the computer, computerized system, software or firmware it covers.

Procedure *(SOP)*

The SOP should contain instructions and procedures that enable the creation of a Performance Qualification protocol document. The Performance Qualification is required in order to provide the performance qualification requirements for the specified software application, computer or computerized system that can be used by a System Owner to qualify that system for use.

The Performance Qualification is prepared as a stand-alone document for all projects. It is not good practice to include it as part of other documents in an attempt to reduce the amount of workload or paperwork. This only causes more work over the length of the project resulting from misunderstandings and confusion. Clear, concise documents that address specific parts of any given project are desirable.

PQ provides a means to test and record the proper performance of the software application, computer or computerized system. This will also provide a baseline record of the system at startup or at its last upgrade that can be used for determining what changes may be required in the future. It ensures that the software functions as purported and intended.

The Performance Qualification protocol provides an overall technical description of the performance qualification requirements of a software application, computer or computerized system,. It includes sufficient technical information on how it is to perform in actual use as purported and intended within a specified range including all users, software, firmware, hardware, peripherals, associated equipment, and operating procedures. It must be based on the User Requirements and Functional Requirements against which the system is tested.

It provides the final information and documentation requirements that are used by the System Owner to test the system for performance.

If the system relies on existing components or infrastructure outside of End-User control, such as a server or network managed by Computer Operations, these are qualified by the appropriate group in accordance with approved procedures. The End-User is not required to duplicate these efforts. However, it is required that the End-User reference and include copies of these documents in the validation package.

Data used in qualification scripts must be identified, justified for its use and simulate actual use.

Boundary testing is performed. That is data that is outside, on and inside the boundaries of a given function to verify that these conditions are properly handled

Post Execution Approval (protocol)

This section contains a place for the QA(U) approval of the executed version of the protocol. The QA(U) signature in this section indicates that the PQ protocol has been properly executed from a compliance standpoint. All sections were completed, all tests and re-tests properly executed, no blank spaces exist and all deviations were properly remediated. It also means that each page of the protocol was reviewed. Should QA(U) determine that the even though the procedure was properly followed but the PQ failed, it is so noted upon signing. The phrase "Disapproved for Use" along with a brief explanation is clearly entered in association with the signature.

Purpose (protocol)

This section should clearly state the computer, computerized system, software or firmware with its performance qualification requirements and information as to why it is considered for use in a regulated system. This short introduction should be concise and present details so that an unfamiliar reader can understand the purpose of the document. For example:

> "The purpose of the Performance Qualification Protocol is to verify and document the proper performance of the XYZ Computerized System. These are provided in the User Requirements, Functional Requirements and Design Specification documents and include:

- Application Software
- Firmware
- Hardware
- Peripherals
- Associated Equipment
- System Standard Operating Procedures"

Scope (protocol)

The Scope section identifies the limitations of a specific Performance Qualification protocol in relationship to validating a specific software application, computer or computerized system. For example, if part of the system consists of a Local area network (LAN) that has already been qualified, then the qualification of the LAN may be excluded by stating that it is being addressed elsewhere.

Equipment / Materials / Tools (protocol)

This section lists any required or recommended equipment, materials, tools, templates or guides. Provide the name and a brief description of the equipment, instruments, materials and/or tools to be used in the procedures or testing covered by the Performance Qualification protocol. This is for reference purposes and requires the user to utilize the specific item or a designated equivalent. The purpose and use is explained elsewhere in the document. Substitution of comparable equipment may be made in accordance with procedures for Change Control.

If the information is included in a job aid or operating manual, a description of where the information may be found with a specific reference to the document number, title, date and revision number should be specified.

Warnings / Notes / General Information / Safety (protocol)

Warnings, notes, general or safety information applicable to the Performance Qualification protocol are listed in this section.

Policies / Guidelines / References (protocol)

This section lists references to appropriate regulatory requirements, SOPs, policies, guidelines, or other relevant documents referred to in the Performance Qualification protocol. The name and a brief description of applicable related SOPs, corporate policies, procedures and guidelines, Quality Control monographs, books, technical papers, publications, specific regulations, batch records, vendor or system manuals, and all other relevant documents that are required by the Performance Qualification protocol should be included. This is for reference purposes and requires the user to use the specific item or a designated equivalent. The purpose and use are explained elsewhere in the document.

Assumptions / Exclusions / Limitations (protocol)

Discuss any assumptions, exclusions or limitations within the context of the Performance Qualification. Specific issues should be directly addressed in this section of the Performance Qualification protocol.

Glossary of Terms (protocol)

The Glossary of Terms should define all of the terms and acronyms that are used in the document or that are specific to the operation being described and may not be generally known to the reader. The reader may also be referred to the FDA document, "Glossary of Computerized System and Software Development Terminology," or other specific glossaries that may be available.

Titles, terms, and definitions that are used in the corporate policies, procedures and guidelines should be defined.

Roles and Responsibilities (protocol)

Validation activities are the result of a cooperative effort involving many people, including the System Owners, End Users, the Computer Validation Group, Subject Matter Experts (SME) and Quality Assurance (Unit).

This section requires clear identification of the roles, responsibilities and individual(s) responsible for the creation and use of the Performance Qualification. It should specify the participants by name, title, department or company affiliation and responsibility in a table. Responsibilities should be specific for the procedures and activities delineated in each specific document.

The System Owners or End-Users should be responsible for the development and understanding of the Performance Qualification. They should acquire and coordinate support from in-house and/or external consultants, the Computer Validation Group, Quality Assurance(Unit) and system developers.

The QA(U) should be responsible to verify that the details in the Performance Qualification are compliant with applicable departmental, interdepartmental and corporate policies and procedures and that the approach is in accordance with cGxP regulations and good business practices, especially 21 CFR Part 11. The QA(U) should review the executed protocol for completeness.

Computer Operations should be responsible for understanding the Computer System Design Specifications document and working with the System Owner/End-Users to develop a Performance Qualification protocol for the software that will satisfy the business needs as indicated in the User and Functional Requirements.

Only qualified Testers execute the protocol.

The System Owner / End-User should write a PQ Summary Report and ensure that it is reviewed and approved by the System Owner, CVG and QA(U).

The Subject Matter Expert (SME) should ensure that changes required as a result of deviations during testing and subsequent remediation are captured in the summary report and incorporated into the Performance Qualification protocol. If the PQ protocol execution is begun prior to the approval of the PQ summary report, then the SME should certify that deviations and remediations have no effect on the PQ. Otherwise, the affected PQ protocol tests need to be reviewed, repeated as necessary, or re-written and regression testing performed.

Procedure (protocol)

Background Information

This section of the Performance Qualification protocol provides an overview of the business operations for the system including workflow, material flow and the control philosophy of what functions, calculations and processes are to be performed by the system, as well as interfaces that must be provided to the system or application. It outlines the activities required for commissioning the equipment.

The execution of the protocol verifies compliance with the Performance Qualification.

Instructions on Testing Procedures

This section describes the procedures that the tester is to follow. It must include how the tests will be documented, how deviations will be handled and other items relevant to the execution of the protocol. Test procedures should be in accordance with a Validation Plan.

PQ testing is done in the qualified production environment.

Execution of each test procedure should observe the following general instructions:

All signatures or initials are written in permanent ink and dated. Ditto marks or other methods to indicate repeated acceptance of a test must not be used (i.e., individual signatures/initials are required).

Each individual involved must be qualified to perform the assigned tasks. *Qualified* means that the person's training, education and experience are appropriate for the task and are suitably documented.

Each test must list the action to be taken, the expected results (also referred to as the acceptance criteria), the observed results, the disposition of the test (e.g., pass or fail), and the signature/date of the person performing the test.

Results must be documented by one individual and reviewed by a second.

The test results are recorded in the spaces provided as each test is executed. When the test listed on a particular page is completed, the person performing the test signs the space labeled "Performed By" on the bottom of the page. If the test was not completed satisfactorily (i.e., a component fails to meet expected criteria), a "Discrepancy Report" is prepared.

Discrepancy Reports are numbered sequentially and noted on the appropriate protocol page. Each Discrepancy Report summarizes the observed problem and indicates the actions taken to resolve it. Once all items on a page are completely resolved and the re-tested procedure passes, the tester signs and dates the bottom of the page. When the overall test is completed, a person other than the tester reviews the work, accepts it as complete by signing on the "Verified By" line at the bottom of the page noting any exceptions.

Screen prints should be used wherever possible as proof of execution, especially when expected and unexpected errors occur.

All pages of the document must have a page number and the total number of pages of the document. This includes attachments, which are also referenced in the primary document. Attachments must be numbered, signed and dated and marked in such a manner that they can be associated with a specific test step.

A signature sheet (where the person's signature, printed name, initials, and affiliation are recorded) is included for each test document to identify all individuals involved in the execution of the test protocol.

Protocol test deviations or failures identified during the testing are addressed on a case-by-case basis. It is the responsibility of the site team to evaluate and resolve these deviations or failures.

Test deviations or failures are initially documented in the test script and then recorded in a Discrepancy / Deviation Report and a Log. This record includes a description of the problem, when it was detected, the circumstances of detection, and who detected it. The site validation team reviews the failures and decides on appropriate actions. The team's review is documented in the Discrepancy / Deviation Report and Log, which contains the identification of the failure, the date reviewed, assessment of the failure, assessment of the impact of the failure on the validation effort, and the team's recommendation for resolution. The log also indicates when and how the deviation or failure was resolved. For each failure, the team determines whether the failure impacts site or core software. Impact on core software must be communicated to all other site teams.

Deviations / Failures are categorized in accordance with a defined scheme. An example of such a scheme:

- Critical

An issue that is critical to cGxP compliance; affects compliance with company SOPs or regulatory requirements; in any way compromises the integrity of the data; or that could directly impact the identity, strength, quality, or purity of a drug product, the operation of a medical device, data that is subject to regulatory review, or the safety of individuals who operate the system.

- Major

An issue for which there is an acceptable workaround solution.

- Minor

Errors in text fields that do not fall into the Criticalor Major category and are unlikely to be misleading.

- Cosmetic

Trivial errors for which no action is required.

Outcomes of the Review Process:

- Change in Software Configuration or Code

This is in accordance with a Change Control and/or Configuration Management SOP.

- Modify System Requirements or Specifications

For major documentation changes, the document is revised, the revision level incremented and the document sent for re-approval. For minor test script changes a deviation report is generated. If such a change is required, the team may elect to escalate the problem to the implementation team for further study or to management.

- Develop a Manual Procedure to Work Around the Problem

The workaround is written into one of the user SOPs

- Modify the Test Script

If after evaluation it is determined that the system is responding to a test as required, but the test script contains either the wrong instructions or an incorrect expected response, the team may recommend a change in the test script. This requires the team to document the change in the test script and retest after the change is approved.

Following execution of the protocols, the completed documents are reviewed for completeness and compliance with the appropriate procedures. If the protocol document is complete and compliant, the reviewer signs and dates it. This includes the review of any attached test outputs.

Tester Identification

This section identifies each person participating in the testing. The following information is entered in a table for identification purposes:

- Full Name – printed
- Signature
- Initials
- Affiliation

Qualification Tasks

This section of the protocol requires a series of processes and tests that must be performed as part of the Performance Qualification. Each procedure and/or test section should consist of the following sections:

- Test Section Number and Title
- Purpose of test
- Execution procedures, if appropriate
- Functional Requirement being tested
- Design Specification being tested
- Procedure or Test

Each Test Step should consist of the following sections:

- Test Step number identification
- Action to be performed
- Expected Results (acceptance criteria)
- Actual Results (observations)
- Pass / Fail Results
- Tester Signature / Date
- Comments
- Reviewer Signature / Date

The Procedure or Test section indicates the actions that the tester is expected to perform. These are specific instructions on how to accomplish each required action or task.

The Expected Results section indicates what the tester is expected to observe as a result of the action.

The Actual Results is an entry of what the tester observes, whether the test performs as expected (passes) or does not (fails).

Screen prints should be required wherever possible as proof of execution, especially when expected and unexpected errors occur.

The Pass / Fails Results section should clearly state the conclusion of the test step by the Tester entering the word "Pass" or "Fail." The use of the letter "P" for pass or the letter "F" for fail *is* acceptable.

The Tester Signature / Date section is where the tester initials and dates the results, indicating that the test was performed by that person.

The Comments section may be used to record comments or explanations as required.

The Reviewer Signature / Date section is where the Reviewer signs or initials, and dates each test or procedure indicating that the item has been reviewed and is acceptable as recorded.

If the procedure is a gathering of information, such as serial numbers, there is no need for Expected Results and Pass / Fail sections. The initials of the Tester and date will suffice.

The procedure for the handling of test deviations and failures is specified in the PQ protocol's Instruction on Testing Procedures section, which is written in accordance with the section Instructions on Testing Procedures of this SOP.

Discrepancy / Deviation Report

This section provides for a form that is used as a Discrepancy / Deviation Report. Whenever a test or procedure results in a discrepancy or deviation, a report must be made detailing the results and explaining the resolution. The form must contain the following sections:

- Report Number
- Discrepancy / Deviation
- Test Section Number and Title
- Test Step Number
- Signature /Date
- Resolution
- Signature / Date
- QA(U) Signature / Date

The Report Number section is a sequential number derived from the Discrepancy / Deviation Log form where the report is to be logged and tracked.

The Discrepancy / Deviation section is used to enter the details of the test or procedure results.

The Test Section Number and Title is used to enter the test section number and title of the test or procedure being reported.

The Test Step Number is used to enter the test or procedure step number being reported.

The Signature / Date section is filled in by the person who issued the report.

The Resolution section is used to enter the details of the resolution or work around employed to remediate the deviation or discrepancy.

The Signature / Date section is filled in by the person who issued the remediation report.

The QA(U) Signature / Date section is used by the QA(U) individual who reviewed and approved the report.

Discrepancy / Deviation Log

This section provides for a form for use as a Discrepancy / Deviation Log. It must contain a table with the following columns:

- Report Number
- PQ
- Test Section Number and Title
- Test Step number
- Deviation / Discrepancy
- Date / Initials

The Report Number is a sequential number used to identify the reports associated with the specific qualification protocol.

The PQ section is used to identify the protocol that the report represents.

The Test Section Number and Title is used to enter the Test Section Number and Title of the test or procedure being reported.

The Test Step Number is used to enter the test or procedure step number being reported.

The Deviation / Discrepancy item is used to enter a brief description of the deviation or discrepancy being logged.

The Date / Initials section is filled in by the person issuing the remediation Report Nunber.

Document Management and Storage (protocol)

A Document Management and Storage section is included in each Performance Qualification protocol. It defines the Document Management and Storage methodology that is followed for issue, revision, replacement, control and management of each Performance Qualification protocol. This methodology is in accordance with an established Change Control SOP and a Document Management and Storage SOP.

Revision History (protocol)

A Revision History section is included in each Performance Qualification protocol. The Performance Qualification protocol Version Number and all Revisions made to the document are recorded in this section. These changes are in accordance with a Change Control Procedure SOP.

The Revision History section consists of a summary of the changes, a brief statement explaining the reasons the document was revised, and the name and affiliation of the person responsible for the entries. It includes the numbers of all items the Performance Qualification protocol replaces and is written so that the original Performance Qualification protocol is not needed to understand the revisions or the revised document being issued.

When the Performance Qualification protocol is a new document, a "000" as the Revision Numb er is entered with a statement such as "Initial Issue" in the comments section.

Regulatory or audit commitments require preservation from being inadvertently lost or modified in subsequent revisions. All entries or modifications made as a result of regulatory or audit commitments require indication

as such in the Revision History and should indicate the organization, date and applicable references. A complete and accurate audit trail must be developed and maintained throughout the project lifecycle.

Attachments (protocol)

If needed, an Attachments section is included in each Performance Qualification protocol in accordance with the SOP on SOPs. An Attachment Index and Attachment section follow the text of a Performance Qualification protocol. The top of each attachment page indicates "Attachment # x," where 'x' is the sequential number of each attachment. The second line of each attachment page contains the attachment title. The number of pages and date of issue corresponding to each attachment are listed in the Attachment Index.

Attachments are placed directly after the Attachment Index and are sequentially numbered (e.g., Attachment # 1, Attachment # 2) with each page of an attachment containing the corresponding document number, page numbering, and total number of pages in the attachment (e.g., Page x of y). Each attachment is labeled consistent with the title of the attachment itself.

Attachments are treated as part of the Performance Qualification protocol. If an attachment needs to be changed, the entire Performance Qualification protocol requires a new revision number and is reissued. New attachments or revisions of existing attachments are in accordance with an established Change Control SOP and a Document Management and Storage SOP.

Document Management and Storage *(SOP)*

A Document Management and Storage section is included in the Performance Qualification SOP. It defines the Document Management and Storage methodology that is followed for issue, revision, replacement, control and management of this SOP. This methodology is in accordance with an established Change Control SOP and a Document Management and Storage SOP.

Revision History *(SOP)*

A Revision History section is included in the Performance Qualification SOP. The Performance Qualification SOP Version Number and all Revisions made to the document are recorded in this section. These changes are in accordance with a Change Control Procedure SOP.

The Revision History section consists of a summary of the changes, a brief statement explaining the reasons the document was revised, and the name and affiliation of the person responsible for the entries. It includes the numbers of all items the Performance Qualification SOP replaces and is written so that the original Performance Qualification SOP is not needed to understand the revisions or the revised document being issued.

When the Performance Qualification SOP is a new document, a "000" as the Revision Number is entered with a statement such as "Initial Issue" in the comments section.

Regulatory or audit commitments require preservation from being inadvertently lost or modified in subsequent revisions. All entries or modifications made as a result of regulatory or audit commitments require indication as such in the Revision History and should indicate the organization, date and applicable references. A complete and accurate audit trail must be developed and maintained throughout the project lifecycle.

Attachments *(SOP)*

If needed, an Attachments section is included in the Performance Qualification SOP in accordance with the SOP on SOPs. An Attachment Index and Attachment section follow the text of the Performance Qualification

SOP. The top of each attachment page indicates "Attachment # x," where 'x' is the sequential number of each attachment. The second line of each attachment page contains the attachment title. The number of pages and date of issue corresponding to each attachment are listed in the Attachment Index.

Attachments are placed directly after the Attachment Index and are sequentially numbered (e.g., Attachment # 1, Attachment # 2) with each page of an attachment containing the corresponding document number, page numbering, and total number of pages in the attachment (e.g., Page x of y). Each attachment is labeled consistent with the title of the attachment itself.

Attachments are treated as part of the Performance Qualification SOP. If an attachment needs to be changed, the entire Performance Qualification SOP requires a new revision number and is reissued. New attachments or revisions of existing attachments are in accordance with an established Change Control SOP and a Document Management and Storage SOP.

Chapter 23 – Electronic Records; Electronic Signatures

In this chapter we will discuss the Standard Operating Procedure (SOP) for requiring and determining compliance of a software application, computer or computerized system with 21 CFR Part 11, *Electronic Records; Electronic Signatures*.

The FDA has determined that in order to maintain the integrity, accuracy, reliability and consistency of operation of computer and computerized systems, the subsequent electronic data and provide a vehicle for the use of electronic signatures that are considered the equivalent of hand written signatures, the application of 21 CFR Part 11 is required in the interest of health and safety. The regulation is intended to prevent fraud and ensure data integrity. It is *not* intended to make systems foolproof.

This SOP is used for all systems that create, modify, maintain, retrieve, transmit or archive records in electronic format. It is used in conjunction with an overlying corporate policy to provide sustainable compliance with the regulation.

Everyone involved in the operation of a software application, computer or computerized system that controls equipment and/or produces electronic data or records in support of all regulated laboratory, manufacturing, clinical or distribution activities uses this SOP.

Title Block, Headers and Footers

The Electronic Records; Electronic Signatures SOP should have a title block, title or cover page along with a header and/or footer on each page of the document that clearly identifies the document. The minimum information that should be displayed for the Electronic Records; Electronic Signatures SOP is as follows:

- Document Title
- Document Number
- Document Revision Number
- Project Name and Number
- Site/Location/Department Identification
- Date Document was issued, last revised or draft status
- Effective Date
- Pagination: page of total pages (e.g., Page x of y)

The Document Title should be in large type font that enables clear identification. It should clearly and concisely describe the contents and intent of the document. Since this document will be part of a set or package of documents within a project, it should be consistent with other documents used for the project and specify that it is the Electronic Records; Electronic Signatures SOP. This information should be on every page of the document.

The Document Number should be in large type font that enables clear identification. Since the Electronic Records; Electronic Signatures SOP will be part of a set or package of documents within a project, it should be consistent with other documents used for the project. This information should appear on every page of the document.

The Document Revision Number, which identifies the revision of the document, should appear on each page of the document. Many documents will go through one or more changes over time, particularly during the draft stage and the review and approval cycle. It is important to maintain accurate revision control because even a subtle change in a document can have an immense impact on the outcomes of activities. The SOP on SOPs should govern the numbering methodology.

See Chapter 6 – The SOP on SOPs.

The Project Name and Number, if applicable, should be indicated on the cover or title page. This information identifies the overall project to which this document belongs.

The Site/Location/Department Identification should be indicated on the cover or title page. It identifies the Site/Location/Department that has responsibility for this document and its associated project.

The Date that the Document was issued, last revised or issued with draft status should be on each page of the document (e.g., 12-Dec-01). This will clearly identify the date on which the document was last modified.

The Effective Date of the Document should be on the title or cover page of the document (e.g., 12-Dec-01). This will clearly identify when to begin using the current version of the document.

Pagination, page of total pages (e.g., Page x of y), should be on each page of the document. This will help identify whether or not all of the pages of the document are present. Pages intentionally left blank should clearly be designated as such.

Approval Section

The Approval Section should contain a table with the Author and Approvers printed names and titles and a place to sign and date the SOP. This SOP on SOPs should govern who the approvers of the document should be. The functional department (e.g., Production / Quality Assurance / Engineering) of the signatory should be placed below each printed name and title along with a statement associated with each approver indicating the person's role or qualification in the approval process. Indicate the significance of the associated signature (e.g., This document meets the requirements of Corporate Policy 9055, Electronic Records; Electronic Signatures, dated 22-Mar-2000).

The completion (signing) of this section indicates that the contents have been reviewed and approved by the listed individuals.

Table of Contents

The SOP should contain a Table of Contents, which appears towards the beginning of the document. This Table of Contents provides a way to easily locate various topics contained in the document.

Purpose

The Purpose section should clearly state that the intention of the document is to provide a Standard Operating Procedure that is used for requiring and determining compliance of a software application, computer or computerized system, or infrastructure with 21 CFR Part 11. This introduction to the SOP concisely describes the document's purpose in sufficient detail so that a non-technical reader can understand its purpose.

Scope

The Scope section identifies the limitations of the Electronic Records; Electronic Signatures SOP. It limits the scope to all GxP systems and is applicable to all computers, computerized systems, infrastructure, LANs and WANs including clinical, laboratory, manufacturing and distribution computer systems. It applies to all Sites, Locations and Departments within the company.

It includes electronic records, electronic signatures and handwritten signatures executed to electronic records. It applies to all records that are created, modified, maintained, archived, retrieved or transmitted in electronic form under any requirements set forth in FDA regulations. Provide the scope or range of control and authority of the SOP. State who, what, where, when, why and how the SOP is to control the processes or procedures specified in the Purpose section.

The Electronic Records; Electronic Signatures SOP is used by everyone involved in the operation of computers or computerized systems that produce electronic data or records in support of all regulated laboratory, manufacturing, clinical or distribution activities.

This SOP applies to records in electronic form that are created, modified, maintained, archived, retrieved, or transmitted, under any records requirements set forth in agency regulations. It also applies to electronic records submitted to the agency under requirements of the Federal Food, Drug, and Cosmetic Act and the Public Health Service Act, even if such records are not specifically identified in agency regulations. However, it does not apply to paper records that are, or have been, transmitted by electronic means.

Equipment / Materials / Tools

This section lists any required or recommended equipment, materials, tools, templates or guides. Provide the name and a brief description of the equipment, instruments, materials and/or tools to be used in the procedures or testing covered by the Electronic Records; Electronic Signatures SOP. This is for reference purposes and requires the user to utilize the specific item or a designated equivalent. The purpose and use is explained elsewhere in the document. Substitution of comparable equipment may be made in accordance with procedures for Change Control

If the information is included in a job aid or operating manual, a description of where the information may be found with a specific reference to the document number, title, date and revision number should be specified.

Warnings / Notes / General Information / Safety

Warnings, notes, general or safety information applicable to the Electronic Records; Electronic Signatures SOP are listed in this section.

Policies, Guidelines, References

This section lists references to appropriate regulatory requirements, SOPs, policies, guidelines, or other relevant documents referred to in the Electronic Records; Electronic Signatures SOP. The name and a brief description of applicable related SOPs, corporate policies, procedures and guidelines, Quality Control monographs, books, technical papers, publications, specific regulations, batch records, vendor or system manuals, and all other relevant documents that are required by the Electronic Records; Electronic Signatures SOP should be included. This is for reference purposes and requires the user to use the specific item or a designated equivalent. The purpose and use are explained elsewhere in the document.

SOPs cite references and incorporate them in whole by reference rather than specifying details that duplicate these items. This is done so that changes in policy, guidelines or reference materials require no changes to SOPs themselves and the most current version of the referenced material is used. Examples are:

- Policy ABC-123, Administrative Policy for Electronic Records; Electronic Signatures
- Corporate, Plan for Compliance with 21 CFR Part 11, Electronic Records; Electronic Signatures
- SOP-456, Electronic Records; Electronic Signatures
- FDA, 21 CFR Part 11, Electronic Records; Electronic Signatures

- FDA, General Principals of Software Validation; Final Guidance for Industry and FDA Staff, January 11, 2002

Assumptions / Exclusions / Limitations

Any and all assumptions, exclusions or limitations within the context of the SOP should be discussed. Specific issues should be directly addressed in this section of the Electronic Records; Electronic Signatures SOP. Examples are:

"Legacy systems are not grandfathered under 21 CFR Part 11, there is no limitation with respect to legacy systems."

Glossary of Terms

A Glossary of Terms should be included that defines all of the terms and acronyms that are used in this document or that are specific to the operation being described and may not be generally known to the reader. The reader may also be referred to the FDA document, "Glossary of Computerized System and Software Development Terminology."

Roles and Responsibilities

The requirements for compliance with 21 CFR Part 11 are, at a minimum, the joint effort of the System Owner, CVG and QAU.

This section clearly specifies the participants by title, department or company affiliation and role in a table.

The System Owner is responsible for the development and understanding of the requirement for compliance with 21 CFR Part 11. He or she acquires and coordinates support from in-house consultants, Quality Assurance(Unit), System Developers and/or contracted external support services.

The QA(U) is responsible to verify that the systems within the scope of this SOP are compliant with the applicable departmental, interdepartmental and corporate practices and policies, appropriate cGxP regulations, good business practices and 21 CFR Part 11.

Procedure

The SOP must contain instructions and procedures that require compliance with the 21 CFR Part 11 for software applications, computer and computerized systems.

Introduction

Provide an introduction to the tasks at hand and explain the background of what is to be performed. For example:

"21 CFR Part 11 went into effect on August 20, 1997. It provides the criteria for FDA acceptance of electronic records to be equivalent to paper records and electronic signatures to be equivalent to handwritten signatures. The rule must be interpreted and applied to both new and legacy systems. Systems in use that predate August 20, 1997 are not exempt from the rule. The FDA recognizes that it will take time for existing systems to attain full compliance. The FDA expects steps to be taken toward achieving full compliance of these systems to 21 CFR Part 11. This document specifies the Standard Operating Procedures that must be used for sustainable compliance with 21 CFR Part 11."

Implementation

All infrastructure, software applications, computer and computerized systems that create, modify, maintain, archive, retrieve or transmit data in electronic form under any requirements set forth in FDA regulations must be brought into sustainable compliance with 21 CFR Part 11. This includes both system and procedural controls. Existing and in-process systems must be assessed. Gaps identified during assessment must be analyzed and a plan created and implemented to remediate them.

New systems must meet all of the requirements of 21 CFR Part 11.

A determination must be made regarding open systems. Open systems are quite difficult to manage and maintain secure. They are not recommended. It is highly recommended that a definitive statement be put in the SOP disallowing open systems. Set a time period for the closing of open systems with a defined process for exception or extension.

Create a 21 CFR Part 11 Master Compliance Plan listing items required to comply with the regulation along with each gap or deficiency to be addressed. The Master Compliance Plan identifies remediation objectives, priorities, timelines and a methodology for completion. Alternately, this information could be cataloged in the Validation Master Plan.

See Chapter 37 – Validation Master Plan.

Inventory

Create an inventory of all infrastructure, software applications, computer and computerized system items regardless of compliance requirement. This will show that nothing was overlooked.

The recommended items to inventory and assess for each software application, computer or computerized system are:

- System
- Description
- Validation Status
- GxP
- Functional Owner
- Administrator
- Corporate System
- IS MGR Contact
- IS Responsibility
- Platform
- Operating System
- Comments
- Last Validation Date
- Criticality
- Recovery Priority
- Backup Frequency
- Backup Type
- Run Frequency

- Recovery Procedures Documented?
- Input Dependencies
- Output Dependencies
- Compliance Gaps

The recommended infrastructure items to inventory and assess are the management of:

- IT/IS Organization
- Infrastructure Installation and Qualification
- Security
- Production Environment Management
- Assets
- Back-up, Archival and Restoration
- Disaster Recovery
- Maintenance
- Record Retention
- Periodic Review
- Change Control
- Configuration Management
- Problem Reporting
- cGxP Procedures, Systems and Training
- Standard Operating Procedures
- Guidelines, Policies, SOPs and Training in Support of SDLC
- GxP Training for IT/IS Staff and Contractors
- Training and Training Documentation
- Vendors and Contractors

FDA Access

Computer systems (including hardware and software), controls, and attendant documentation subject to the requirements of 21 CFR Part 11 must be readily available for FDA inspection. Electronic records must be available to the FDA in both electronic and human readable form. The FDA, at their sole discretion and request, is entitled to copies of any records, paper or electronic, that are subject to the regulation or other GxP requirements.

It is highly recommended that exact copies of manual and electronic records supplied to the FDA be retained in accordance with a Document Management and Storage SOP.

Record and System Security

Systems used to create, modify, maintain, retrieve, transmit or archive records in electronic format must use procedures and controls that ensure authenticity, integrity and security of the electronic records they create. The term "systems" means people, hardware, software and infrastructure. The method must ensure that the signer cannot disclaim the authenticity of the electronically signed record.

Physical Security

All core network infrastructure equipment including, but not limited to, servers, hubs, routers, switches, cabling, firewalls and wiring closets must be physically secure. Controlled access is required to prevent unauthorized access. All activities in these areas are logged either manually or electronically in accordance with one or more SOPs.

It is highly recommended that security breeches immediately cause an alarm indication that notifies a security department.

It is important to have and demonstrate control of the physical environment as well as the logical computing environment.

Logical Security

Accidental or intentional modification or deletion of data and metadata along with associated programs that affect it must be prevented. This includes, but is not limited to, programming code, algorithms, data, metadata, data schema and structures, audit trails, electronic signature data, relational metadata, and logs.

A method of access based on user-required functionality must be implemented. The user authority determines the access level to system, application, functions and data. There should be at least six levels of access based on the User or device authority and functionality: None, Read, Deposit, Write, Edit, System Administrator. This includes internal accounts set up by vendors of commercial software. Only Read-only access accounts may be shared by more than one user.

Persons having administrator access to applications, data, and metadata must *not* be the owners of the data. Each administrator account must be restricted to one assigned individual.

Accounts must be unique to one individual and must not be reused by, or reassigned to, anyone else. Name changes are *not* an exception. Requests to add, change or disable an account must be authorized by the System Owner and reviewed on a periodic basis. All accounts must be maintained for the life of the data with which they are associated for identification purposes. Account deletion is not permitted.

Verification of the identity of each person requesting and acquiring access must be made prior to account creation and authorization for use.

When an electronically signed electronic record is viewed or printed, it must be clearly determinable that it has been electronically signed. The electronic signature identifying information must be displayed on the screen or printouts. If the record has been altered after signing, there must be a clear indication that a change has occurred and that the electronic signature is no longer valid.

Electronic records that have been changed constitute new records for signing purposes.

When an electronic record is copied or moved to another system, the link between it and any associated manual and/or electronic signatures must be maintained in the original system. Although it is unnecessary to copy the manual or electronic signature with the electronic record, there must be a mechanism to trace the source of the record and the corresponding signatures to their origin.

It is highly recommended that the following be implemented in accordance with one or more SOPs:

- Automatic logoff after a period of inactivity
- Electronic signatures not to be stored in human readable form
- Firewall protection
- Password management
- Periodic account verification
- User changeable passwords
- Virus protection
- Back-up and Restore

Electronic data, metadata, software applications, and supporting files are backed up on a routine basis, preferably daily, in accordance with a Backup and Restore SOP. Each software application, computer and/or computerized system is governed by SOPs specific to that system.

Restoration procedures in accordance with these SOPs must be validated and periodically verified and documented as functioning properly.

A recovery plan should be in place in the event a restoration fails.

Backup media is stored in a secure location in accordance with an SOP.

Electronic Signature information must be backed up and restored with their corresponding electronic records.

Disaster Recovery

A disaster recovery plan should be in place, validated, and periodically verified and documented as being functional.

Audit Trail

Audit trails must record the local[10] date, time, and the information affected by all user and administrative activity that creates, modifies, or marks as inactive any electronic record. Electronic data must *not* be deleted. Audit Trails must be automatically computer generated, secure, independent, outside of operator control, and protected against accidental or intentional alteration or deletion.

Audit Trails must not have an inactive or inhibited state.

Audit Trails must be available for FDA review, print and copy.

Audit Trails must be backed up and restored with their associated data.

Each Audit Trail must contain, at a minimum, the following information:

- Actual user name
- Action taken (e.g.- add, change, set as inactive)
- Local date and time of the action
- History of previous data values for modified items
- Reason for the action or change

The Audit Trail is recorded at the time an action is completed. Audit trails are not required for draft documents but must begin concurrent with the first signature and/or official use.

Validation

Software applications, computer and computerized systems subject to the requirements of 21 CFR Part 11 must be validated to ensure accuracy, reliability, consistent intended performance, and the ability to discern invalid or altered records. This means both system (technical) and procedural controls including evidence of

[10] Specify the local time zone. Indicate daylight savings time, if it is in effect.

sustainable compliance. That is, proof that the procedures are actually being used and followed on an ongoing basis.

An overlying policy requiring validation and sustainable compliance in accordance with the requirements of 21 CFR Part 11is mandated by the validation process (current best practice) and 21 CFR Part 11. In theory this policy should be in place prior to the implementation of SOPs and the System Development Life Cycle (SDLC) methodology. Reality is generally the accomplishment of the minimum requirement to satisfy the condition in the near term. Be sure that required policies are in place within a reasonable amount time or preferably concurrent with the implementation of SDLC methods that satisfy the requirements of 21 CFR Part 11.

System Checks

Operational system checks must be used to enforce permitted sequencing of steps and events, as appropriate. Device (e.g., terminal) checks must be used to determine, as appropriate, the validity of the source of data input or operational instruction. These are specified as part of the User and Functional Requirements, and the Design Specifications. If it is not possible or feasible to provide these checks and controls programmatically, then they must be provided by means of SOPs.

Education and Experience

Persons who develop, maintain, or use electronic record/electronic signature systems must have the education, training, and experience to perform their assigned tasks. SOPs that govern training in the areas of curriculum, training manuals, assessment, job aids and records for all software applications, computers and computerized systems governed by 21 CFR Part 11 and GxP requirements must be in place. This applies to employees, consultants, vendors, and contractors.

Systems Documentation

The System Owner is responsible for the distribution, access, and use of documentation for system operation and maintenance. He or she is also responsible for revision and change control procedures to maintain an audit trail that documents time-sequenced development and modification of systems documentation.

Systems documentation explains how a software application, computer or computerized system functions, is used, and maintained. This includes SOPs, administration and user manuals, SDLC documentation, maintenance manuals and logs, periodic validation and qualification reports, documentation storage, and change control procedures. The may be in paper or electronic form. Electronic versions must comply with all 21 CFR Part 11and cGxP requirements.

Only the current version of a given document can be in use. This must correspond to the configuration management version of the associated system in use. Documentation, by version, must be restricted, distributed, tracked, and controlled in accordance with a methodology set forth in an SOP.

All documentation must be retained for the same period of time as the data its system collects based on Predicate Rules.

Open Systems

System access must be limited to authorized individuals.

Persons using open systems to create, modify, maintain, or transmit electronic records must employ procedures and controls designed to ensure the authenticity, integrity, and, as appropriate, the confidentiality of electronic records from the point of their creation to the point of their receipt. Such procedures and controls must include those identified in 21 CFR 11.10, as appropriate, and additional measures such as document encryption and use of appropriate digital signature standards to ensure, as necessary under the circumstances, record authenticity, integrity, and confidentiality.

An open system is an environment in which system access is not controlled by persons who are responsible for the content of electronic records that are on the system.

For example, the Internet or World Wide Web is an open system. Virtual Private Networks operating over the Internet are open systems and require controls, such as encryption. Dial-in connectivity, Local Area Networks (LAN), Wide Area Networks (WAN) and local company intranets are closed systems provided they have a secure firewall separation from the Internet, if connected.

Where open systems are used to create, modify, maintain, retrieve, transmit or archive records in electronic format as regulated by 21 CFR Part 11, all of the requirements of closed systems apply. These include, but are not limited to, the use of encryption, electronic signatures and system checks.

Keep in mind that the need or requirement for confidentiality is not a 21 CFR Part 11 requirement. Confidentiality is required by other regulation for patient confidentiality and other situations.

SOPs must be in place that define these controls and specify how they are to be implemented and used to maintain sustainable compliance with these requirements.

Electronic Signatures

An electronic signature is a computer data compilation of any symbol or series of symbols executed, adopted, or authorized by an individual to be the legally binding equivalent of the individual's handwritten signature.

A digital signature is an electronic signature based upon cryptographic methods of originator authentication, computed by using a set of rules and a set of parameters such that the identity of the signer and the integrity of the data can be verified.[11]

Where the signing of electronic records is regulated, all Electronic Signatures, whether biometric or non-biometric, need to comply with the requirements of 21 CFR Part 11.

Electronic signatures and handwritten signatures executed to electronic records must be linked to their respective electronic records to ensure that the signatures cannot be excised, copied, or otherwise transferred to falsify an electronic record by ordinary means.

General Requirements

Each Electronic Signature for a given system must be unique to one person for the duration of the retention period of the data. It must not be reused or shared. At a minimum it consists of a User ID and password. It must clearly identify its owner by his or her legal name. Abbreviation or nicknames must not be used. Where more than one person may have the same legal name, a means of unique identification must be provided.

---

[11] FDA, 21 CFR Part 11, Electronic Records; Electronic Signatures

More than one electronic signature may be given a user over time as a result of a change in legal name. However, only one electronic signature for an individual can be active at a time in any given system or environment. Name changes must *not* be made to an electronic signature. When a person's legal name changes, a new, unique electronic signature is issued.

Electronic records signed with an electronic signature must, at a minimum, indicate the legal name of the signer, the local date and time of the signing, and what the signature represents (e.g.- approval, disapproval, author, data entry, verification, review).

For multiple electronic signatures during the same working session, each electronic signature must comply with all of the requirements and contain all of the required electronic signature components. Since the user has already been identified at the initial login, only the password is required for successive signings.

When an individual executes one or more signings not performed during a single, continuous period of controlled system access, each signing must be executed using all of the electronic signature components and identification requirements.

People using electronic signatures must be held accountable and responsible for their actions by means of SOPs. They are trained in the use of these SOPs and provide a "wet ink signature" indicating their agreement and understanding that electronic signatures are legally binding upon them. A means is implemented to ensure sustainable compliance. Training documentation should be maintained for the life of the data.

There must be a means in place that prevents copying and pasting an electronic signature from one record to another.

Based on business requirements as documented in the User Requirements, Functional Requirements, Design Specifications or by the System Owner, it is permissible for one person to sign on behalf of another providing the signer uses their own electronic signature and there is an indication that the electronic signature is made on behalf of a specified individual.

A means must be implemented that prevents and detects unauthorized attempts to use electronic signatures and reports the activity to management.

Devices such as tokens or cards must be initially and periodically tested to ensure that they have not been altered.

Non-Biometric Electronic Signatures

Non-biometric electronic signatures must employ at least two distinct identification components such as an identification code and password. This combination must be unique with no two people having the same identification code and password.

Non-biometric electronic signatures must be used only by their genuine owners.

Non-biometric electronic signatures must be administered and executed to ensure that attempted use of an individual's electronic signature by anyone other than its genuine owner requires collaboration of two or more individuals.

Biometric Signatures

Electronic signatures based upon biometrics must be designed to ensure that they cannot be used by anyone other than their genuine owners.

Biometrics means a method of verifying an individual's identity based on measurement of the individual's physical feature(s) or repeatable action(s) where those features and/or actions are both unique to that individual and measurable.

An electronic signature based upon biometrics must not be able to be used by anyone other than its genuine owner (e.g.- retinal or fingerprint pattern recognition).

See Record and System Security - Logical Security.

Hybrid Systems

A hybrid system consists of a combination of electronic records with paper records and/or hand written signatures.

When a handwritten "wet ink signature" is used for signing electronic records, it must include the handwritten signature, the printed name of the signer, the local date and time of the signing, what the signing represents (e.g.- approval, disapproval, author, data entry, verification, review) and clear unique identification of the linked electronic record or records.

Electronic records that have been changed constitute new records for signing purposes.

See Record and System Security - Logical Security.

Record Retention / System Retirement

Record Retention

Electronic records must be protected to enable their accurate and ready retrieval throughout the records retention period. This includes the hardware, software, attendant documentation, data, metadata, audit trails and electronic signatures that constitute the computer or computerized system.

See FDA Access.

Copies

The systems must have the ability to generate accurate and complete copies of readable electronic records. This includes documentation, data, metadata, audit trails and electronic signatures. The ability to reproduce both electronically and in printed form any information that was normally available during the live operation of the system must be available during the retention period. Paper records alone are not acceptable as there is more information in the electronic data than in a printout.

The ability to select specific data or information for copying must be provided.

See FDA Access.

Media Refresh

Electronic storage media used for backup and archival purposes must be periodically refreshed in accordance with an SOP specific to its system to ensure readability. All media deteriorates over time.

Conversion / Migration Policy

When electronic records are migrated to another system, whether or not the data is converted for compatibility with the new system, all data, metadata, audit trails and electronic signatures that constitute the electronic records must be retained, properly linked and retrievable throughout the records retention period (Predicate Rule).

Obsolete Systems

Obsolete systems do not need to be maintained providing all data, metadata, audit trails and electronic signatures that constitute the electronic records are retained, properly linked, and retrievable throughout the records retention period. All documentation must be maintained and retrievable throughout the records retention period (Predicate Rule).

System retirement must provide a means in accordance with an SOP or plan for the migration of data.

See Conversion / Migration policy.

Time

All entries and transactions must be date and time stamped with the local date and time including time zone. This includes electronic records, electronic signatures, and audit trails. Where a transaction crosses time zones, there must be a means to reconstruct the events in their original order.

Regulatory Compliance Assessment

An assessment identifying areas that require change to achieve full compliance is performed for all infrastructure, software applications, computer and computerized systems that create, modify, maintain, archive, retrieve or transmit data in electronic form under any requirements set forth in FDA regulations. A Standard Operating Procedure is used to assess the level of compliance of infrastructure, all software applications, computer and computerized systems with 21 CFR Part 11.

The assessment is, at a minimum, the joint effort of a Lead Assessor who is independent with respect to the System Owner, the System Owner, a Reviewer and QA(U). A Summary Report of the findings and conclusions of the assessment are prepared.

The System Owner is responsible for the development and understanding of the assessment. He or she acquires and coordinates support from in-house consultants, Quality Assurance (Unit), System Developers and/or contracted external support services.

QA(U) is responsible to verify that the details in the assessment protocol are compliant with the applicable departmental, interdepartmental and corporate practices and policies. They also assure that the approach meets appropriate cGxP regulations, good business practices and compliance with 21 CFR Part 11.

The qualifications of the individuals involved in the project are kept on file and made available as needed in accordance with cGxP. The qualifications are kept with the project documentation or are available from the Human Resources Department.

Document Management and Storage

A Document Management and Storage section is included in the Electronic Records; Electronic Signatures SOP. It defines the Document Management and Storage methodology that is followed for issue, revision, replacement, control and management of this SOP. This methodology is in accordance with an established Change Control SOP and a Document Management and Storage SOP.

Revision History

A Revision History section is included in the Electronic Records; Electronic Signatures SOP. The Electronic Records; Electronic Signatures SOP Version Number and all Revisions made to the document are recorded in this section. These changes are in accordance with a Change Control Procedure SOP.

The Revision History section consists of a summary of the changes, a brief statement explaining the reasons the document was revised, and the name and affiliation of the person responsible for the entries. It includes the numbers of all items the Electronic Records; Electronic Signatures SOP replaces and is written so that the original Electronic Records; Electronic Signatures SOP is not needed to understand the revisions or the revised document being issued.

When the Electronic Records; Electronic Signatures SOP is a new document, a "000" as the Revision Number is entered with a statement such as "Initial Issue" in the comments section.

Regulatory or audit commitments require preservation from being inadvertently lost or modified in subsequent revisions. All entries or modifications made as a result of regulatory or audit commitments require indication as such in the Revision History and should indicate the organization, date and applicable references. A complete and accurate audit trail must be developed and maintained throughout the project lifecycle.

Attachments

If needed, an Attachments section is included in the Electronic Records; Electronic Signatures SOP in accordance with the SOP on SOPs. An Attachment Index and Attachment section follow the text of the Electronic Records; Electronic Signatures SOP. The top of each attachment page indicates "Attachment # x," where 'x' is the sequential number of each attachment. The second line of each attachment page contains the attachment title. The number of pages and date of issue corresponding to each attachment are listed in the Attachment Index.

Attachments are placed directly after the Attachment Index and are sequentially numbered (e.g., Attachment # 1, Attachment # 2) with each page of an attachment containing the corresponding document number, page numbering, and total number of pages in the attachment (e.g., Page x of y). Each attachment is labeled consistent with the title of the attachment itself.

Attachments are treated as part of the Electronic Records; Electronic Signatures SOP. If an attachment needs to be changed, the entire Electronic Records; Electronic Signatures SOP requires a new revision number and is reissued. New attachments or revisions of existing attachments are in accordance with an established Change Control SOP and a Document Management and Storage SOP.

Chapter 24 - 21 CFR Part 11 Assessment

In this chapter we will discuss the Standard Operating Procedure (SOP) for determining the compliance of a software application, computer or computerized system with 21 CFR Part 11, *Electronic Records; Electronic Signatures.*

The FDA has determined that in order to maintain the integrity, accuracy, reliability and consistency of operation of computer and computerized systems, the subsequent electronic data and provide a vehicle for the use of electronic signatures that are considered the equivalent of hand written signatures, the application of 21 CFR Part 11 is required in the interest of health and safety. It is needed to determine compliance with 21 CFR Part 11.

This SOP is used for all systems that create, modify, maintain, retrieve, transmit or archive records in electronic format. It is used in conjunction with an overlying corporate policy to provide sustainable compliance with the regulation.

Everyone involved in the operation of a software application, computer or computerized system that controls equipment and/or produces electronic data or records in support of all regulated laboratory, manufacturing, clinical or distribution activities uses this SOP.

Title Block, Headers and Footers

The 21 CFR Part 11 Assessment SOP should have a title block, title or cover page along with a header and/or footer on each page of the document that clearly identifies the document. The minimum information that should be displayed for the 21 CFR Part 11 Assessment SOP is as follows:

- Document Title
- Document Number
- Document Revision Number
- Project Name and Number
- Site/Location/Department Identification
- Date Document was issued, last revised or draft status
- Effective Date
- Pagination: page of total pages (e.g., Page x of y)

The Document Title should be in large type font that enables clear identification. It should clearly and concisely describe the contents and intent of the document. Since this document will be part of a set or package of documents within a project, it should be consistent with other documents used for the project and specify that it is the 21 CFR Part 11 Assessment SOP. This information should be on every page of the document.

The Document Number should be in large type font that enables clear identification. Since the 21 CFR Part 11 Assessment SOP will be part of a set or package of documents within a project, it should be consistent with other documents used for the project. This information should appear on every page of the document.

The Document Revision Number, which identifies the revision of the document, should appear on each page of the document. Many documents will go through one or more changes over time, particularly during the draft stage and the review and approval cycle. It is important to maintain accurate revision control because even a subtle change in a document can have an immense impact on the outcomes of activities. The SOP on SOPs should govern the numbering methodology.

See Chapter 6 – "The SOP on SOPs

The Project Name and Number, if applicable, should be indicated on the cover or title page. This information identifies the overall project to which this document belongs.

The Site/Location/Department Identification should be indicated on the cover or title page. It identifies the Site/Location/Department that has responsibility for this document and its associated project.

The Date that the Document was issued, last revised or issued with draft status should be on each page of the document (e.g., 12-Dec-01). This will clearly identify the date on which the document was last modified.

The Effective Date of the Document should be on the title or cover page of the document (e.g., 12-Dec-01). This will clearly identify when to begin using the current version of the document.

Pagination, page of total pages (e.g., Page x of y), should be on each page of the document. This will help identify whether or not all of the pages of the document are present. Pages intentionally left blank should clearly be designated as such.

Approval Section

The Approval Section should contain a table with the Author and Approvers printed names and titles and a place to sign and date the SOP. This SOP on SOPs should govern who the approvers of the document should be. The functional department (e.g., Production / Quality Assurance / Engineering) of the signatory should be placed below each printed name and title along with a statement associated with each approver indicating the person's role or qualification in the approval process. Indicate the significance of the associated signature (e.g., This document meets the requirements of Corporate Policy 9055, Electronic Records; Electronic Signatures, dated 22-Mar-2000).

The completion (signing) of this section indicates that the contents have been reviewed and approved by the listed individuals.

Table of Contents

The SOP should contain a Table of Contents, which appears towards the beginning of the document. This Table of Contents provides a way to easily locate various topics contained in the document.

Purpose *(SOP)*

The Purpose section should clearly state that the intention of the document is to provide a Standard Operating Procedure to be used to assess the level of compliance of a software application, computer or computerized system or infrastructure with 21 CFR Part 11. This introduction to the SOP concisely describes the document's purpose in sufficient detail so that a non-technical reader can understand its purpose.

Scope *(SOP)*

The Scope section identifies the limitations of the 21 CFR Part 11 Assessment SOP. It limits the scope to all GxP systems and is applicable to all computers, computerized systems, infrastructure, LANs and WANs including clinical, laboratory, manufacturing and distribution computer systems. It applies to all Sites, Locations and Departments within the company.

It includes electronic records, electronic signatures and handwritten signatures executed to electronic records. It applies to all records that are created, modified, maintained, archived, retrieved or transmitted in electronic form under any requirements set forth in FDA regulations. Provide the scope or range of control and authority of the SOP. State who, what, where, when, why and how the SOP is to control the processes or procedures specified in the Purpose section.

Equipment / Materials / Tools

This section lists any required or recommended equipment, materials, tools, templates or guides. Provide the name and a brief description of the equipment, instruments, materials and/or tools to be used in the procedures or testing covered by the 21 CFR Part 11Assessment SOP. This is for reference purposes and requires the user to utilize the specific item or a designated equivalent. The purpose and use is explained elsewhere in the document. Substitution of comparable equipment may be made in accordance with procedures for Change Control.

If the information is included in a job aid or operating manual, a description of where the information may be found with a specific reference to the document number, title, date and revision number should be specified.

Warnings / Notes / General Information / Safety

Warnings, notes, general or safety information applicable to the 21 CFR Part 11 Assessment SOP are listed in this section.

Policies, Guidelines, References

This section lists references to appropriate regulatory requirements, SOPs, policies, guidelines, or other relevant documents referred to in the 21 CFR Part 11 Assessment SOP. The name and a brief description of applicable related SOPs, corporate policies, procedures and guidelines, Quality Control monographs, books, technical papers, publications, specific regulations, batch records, vendor or system manuals, and all other relevant documents that are required by the 21 CFR Part 11 Assessment SOP should be included. This is for reference purposes and requires the user to use the specific item or a designated equivalent. The purpose and use are explained elsewhere in the document.

SOPs cite references and incorporate them in whole by reference rather than specifying details that duplicate these items. This is done so that changes in policy, guidelines or reference materials require no changes to SOPs themselves and the most current version of the referenced material is used. Examples are:

- Policy ABC-123, Administrative Policy for Electronic Records; Electronic Signatures
- Corporate, Plan for Compliance with 21 CFR Part 11, Electronic Records; Electronic Signatures
- SOP-456, Electronic Records; Electronic Signatures
- FDA, 21 CFR Part 11, Electronic Records; Electronic Signatures
- FDA, General Principals of Software Validation; Final Guidance for Industry and FDA Staff, January 11, 2002

Assumptions / Exclusions / Limitations

Any and all assumptions, exclusions or limitations within the context of the SOP should be discussed. Specific issues should be directly addressed in this section of the 21 CFR Part 11 Assessment SOP. Examples are:

"Legacy systems are not grandfathered under 21 CFR Part 11, there is no limitation with respect to legacy systems."

"Networks that have been previously shown to be sustainedly compliant with 21 CFR Part 11 may be excluded from re-assessment when a new or existing system is evaluated."

Glossary of Terms

A Glossary of Terms should be included that defines all of the terms and acronyms that are used in this document or that are specific to the operation being described and may not be generally known to the reader. The reader may also be referred to the FDA document, "Glossary of Computerized System and Software Development Terminology."

Roles and Responsibilities

The assessment for compliance with 21 CFR Part 11 should be, at a minimum, the joint effort of a Lead Assessor, the System Owner and a Reviewer with a member of QA(U) being the final signatory of both the protocol and the Summary Report.

This section should clearly specify the participants by name, title, department or company affiliation and role in a table. The information in this section should conform to the corresponding information in the project Validation Plan.

The System Owner should be responsible for the development and understanding of the assessment. They should acquire and coordinate support from in-house consultants, Quality Assurance (Unit), system developers and/or contracted external support services.

The QA(U) should be responsible to verify that the details in the assessment protocol are compliant with the applicable departmental, interdepartmental and corporate practices and policies. They should also assure that the approach meets appropriate cGxP regulations, good business practices and compliance with 21 CFR Part 11.

The qualifications of the individuals involved in the project should be kept on file and made available as needed in accordance with cGxP. The qualifications should be kept with the project documentation or be available from the Human Resources Department. The location of the qualification documentation should be stated in this section.

Procedure

The protocol should contain instructions and procedures that enable the assessment of a software application, computer or computerized system. These include, but are not limited to the following items.

Title (protocol)

The title identifies the system to which assessment applies and states that it is an assessment in accordance with 21 CFR Part 11(e.g., 21 CFR Part 11 Assessment for the ACX-043 Mass Spectrometry Computerized System). It should be on the every page of the document.

Introduction (protocol)

Provide an introduction to the protocol that is an overview of the tasks at hand and explains the background of what is to be performed. For example:

Chapter 24 – 21 CFR Part 11 Assessment

"21 CFR Part 11 went into effect on August 20, 1997. It provides the criteria for FDA acceptance of electronic records to be equivalent to paper records and electronic signatures to be equivalent to handwritten signatures. The rule must be interpreted and applied to both new and legacy systems. Systems in use that predate August 20, 1997 are not exempt from the rule. The FDA recognizes that it will take time for existing systems to attain full compliance. The FDA expects steps to be taken toward achieving full compliance of these systems to 21 CFR Part 11. This document is an assessment method that will identify areas in the system and system operation that require change to achieve full compliance."

Purpose (protocol)

The Purpose section should clearly state that the intention of the document is to provide a means for assessing the level of compliance of a software application, computer or computerized system, or infrastructure with 21 CFR Part 11. This introduction to the protocol should be concise and describe the document's purpose in sufficient detail so that a non-technical reader can understand its purpose. For example:

"The purpose of this document is to evaluate the ACX-043 Mass Spectrometry Computerized System to determine the requirements that will be necessary to allow sustainable compliance with 21 CFR Part 11."

Scope (protocol)

The Scope section identifies the limitations of a specific 21 CFR Part 11 Assessment protocol with respect to the procedure to be performed in the assessment. It limits the scope to a specific system or software application. It applies to a specific Site, Location and/or Department within the company. It includes electronic records, electronic signatures and handwritten signatures executed to electronic records. It applies to all records that are created, modified, maintained, archived, retrieved or transmitted in electronic form under any requirements set forth in FDA regulations. For example:

"The scope of this document covers regulations pertaining to 21 CFR Part 11, Electronic Records; Electronic Signatures. This protocol is designed to assist with the assessment of computerized systems that create, modify, maintain, archive, retrieve, or transmit electronic records(s) that are required to demonstrate compliance with FDA regulations, and that are used in lieu of paper records. It is applicable to the XYZ Mass Spectrometer System 985 located in Department 600, Bioanlytical Laboratory located in building 35, Anywhere, US."

Applicable Requirements

The protocol identifies all applicable requirements that are necessary to perform the assessment. These are generally the same as those specified in the governing SOP and any additional items deemed necessary.

It specifies the qualifications of the people who are to perform the assessment.

Overview

The Overview provides a general description of the system, associated software applications and system interfaces. The execution of the protocol assesses compliance to 21 CFR Part 11.

Roles and Responsibilities

The assessment for compliance with 21 CFR Part 11 should be, at a minimum, the joint effort of a Lead Assessor who is independent of the System Owner, the System Owner, and a Reviewer with a member of QA(U) being the final signatory of both the protocol and the Summary Report.

This section clearly specifies the participants by name, title, department or company affiliation and role in a table. The information in this section should conform to the corresponding information in the project Validation Plan.

The System Owner is responsible for the development and understanding of the assessment. He or she acquires and coordinates support from in-house consultants, QA(U), System Developers and/or contracted external support services.

The QA(U) is responsible for verifying that the details in the assessment protocol are compliant with the applicable departmental, interdepartmental and corporate practices and policies. They also assure that the approach meets appropriate cGxP regulations, good business practices and compliance with 21 CFR Part 11.

The qualifications of the individuals involved in the project are kept on file and made available as needed in accordance with cGxP. The qualifications are kept with the project documentation or are available from the Human Resources Department. The location of the qualification documentation is stated in this section.

Instructions on Testing Procedures

This section describes the procedures that the Assessor is expected to follow. It includes how the tests will be documented, how deviations will be handled and other administrative items relevant to the execution of the protocol. The test procedures should be in accordance with the SOP governing 21 CFR Part 11 compliance assessments.

Assessor Identification

This section identifies each person participating in the testing. The following information is entered in a table for identification purposes:

- Full Name – printed
- Signature
- Initials
- Affiliation

Inventory

Identify what is being assessed. In a table or other suitable format, inventory items, such as, the System Name, System ID, Start Date, System Location, System Owner, End Date, System Type, Dial-in Access and Biometric Method.

Regulatory Compliance Checklist

This is the main section of the protocol and lists a series of tests that are to be performed as part of the assessment. Each test section should consist of the following:

- Identification of the Subpart(s) of the Regulation being assessed
- Requirements of the Subpart(s) of the Regulation being assessed
- Assessment Question
- Response - Yes / No / Not Applicable
- Observation (with source of information)
- Assessor Section with Signature and Date
- Reviewer Section with Signature and Date

- Approval Section with Signature and Date

The following Subparts should be assessed:

- 21 CFR 11.10 (a), 11.30, Systems used to create, modify, maintain, or transmit electronic records are validated.
- 21 CFR 11.10 (b), 11.30, Systems have the ability to generate accurate and complete copies of readable electronic records.
- 21 CFR 11.10 (c), 11.30, Protection of records to enable their accurate and ready retrieval throughout the records retention periods.
- 21 CFR 11.10 (d), 11.30, Limiting system access to authorized individuals.
- 21 CFR 11.10 (e), 11.30, Electronic records have audit trails.
- 21 CFR 11.10 (f), 11.30, Use of operational system checks to enforce permitted sequencing of steps and events, as appropriate.
- 21 CFR 11.10 (g), 11.30, Use of authority checks to ensure that only authorized individuals can use the system, electronically sign a record, access the operation or computer system input or output device, alter a record, or perform the operation at hand.
- 21 CFR 11.10 (h), 11.30, Device checks are in place to verify the source of data input or operational instruction.
- 21 CFR 11.10 (i), 11.30, Training is provided for those persons developing, maintaining, or using electronic record/electronic signature systems.
- 21 CFR 11.10 (j), 11.30, Written policies are in place that hold individuals accountable for and responsible for actions initiated under their electronic signatures.
- 21 CFR 11.10 (k), 11.30, Adequate control is in place for distribution, access, and use of electronic records. Revision and change control procedures maintain an audit trail of documentation development and modification.
- 21 CFR 11.30, Document encryption and use of appropriate digital signature standards must be in place. APPLICABLE TO OPEN SYSTEMS ONLY.
- 21 CFR 11.50, Signed electronic records contain: Signer's printed name, Date and time of signature, Signature meaning (e.g., review, approval)
- 21 CFR 11.70, Electronic signatures and handwritten signatures executed to electronic records are linked to their respective electronic records.
- 21 CFR 11.100 (a), Electronic signatures are unique to one individual and used only by that same individual.
- 21 CFR 11.100 (b), The identities of individuals are verified before electronic signatures are assigned.
- 21 CFR 11.100 (c), Electronic signatures used by individuals on or after August 20, 1997 are the legally binding equivalent of traditional handwritten signatures.
- 21 CFR 11.200 (a and b), Electronic signatures are designed to ensure they cannot be used by anyone other than their genuine owners. (NON-BIOMETRIC)
- 21 CFR 11.200 (a and b), Electronic signatures are designed to ensure they cannot be used by anyone other than their genuine owners. (BIOMETRIC)
- 21 CFR 11.300, Persons who use electronic signatures based upon use of identification codes in combination with passwords shall employ controls to ensure their security and integrity.

Document Management and Storage (protocol)

A Document Management and Storage section is included in each 21 CFR Part 11 Assessment. It defines the Document Management and Storage methodology that is followed for issue, revision, replacement, control and management of each 21 CFR Part 11 Assessment. This methodology is in accordance with an established Change Control SOP and a Document Management and Storage SOP.

Revision History (protocol)

A Revision History section is included in each 21 CFR Part 11 Assessment. The 21 CFR Part 11 Assessment Version Number and all Revisions made to the document are recorded in this section. These changes are in accordance with a Change Control Procedure SOP.

The Revision History section consists of a summary of the changes, a brief statement explaining the reasons the document was revised, and the name and affiliation of the person responsible for the entries. It includes the numbers of all items the 21 CFR Part 11 Assessment replaces and is written so that the original 21 CFR Part 11 Assessment is not needed to understand the revisions or the revised document being issued.

When the 21 CFR Part 11 Assessment is a new document, a "000" as the Revision Number is entered with a statement such as "Initial Issue" in the comments section.

Regulatory or audit commitments require preservation from being inadvertently lost or modified in subsequent revisions. All entries or modifications made as a result of regulatory or audit commitments require indication as such in the Revision History and should indicate the organization, date and applicable references. A complete and accurate audit trail must be developed and maintained throughout the project lifecycle.

Attachments (protocol)

If needed, an Attachments section is included in each 21 CFR Part 11 Assessment in accordance with the SOP on SOPs. An Attachment Index and Attachment section follow the text of a 21 CFR Part 11 Assessment. The top of each attachment page indicates "Attachment # x," where 'x' is the sequential number of each attachment. The second line of each attachment page contains the attachment title. The number of pages and date of issue corresponding to each attachment are listed in the Attachment Index.

Attachments are placed directly after the Attachment Index and are sequentially numbered (e.g., Attachment # 1, Attachment # 2) with each page of an attachment containing the corresponding document number, page numbering, and total number of pages in the attachment (e.g., Page x of y). Each attachment is labeled consistent with the title of the attachment itself.

Attachments are treated as part of the 21 CFR Part 11 Assessment. If an attachment needs to be changed, the entire 21 CFR Part 11 Assessment requires a new revision number and is reissued. New attachments or revisions of existing attachments are in accordance with an established Change Control SOP and a Document Management and Storage SOP.

Document Management and Storage (SOP)

A Document Management and Storage section is included in the 21 CFR Part 11 Assessment SOP. It defines the Document Management and Storage methodology that is followed for issue, revision, replacement, control and management of this SOP. This methodology is in accordance with an established Change Control SOP and a Document Management and Storage SOP.

Revision History (SOP)

A Revision History section is included in the 21 CFR Part 11 Assessment SOP. The 21 CFR Part 11 Assessment SOP Version Number and all Revisions made to the document are recorded in this section. These changes are in accordance with a Change Control Procedure SOP.

The Revision History section consists of a summary of the changes, a brief statement explaining the reasons the document was revised, and the name and affiliation of the person responsible for the entries. It includes the numbers of all items the 21 CFR Part 11 Assessment SOP replaces and is written so that the original 21 CFR Part 11 Assessment SOP is not needed to understand the revisions or the revised document being issued.

When the 21 CFR Part 11 Assessment SOP is a new document, a "000" as the Revision Number is entered with a statement such as "Initial Issue" in the comments section.

Regulatory or audit commitments require preservation from being inadvertently lost or modified in subsequent revisions. All entries or modifications made as a result of regulatory or audit commitments require indication as such in the Revision History and should indicate the organization, date and applicable references. A complete and accurate audit trail must be developed and maintained throughout the project lifecycle.

Attachments *(SOP)*

If needed, an Attachments section is included in the 21 CFR Part 11 Assessment SOP in accordance with the SOP on SOPs. An Attachment Index and Attachment section follow the text of the 21 CFR Part 11 Assessment SOP. The top of each attachment page indicates "Attachment # x," where 'x' is the sequential number of each attachment. The second line of each attachment page contains the attachment title. The number of pages and date of issue corresponding to each attachment are listed in the Attachment Index.

Attachments are placed directly after the Attachment Index and are sequentially numbered (e.g., Attachment # 1, Attachment # 2) with each page of an attachment containing the corresponding document number, page numbering, and total number of pages in the attachment (e.g., Page x of y). Each attachment is labeled consistent with the title of the attachment itself.

Attachments are treated as part of the 21 CFR Part 11 Assessment SOP. If an attachment needs to be changed, the entire 21 CFR Part 11 Assessment SOP requires a new revision number and is reissued. New attachments or revisions of existing attachments are in accordance with an established Change Control SOP and a Document Management and Storage SOP.

Chapter 25 – Summary Reports

In this chapter we will discuss the Standard Operating Procedure (SOP) for creating a Summary Report for Design Qualification, Unit Testing, Installation Qualification (IQ), Operational Qualification (OQ), Performance Qualification (PQ) testing, the 21 CFR Part 11 Assessment or Retrospective Validation and Gap Analysis for a validation project. It specifies the requirements for preparing a Summary Report for or in support incremental qualification of a Good Clinical, Laboratory, Manufacturing or Distribution Practices (GxP) project.

Best industry practices and FDA Guidance documents indicate that in order to demonstrate that a computer or computerized system performs as it purports and is intended to perform, interim Summary Reports are required to demonstrate, in a summarized format, the outcomes of qualification testing.

This SOP is used for all systems that create, modify, maintain, retrieve, transmit or archive records in electronic format. It is used in conjunction with an overlying corporate policy to provide sustainable compliance with the regulation. All Unit Testing, IQ, OQ, PQ procedures and Retrospective Validation and Gap Analyses must have their outcomes summarized in accordance with this SOP.

This SOP is used as a means to summarize the outcomes of qualification testing by everyone involved in projects for a software application, computer or computerized system that controls equipment and/or produces electronic data or records in support of all regulated laboratory, manufacturing, clinical or distribution activities.

Title Block, Headers and Footers

The Summary Reports SOP should have a title block, title or cover page along with a header and/or footer on each page of the document that clearly identifies the document. The minimum information that should be displayed for the Summary Reports SOP is as follows:

- Document Title
- Document Number
- Document Revision Number
- Project Name and Number
- Site/Location/Department Identification
- Date Document was issued, last revised or draft status
- Effective Date
- Pagination: page of total pages (e.g., Page x of y)

The Document Title should be in large type font that enables clear identification. It should clearly and concisely describe the contents and intent of the document. Since this document will be part of a set or package of documents within a project, it should be consistent with other documents used for the project and specify that it is the Summary Reports SOP. This information should be on every page of the document.

The Document Number should be in large type font that enables clear identification. Since the Summary Reports SOP will be part of a set or package of documents within a project, it should be consistent with other documents used for the project. This information should appear on every page of the document.

The Document Revision Number, which identifies the revision of the document, should appear on each page of the document. Many documents will go through one or more changes over time, particularly during the draft stage and the review and approval cycle. It is important to maintain accurate revision control because even a

subtle change in a document can have an immense impact on the outcomes of activities. The SOP on SOPs should govern the numbering methodology.

See Chapter 6 – "The SOP on SOPs."

The Project Name and Number, if applicable, should be indicated on the cover or title page. This information identifies the overall project to which this document belongs.

The Site/Location/Department Identification should be indicated on the cover or title page. It identifies the Site/Location/Department that has responsibility for this document and its associated project.

The Date that the Document was issued, last revised or issued with draft status should be on each page of the document (e.g., 12-Dec-01). This will clearly identify the date on which the document was last modified.

The Effective Date of the Document should be on the title or cover page of the document (e.g., 12-Dec-01). This will clearly identify when to begin using the current version of the document.

Pagination, page of total pages (e.g., Page x of y), should be on each page of the document. This will help identify whether or not all of the pages of the document are present. Pages intentionally left blank should clearly be designated as such.

Approval Section *(SOP)*

The Approval Section should contain a table with the Author and Approvers printed names, titles and a place to sign and date the SOP. The SOP on SOPs should govern who the approvers of the document should be. The functional department (e.g., Production / Quality Assurance / Engineering) of the signatory should be placed below each printed name and title along with a statement associated with each approver indicating the person's role or qualification in the approval process. Indicate the significance of the associated signature (e.g., This document meets the requirements of Corporate Policy 9055, Electronic Records; Electronic Signatures, dated 22-Mar-2000).

The completion (signing) of this section indicates that the contents have been reviewed and approved by the listed individuals.

Table of Contents *(SOP)*

The SOP should contain a Table of Contents, which appears towards the beginning of the document. This Table of Contents provides a way to easily locate various topics contained in the document.

Purpose *(SOP)*

The Purpose section should clearly state that the intention of the document is to provide a Standard Operating Procedure to be used to create Summary Reports for use in the incremental qualification of new or upgraded computer or computerized systems.

This introduction to the SOP concisely describes the document's purpose in sufficient detail so that a non-technical reader can understand its purpose.

Scope (SOP)

The Scope section identifies the limitations of the Summary Reports SOP. It limits the scope to all GxP systems and is applicable to all computers, computerized systems, infrastructure, LANs and WANs including clinical, laboratory, manufacturing and distribution computer systems. It applies to all Sites, Locations and Departments within the company.

All Unit Testing, IQ, OQ and PQ procedures must have their outcomes summarized in accordance with this SOP.

Equipment / Materials / Tools)

This section lists any required or recommended equipment, materials, tools, templates or guides. Provide the name and a brief description of the equipment, instruments, materials and/or tools to be used in the procedures or testing covered by the Summary Report SOP. This is for reference purposes and requires the user to utilize the specific item or a designated equivalent. The purpose and use is explained elsewhere in the document. Substitution of comparable equipment may be made in accordance with procedures for Change Control.

If the information is included in a job aid or operating manual, a description of where the information may be found with a specific reference to the document number, title, date and revision number should be specified.

Warnings / Notes / General Information / Safety

Warnings, notes, general or safety information applicable to the Summary Reports SOP are listed in this section.

Policies, Guidelines, References

This section lists references to appropriate regulatory requirements, SOPs, policies, guidelines, or other relevant documents referred to in the Summary Reports SOP. The name and a brief description of applicable related SOPs, corporate policies, procedures and guidelines, Quality Control monographs, books, technical papers, publications, specific regulations, batch records, vendor or system manuals, and all other relevant documents that are required by the Summary Reports SOP should be included. This is for reference purposes and requires the user to use the specific item or a designated equivalent. The purpose and use are explained elsewhere in the document.

SOPs cite references and incorporate them in whole by reference rather than specifying details that duplicate these items. This is done so that changes in policy, guidelines or reference materials require no changes to SOPs themselves and the most current version of the referenced material is used. Examples are:

- Policy ABC-123, Administrative Policy for Electronic Records; Electronic Signatures
- Corporate, Plan for Compliance with 21 CFR Part 11, Electronic Records; Electronic Signatures
- SOP-456, Electronic Records; Electronic Signatures
- FDA, 21 CFR Part 11, Electronic Records; Electronic Signatures
- FDA, General Principals of Software Validation; Final Guidance for Industry and FDA Staff, January 11, 2002
- GAMP, Supplier Guide for Validation of Automated Systems in Pharmaceutical Manufacture

Assumptions / Exclusions / Limitations (SOP)

Any and all assumptions, exclusions or limitations within the context of the SOP should be discussed. Specific issues should be directly addressed in this section of the Summary Reports SOP.

Glossary of Terms

A Glossary of Terms should be included that defines all of the terms and acronyms that are used in this document or that are specific to the operation being described and may not be generally known to the reader. The reader may also be referred to the FDA document, "Glossary of Computerized System and Software Development Terminology."

Roles and Responsibilities

The System Owner / End-User is responsible for preparing Summary Reports for each Unit Test, IQ, OQ or PQ protocol performed and ensure that it is reviewed and approved by the System Owner, CVG and QA(U).

The Subject Matter Expert (SME) ensures that changes required as a result of deviations during testing and subsequent remediation are captured in the Summary Report. If a qualification protocol execution was begun prior to the approval of the summary report, then the SMEshould certify that deviations and remediations have no effect on the qualification. If the deviations or remediations have an affect, then the Summary Report should state that affect.

Procedure

The Summary Reports SOP should contain instructions and procedures that enable the creation of a Summary Report for Design Qualification, Unit Testing, Installation Qualification (IQ), Operational Qualification (OQ), Performance Qualification (PQ) testing, the 21 CFR Part 11 Assessment or Retrospective Validation and Gap Analysis.

Summary Reports are prepared as stand-alone documents for all projects. It is not good practice to include them as part of other documents in an attempt to reduce the amount of workload or paperwork. This only causes more work over the length of the project resulting from misunderstandings and confusion. Clear, concise documents that address specific parts of any given project are desirable.

Purpose (summary report)

This section clearly states that the purpose of the summary report is to summarize the outcome of a specified qualification procedure. This short introduction should be concise and present details so that an unfamiliar reader can understand the purpose of the document with regard to its use.

Scope (summary report)

The Scope section identifies the limitations of a specific summary report in relationship to the purpose of the qualification procedure. The objective of a summary report is to summarize the results of a qualification procedure.

Assumptions / Exclusions / Limitations (summary report)

Discuss any assumptions, exclusions or limitations within the context of the report. Specific issues should be directly addressed in this section of the Summary Report.

Overview of the Process

Provide an overview of the business operations for the system including workflow, material flow and the control philosophy of what functions, calculations and processes are to be performed by the system, as well as interfaces that the system or application must be provided. For Unit Testing, IQ, OQ and PQ, outline the activities required for acceptance of the work performed

Participants

List the participants in the procedure by name, title, department or company affiliation and responsibility in a table.

Accomplishments and Impact

Discuss the accomplishments of the procedure and the impact on the overall project.

Key Learning

List the key items learned from the experience with recommendations, as appropriate.

Rework & Delay Occurrences

Explain any rework that was required. What caused it. Discuss any delays that it may have caused and their impact on the project.

Summary

Summarize the overall work performed and the outcome of the procedure.

Conclusions

Provide the conclusions drawn from the results of the procedure. Specify whether the results of the process are acceptable. Does the outcome of the testing warrant approval and/or qualification for use of the item(s) tested? Is it permissible to proceed with the next qualification task or part of the process or project?

Document Management and Storage

A Document Management and Storage section is included in the Summary Report SOP. It defines the Document Management and Storage methodology that is followed for issue, revision, replacement, control and management of this SOP and the Summary Report. This methodology is in accordance with an established Change Control SOP and a Document Management and Storage SOP.

Revision History

A Revision History section is included in the Summary Report SOP. The Summary Report SOP Version Number and all Revisions made to the document are recorded in this section. These changes are in accordance with a Change Control Procedure SOP.

The Revision History section consists of a summary of the changes, a brief statement explaining the reasons the document was revised, and the name and affiliation of the person responsible for the entries. It includes the numbers of all items the Summary Report SOP replaces and is written so that the original Summary Report SOP is not needed to understand the revisions or the revised document being issued.

When the Summary Report SOP is a new document, a "000" as the Revision Number is entered with a statement such as "Initial Issue" in the comments section.

Regulatory or audit commitments require preservation from being inadvertently lost or modified in subsequent revisions. All entries or modifications made as a result of regulatory or audit commitments require indication as such in the Revision History and should indicate the organization, date and applicable references. A complete and accurate audit trail must be developed and maintained throughout the project lifecycle.

Attachments

If needed, an Attachments section is included in the Summary Report SOP in accordance with the SOP on SOPs. An Attachment Index and Attachment section follow the text of the Summary Report SOP. The top of each attachment page indicates "Attachment # x," where 'x' is the sequential number of each attachment. The second line of each attachment page contains the attachment title. The number of pages and date of issue corresponding to each attachment are listed in the Attachment Index.

Attachments are placed directly after the Attachment Index and are sequentially numbered (e.g., Attachment # 1, Attachment # 2) with each page of an attachment containing the corresponding document number, page numbering, and total number of pages in the attachment (e.g., Page x of y). Each attachment is labeled consistent with the title of the attachment itself.

Attachments are treated as part of the Summary Report SOP. If an attachment needs to be changed, the entire Summary Report SOP requires a new revision number and is reissued. New attachments or revisions of existing attachments are in accordance with an established Change Control SOP and a Document Management and Storage SOP.

Chapter 26 – Traceability Matrix

In this chapter we will discuss the Standard Operating Procedure (SOP) for creating a Traceability Matrix that cross-references User Requirements, Functional Requirements, Design Specifications, Installation Qualification (IQ), Operational Qualification (OQ) and Performance Qualification (PQ) for a validation project. It specifies the requirements for preparing a Traceability Matrix to track and determine the compliance of a project with the business needs and requirements of the system owner in support of a Good Clinical, Laboratory, Manufacturing or Distribution Practices (GxP) project.

Best industry practices and FDA Guidance documents indicate that in order to demonstrate that a computer or computerized system performs as it purports and is intended to perform, a Traceability Matrix is required to demonstrate, in a summarized format, that the system is being built in accordance with the business needs and requirements of the project. It aids in demonstrating that the correct system was built.

This SOP is used for all systems that create, modify, maintain, retrieve, transmit or archive records in electronic format. It is used in conjunction with an overlying corporate policy to provide sustainable compliance with the regulation. All requirements, specifications and qualification protocols must be cross-referenced with each other in accordance with this SOP.

Everyone involved in projects for a software application, computer or computerized system that controls equipment and/or produces electronic data or records in support of all regulated laboratory, manufacturing, clinical or distribution activities uses this SOP.

Title Block, Headers and Footers

The Traceability Matrix SOP should have a title block, title or cover page along with a header and/or footer on each page of the document that clearly identifies the document. The minimum information that should be displayed for the Traceability Matrix SOP is as follows:

- Document Title
- Document Number
- Document Revision Number
- Project Name and Number
- Site/Location/Department Identification
- Date Document was issued, last revised or draft status
- Effective Date
- Pagination: page of total pages (e.g., Page x of y)

The Document Title should be in large type font that enables clear identification. It should clearly and concisely describe the contents and intent of the document. Since this document will be part of a set or package of documents within a project, it should be consistent with other documents used for the project and specify that it is the Traceability Matrix SOP. This information should be on every page of the document.

The Document Number should be in large type font that enables clear identification. Since the Traceability Matrix SOP will be part of a set or package of documents within a project, it should be consistent with other documents used for the project. This information should appear on every page of the document.

The Document Revision Number, which identifies the revision of the document, should appear on each page of the document. Many documents will go through one or more changes over time, particularly during the draft stage and the review and approval cycle. It is important to maintain accurate revision control because even a

subtle change in a document can have an immense impact on the outcomes of activities. The SOP on SOPs should govern the numbering methodology.

See Chapter 6 – "The SOP on SOPs."

The Project Name and Number, if applicable, should be indicated on the cover or title page. This information identifies the overall project to which this document belongs.

The Site/Location/Department Identification should be indicated on the cover or title page. It identifies the Site/Location/Department that has responsibility for this document and its associated project.

The Date that the Document was issued, last revised or issued with draft status should be on each page of the document (e.g., 12-Dec-01). This will clearly identify the date on which the document was last modified.

The Effective Date of the Document should be on the title or cover page of the document (e.g., 12-Dec-01). This will clearly identify when to begin using the current version of the document.

Pagination, page of total pages (e.g., Page x of y), should be on each page of the document. This will help identify whether or not all of the pages of the document are present. Pages intentionally left blank should clearly be designated as such.

Approval Section *(SOP)*

The Approval Section should contain a table with the Author and Approvers printed names, titles and a place to sign and date the SOP. The SOP on SOPs should govern who the approvers of the document should be. The functional department (e.g., Production / Quality Assurance / Engineering) of the signatory should be placed below each printed name and title along with a statement associated with each approver indicating the person's role or qualification in the approval process. Indicate the significance of the associated signature (e.g., This document meets the requirements of Corporate Policy 9055, Electronic Records; Electronic Signatures, dated 22-Mar-2000).

The completion (signing) of this section indicates that the contents have been reviewed and approved by the listed individuals.

Table of Contents *(SOP)*

The SOP should contain a Table of Contents, which appears towards the beginning of the document. This Table of Contents provides a way to easily locate various topics contained in the document.

Purpose

The Purpose section should clearly state that the intention of the document is to provide a Standard Operating Procedure to be used to create a Traceability Matrix that cross references User Requirements, Functional Requirements, Design Specifications, Installation Qualification (IQ), Operational Qualification (OQ) and Performance Qualification (PQ) for a validation project of new or upgraded computer or computerized systems. It specifies the requirements for preparing a Traceability Matrix to track and determine the compliance of a project with the business needs of the system owner in support of a Good Clinical, Laboratory, Manufacturing or Distribution Practices (GxP) project.

This introduction to the SOP concisely describes the document's purpose in sufficient detail so that a non-technical reader can understand its purpose.

Scope

The Scope section identifies the limitations of the Traceability Matrix SOP. It limits the scope to all GxP systems and is applicable to all computers, computerized systems, infrastructure, LANs and WANs including clinical, laboratory, manufacturing and distribution computer systems. It applies to all Sites, Locations and Departments within the company.

All requirements, specifications, IQ, OQ and PQ procedures must have their items cross-referenced in accordance with this SOP to demonstrate that each of the User Requirements, Functional Requirements and Design Specifications has been addressed.

Equipment / Materials / Tools

This section lists any required or recommended equipment, materials, tools, templates or guides. Provide the name and a brief description of the equipment, instruments, materials and/or tools to be used in the procedures or testing covered by the Traceability Matrix SOP. This is for reference purposes and requires the user to utilize the specific item or a designated equivalent. The purpose and use is explained elsewhere in the document. Substitution of comparable equipment may be made in accordance with procedures for Change Control.

If the information is included in a job aid or operating manual, a description of where the information may be found with a specific reference to the document number, title, date and revision number should be specified.

Warnings / Notes / General Information / Safety

Warnings, notes, general or safety information applicable to the Traceability Matrix SOP are listed in this section.

Policies, Guidelines, References

This section lists references to appropriate regulatory requirements, SOPs, policies, guidelines, or other relevant documents referred to in the Traceability Matrix SOP. The name and a brief description of applicable related SOPs, corporate policies, procedures and guidelines, Quality Control monographs, books, technical papers, publications, specific regulations, batch records, vendor or system manuals, and all other relevant documents that are required by the Traceability Matrix SOP should be included. This is for reference purposes and requires the user to use the specific item or a designated equivalent. The purpose and use are explained elsewhere in the document.

SOPs cite references and incorporate them in whole by reference rather than specifying details that duplicate these items. This is done so that changes in policy, guidelines or reference materials require no changes to SOPs themselves and the most current version of the referenced material is used. Examples are:

- Policy ABC-123, Administrative Policy for Electronic Records; Electronic Signatures
- Corporate, Plan for Compliance with 21 CFR Part 11, Electronic Records; Electronic Signatures
- SOP-456, Electronic Records; Electronic Signatures
- FDA, 21 CFR Part 11, Electronic Records; Electronic Signatures
- FDA, General Principals of Software Validation; Final Guidance for Industry and FDA Staff, January 11, 2002

Assumptions / Exclusions / Limitations

Any and all assumptions, exclusions or limitations within the context of the SOP should be discussed. Specific issues should be directly addressed in this section of the Traceability Matrix SOP.

Glossary of Terms

A Glossary of Terms should be included that defines all of the terms and acronyms that are used in this document or that are specific to the operation being described and may not be generally known to the reader. The reader may also be referred to the FDA document, "Glossary of Computerized System and Software Development Terminology."

Roles and Responsibilities

The System Owner / End-User is responsible for preparing and maintaining the Traceability Matrix for each the validation project and ensure that it is reviewed and approved by the System Owner, CVG and QA(U).

Procedure

The Traceability Matrix SOP should contain instructions and procedures that enable the creation of a Traceability Matrix to track all requirements, specifications, IQ, OQ, and PQ procedures.

The Traceability Matrix is prepared as stand-alone documents for all projects. It is not good practice to include them as part of other documents in an attempt to reduce the amount of workload or paperwork. This only causes more work over the length of the project resulting from misunderstandings and confusion. Clear, concise documents that address specific parts of any given project are desirable.

This document will be a living document until the end of the System Development Life Cycle. Because it will not be signed off until the end of the project, when it will be essentially too late to fix it, someone other than the preparer must review it on a regular basis.

Traceability Matrix

The Traceability Matrix is a spreadsheet that cross-references the individual Functional Requirements, Design Specifications, and qualification tasks from the IQ, OQ and PQ back to their predecessor requirements or specifications and ultimately back to the User Requirements. Each row represents an item that requires accountability. Each column represents the document items being used to fulfill the validation requirements.

Each item in the User Requirements must be shown to have been satisfied in the Functional Requirements. Each item in the Functional Requirements must be shown to have been satisfied in either the Hardware or Software Design Specification. It must be shown where each item in each design specification is tested in the IQ, OQ and/or PQ.

Where a requirement is not met, an explanation must be made explaining what wasn't met, why it wasn't met and the workaround, if any, that is to be followed to provide for the functionality not supplied by the system. This information is entered in the Exceptions column.

The first column in the table is the User Requirements column. It contains a listing, by item number and description, of the individual User Requirements that must be met for the system to be considered built as required.

For each column after the User Requirements column, simply place all of the document's item reference numbers that satisfy the predecessor requirement in the same cell on that row. It is not necessary to enter the descriptions for other than the User Requirements. Where an item is applicable to more than one predecessor, place it's listing in all of the appropriate locations in that column. Provide a column in the table for notes that can be associated with each row in the table. The columns and their order are as follows:

- User Requirements Column
- Functional Requirements Column
- Design Specification Column(s)
- Installation Qualification Column(s)
- Operational Qualification Column(s)
- Performance Qualification Column
- Exceptions Column
- Notes Column

There may be more than one Design Specification column (e.g.- hardware, software and/or firmware). The same applies for IQ and OQ.

Document Management and Storage (matrix)

A Document Management and Storage section is included in each Traceability Matrix. It defines the Document Management and Storage methodology that is followed for issue, revision, replacement, control and management of each Traceability Matrix. This methodology is in accordance with an established Change Control SOP and a Document Management and Storage SOP.

Revision History (matrix)

A Revision History section is included in each Traceability Matrix. The Traceability Matrix Version Number and all Revisions made to the document are recorded in this section. These changes are in accordance with a Change Control Procedure SOP.

The Revision History section consists of a summary of the changes, a brief statement explaining the reasons the document was revised, and the name and affiliation of the person responsible for the entries. It includes the numbers of all items the Traceability Matrix replaces and is written so that the original Traceability Matrix is not needed to understand the revisions or the revised document being issued.

When the Traceability Matrix is a new document, a "000" as the Revision Number is entered with a statement such as "Initial Issue" in the comments section.

Regulatory or audit commitments require preservation from being inadvertently lost or modified in subsequent revisions. All entries or modifications made as a result of regulatory or audit commitments require indication as such in the Revision History and should indicate the organization, date and applicable references. A complete and accurate audit trail must be developed and maintained throughout the project lifecycle.

Attachments (matrix)

If needed, an Attachments section is included in each Traceability Matrix in accordance with the SOP on SOPs. An Attachment Index and Attachment section follow the text of a Traceability Matrix. The top of each attachment page indicates "Attachment # x," where 'x' is the sequential number of each attachment. The second line of each attachment page contains the attachment title. The number of pages and date of issue corresponding to each attachment are listed in the Attachment Index.

Attachments are placed directly after the Attachment Index and are sequentially numbered (e.g., Attachment # 1, Attachment # 2) with each page of an attachment containing the corresponding document number, page numbering, and total number of pages in the attachment (e.g., Page x of y). Each attachment is labeled consistent with the title of the attachment itself.

Attachments are treated as part of the Traceability Matrix. If an attachment needs to be changed, the entire Traceability Matrix requires a new revision number and is reissued. New attachments or revisions of existing attachments are in accordance with an established Change Control SOP and a Document Management and Storage SOP.

Document Management and Storage *(SOP)*

A Document Management and Storage section is included in the Traceability Matrix SOP. It defines the Document Management and Storage methodology that is followed for issue, revision, replacement, control and management of this SOP. This methodology is in accordance with an established Change Control SOP and a Document Management and Storage SOP.

Revision History *(SOP)*

A Revision History section is included in the Traceability Matrix SOP. The Traceability Matrix SOP Version Number and all Revisions made to the document are recorded in this section. These changes are in accordance with a Change Control Procedure SOP.

The Revision History section consists of a summary of the changes, a brief statement explaining the reasons the document was revised, and the name and affiliation of the person responsible for the entries. It includes the numbers of all items the Traceability Matrix SOP replaces and is written so that the original Traceability Matrix SOP is not needed to understand the revisions or the revised document being issued.

When the Traceability Matrix SOP is a new document, a "000" as the Revision Number is entered with a statement such as "Initial Issue" in the comments section.

Regulatory or audit commitments require preservation from being inadvertently lost or modified in subsequent revisions. All entries or modifications made as a result of regulatory or audit commitments require indication as such in the Revision History and should indicate the organization, date and applicable references. A complete and accurate audit trail must be developed and maintained throughout the project lifecycle.

Attachments *(SOP)*

If needed, an Attachments section is included in the Traceability Matrix SOP in accordance with the SOP on SOPs. An Attachment Index and Attachment section follow the text of the Traceability Matrix SOP. The top of each attachment page indicates "Attachment # x," where 'x' is the sequential number of each attachment. The second line of each attachment page contains the attachment title. The number of pages and date of issue corresponding to each attachment are listed in the Attachment Index.

Attachments are placed directly after the Attachment Index and are sequentially numbered (e.g., Attachment # 1, Attachment # 2) with each page of an attachment containing the corresponding document number, page numbering, and total number of pages in the attachment (e.g., Page x of y). Each attachment is labeled consistent with the title of the attachment itself.

Attachments are treated as part of the Traceability Matrix SOP. If an attachment needs to be changed, the entire Traceability Matrix SOP requires a new revision number and is reissued. New attachments or revisions of existing attachments are in accordance with an established Change Control SOP and a Document Management and Storage SOP.

Chapter 27 – Change Control

In this chapter we will discuss the Standard Operating Procedure (SOP) for performing changes to a software application, computer, or computerized system in accordance with a change control methodology for a validation project. It specifies the requirements for preparing a change control request to track and determine the compliance of a project with the business needs and requirements of the System Owner in support of a Good Clinical, Laboratory, Manufacturing or Distribution Practices (GxP) project.

Best industry practices and FDA Guidance documents indicate that in order to demonstrate that a computer or computerized system performs as it purports and is intended to perform after modification, a change control procedure is required to demonstrate, that the system has been modified in accordance with the business needs of the project.

This SOP is used for all systems that create, modify, maintain, retrieve, transmit or archive records in electronic format. It is used in conjunction with an overlying corporate policy to provide sustainable compliance with the regulations.

This SOP is used by everyone involved in projects for a software application, computer or computerized system that controls equipment and/or produces electronic data or records in support of all regulated laboratory, manufacturing, clinical or distribution activities where modification of the software, computer or computerized system, process, workflow or procedure is involved.

Title Block, Headers and Footers

The Change Control SOP should have a title block, title or cover page along with a header and/or footer on each page of the document that clearly identifies the document. The minimum information that should be displayed for the Change Control SOP is as follows:

- Document Title
- Document Number
- Document Revision Number
- Project Name and Number
- Site/Location/Department Identification
- Date Document was issued, last revised or draft status
- Effective Date
- Pagination: page of total pages (e.g., Page x of y)

The Document Title should be in large type font that enables clear identification. It should clearly and concisely describe the contents and intent of the document. Since this document will be part of a set or package of documents within a project, it should be consistent with other documents used for the project and specify that it is the Change Control SOP. This information should be on every page of the document.

The Document Number should be in large type font that enables clear identification. Since the Change Control SOP will be part of a set or package of documents within a project, it should be consistent with other documents used for the project. This information should appear on every page of the document.

The Document Revision Number, which identifies the revision of the document, should appear on each page of the document. Many documents will go through one or more changes over time, particularly during the draft stage and the review and approval cycle. It is important to maintain accurate revision control because even a subtle change in a document can have an immense impact on the outcomes of activities. The SOP on SOPs should govern the numbering methodology.

See Chapter 6 – The SOP on SOPs.

The Project Name and Number, if applicable, should be indicated on the cover or title page. This information identifies the overall project to which this document belongs.

The Site/Location/Department Identification should be indicated on the cover or title page. It identifies the Site/Location/Department that has responsibility for this document and its associated project.

The Date that the Document was issued, last revised or issued with draft status should be on each page of the document (e.g., 12-Dec-01). This will clearly identify the date on which the document was last modified.

The Effective Date of the Document should be on the title or cover page of the document (e.g., 12-Dec-01). This will clearly identify when to begin using the current version of the document.

Pagination, page of total pages (e.g., Page x of y), should be on each page of the document. This will help identify whether or not all of the pages of the document are present. Pages intentionally left blank should clearly be designated as such.

Approval Section

The Approval Section should contain a table with the Author and Approvers printed names, titles and a place to sign and date the SOP. The SOP on SOPs should govern who the approvers of the document should be. The functional department (e.g., Production / Quality Assurance / Engineering) of the signatory should be placed below each printed name and title along with a statement associated with each approver indicating the person's role or qualification in the approval process. Indicate the significance of the associated signature (e.g., This document meets the requirements of Corporate Policy 9055, Electronic Records; Electronic Signatures, dated 22-Mar-2000).

The completion (signing) of this section indicates that the contents have been reviewed and approved by the listed individuals.

Table of Contents

The SOP should contain a Table of Contents, which appears towards the beginning of the document. This Table of Contents provides a way to easily locate various topics contained in the document.

Purpose

The Purpose section should clearly state that the intention of the document is to provide a Standard Operating Procedure to be used to create a Traceability Matrix that cross references User Requirements, Functional Requirements, Design Specifications, Installation Qualification (IQ), Operational Qualification (OQ) and Performance Qualification (PQ) for a validation project of new or upgraded computer or computerized systems. It specifies the requirements for preparing a Traceability Matrix to track and determine the compliance of a project with the business needs of the system owner in support of a Good Clinical, Laboratory, Manufacturing or Distribution Practices (GxP) project.

This introduction to the SOP concisely describes the document's purpose in sufficient detail so that a non-technical reader can understand its purpose.

Scope

The Scope section identifies the limitations of the Change Control SOP. It limits the scope to all GxP systems and is applicable to all computers, computerized systems, infrastructure, LANs and WANs including clinical, laboratory, manufacturing and distribution computer systems. It applies to all Sites, Locations and Departments within the company.

This SOP applies to the modification and change of all software, computer and computerized systems governed by GxP regulations.

Equipment / Materials / Tools

This section lists any required or recommended equipment, materials, tools, templates or guides. Provide the name and a brief description of the equipment, instruments, materials and/or tools to be used in the procedures or testing covered by the Change Control SOP. This is for reference purposes and requires the user to utilize the specific item or a designated equivalent. The purpose and use is explained elsewhere in the document. Substitution of comparable equipment may be made in accordance with procedures for Change Control.

If the information is included in a job aid or operating manual, a description of where the information may be found with a specific reference to the document number, title, date and revision number should be specified.

Warnings / Notes / General Information / Safety

Warnings, notes, general or safety information applicable to the Change Control SOP are listed in this section.

Policies, Guidelines, References

This section lists references to appropriate regulatory requirements, SOPs, policies, guidelines, or other relevant documents referred to in the Change Control SOP. The name and a brief description of applicable related SOPs, corporate policies, procedures and guidelines, Quality Control monographs, books, technical papers, publications, specific regulations, batch records, vendor or system manuals, and all other relevant documents that are required by the Change Control SOP should be included. This is for reference purposes and requires the user to use the specific item or a designated equivalent. The purpose and use are explained elsewhere in the document.

SOPs cite references and incorporate them in whole by reference rather than specifying details that duplicate these items. This is done so that changes in policy, guidelines or reference materials require no changes to SOPs themselves and the most current version of the referenced material is used. Examples are:

- Policy ABC-123, Administrative Policy for Electronic Records; Electronic Signatures
- Corporate, Plan for Compliance with 21 CFR Part 11, Electronic Records; Electronic Signatures
- SOP-456, Electronic Records; Electronic Signatures
- FDA, 21 CFR Part 11, Electronic Records; Electronic Signatures
- FDA, General Principals of Software Validation; Final Guidance for Industry and FDA Staff, January 11, 2002

Assumptions / Exclusions / Limitations

Any and all assumptions, exclusions or limitations within the context of the SOP should be discussed. Specific issues should be directly addressed in this section of the Change Control SOP.

Glossary of Terms

A Glossary of Terms should be included that defines all of the terms and acronyms that are used in this document or that are specific to the operation being described and may not be generally known to the reader. The reader may als o be referred to the FDA document, "Glossary of Computerized System and Software Development Terminology."

Roles and Responsibilities

The System Owner/ End-User is responsible for preparing and maintaining the change control documentation for each the validation project and ensure that it is reviewed and approved by the System Owner, CVG and QA(U) and others as appropriate.

Procedure

The SOP should contain instructions and procedures that enable the creation and processing of Change Control documentation that track and governs all change requests and procedures.

Change Control documents are prepared as stand-alone documents for all projects. It is not good practice to include them as part of other documents in an attempt to reduce the amount of workload or paperwork. This only causes more work over the length of the project resulting from misunderstandings and confusion. Clear, concise documents that address specific parts of any given project are desirable.

Hardware and Software Change Control

Formal change control ensures that a software application, computer, or computerized system continues to operate as expected following a change and is maintained in a sustainable compliant manner. Assessment of the impact a change makes on a system is important. The system must perform as originally intended and purported and in accordance with any specified changes in the requirements and design specifications. Regression analysis and testing is performed on those areas of the system that are impacted by the change. All changes and their associated support activities are documented prior to imp lementation of the change into the production environment.

Create a plan for making and implementing the requested changes. During the execution of the plan, follow the same procedures that are used for System Development Life Cycle methods. This includes the documentation and procedures.

As the change plan is executed, documentation that would normally be produced as part of a System Development Life Cycle is also produced to support the changed system and provide evidence of sustained compliance. Impacted items may include requirements, designs, IQs, OQs or PQs with results, database conversion plans and results, SOPs, manuals, and training.

The process is under constant review of the System Owner, QA(U), CVG, SMEs and other groups as needed.

All changes are first implemented and validated in a development environment prior to entering the production environment. If this is not possible, then an alternative approach must be defined and approved by, at a minimum, the System Owner, QA(U) and CVG.

Chapter 27 – Change Control

Change Control Steps

Identify the Problem or Reason for Change

A request for change is usually the result of something not working as it was intended, identification of an enhancement, or the result of a change in how a process or workflow is performed. Clearly identify the reason for the change and obtain as much information as possible surrounding the need. Obtain the "who, what, when, where, why and how" of the situation in order to properly prepare the Change Control Form and plan the change.

Identify the System

The affects of a Change Request are not always apparent. Systems generally have complex interconnectivity both internally and externally. Data and workflow that transitions interfaces may not be apparent at first. The system, *all* of its components and sub-components, mu st be carefully identified prior to preparing a Change Request.

Propose a Solution

Formulate a solution than is non-technical in nature. Specify what is to be changed with a synopsis of how to accomplish it.

Determine the Impact of Change

Perform regression analysis to determine the extent of testing that must be repeated after the proposed changes are made. Determine the impact on SOPs, workflow, manuals, training and other documentation. Regression analysis and testing involves identifying and testing only those items impacted by a change.

Prepare the Change Control Form

Fill in the initial information to formalize the change request. Since this is a living document for the duration of the process, it will need to be updated as work progresses.

Submission of Change

The System Owner or designee submits the change request (Change Control Form) to the appropriate group(s) for approval. The groups are based on the Roles and Responsibilities as outlined in Chapter 4 and as determined by the System Owner, CVG and QA(U) and other members of the project team.

Analysis and Decision of the Project Team

The project team determines the adequacy of the impact analysis. It then decides, based on the proposed information in the request , whether or not to approve the request and allow the work to proceed.

Create Project Plan

Major changes to a system generally follow the full System Development Life Cycle methodology, which is the overall discussion of this book. Depending on the complexity of the change and its impact, the System Owner, CVG and QA(U) determine the extent of documentation and process required to perform and manage the project. The process follows the overall method discussed in Chapter 7 – Project Plan.

If the changes are minor, it may be possible to eliminate the Project Plan entirely and work directly with the Change Control Form.

Implement Project Plan

The change(s) are made to the system as specified in the Change Control Form and project plan. Be careful to follow all of the procedures in order to maintain sustainable compliance with cGxP regulations and requirements.

Testing / Retesting

All verification and validation testing is performed in a qualified pre -production environment following the rules for IQ and OQ. *All* changes to the software, computer and/or computerized system must be tested. Additionally, regression testing based on regression analysis must be performed on all affected items.

Implementation to Production

Once all of the processes and procedures have been performed, all testing has been satisfactorily completed and passed with all deviations having been remediated, the documentation has been updated, and all required training has been performed, the system is approved for use in production. The system is then installed in the production environment, PQ is performed, and the updated system is approved for production use.

Emergency Changes

Emergency changes, as determined by the System Owner or designee with the concurrence of the CVG and QA(U), may be implemented prior to obtaining formal approval. After completing emergency changes, the formal documentation and review process is completed within a specified period of time. Generally not more than five working days.

Change Control Documentation

Each Change Control Form contains a revision history section. This is used to document the updates to the form as the project progresses. Modifications to the Change Control Form follow the same Document Management and Storage procedures as other GxP documentation.

The staff performing the changes to the system must be notified of the events in a timely manner.

Change Control Form

The Change Control Form should contain, at a minimum, the following fields:

- Form Title – (e.g.- Change Control Form)
- Request Title – for the requestor of the change to enter the name by which the change is to be known.
- System Name – the name of the software, computer or computerized system or systems to be changed.
- Description of Change – for the requestor of the change to enter a detailed description of the change requested.
- Requested By Name & Date – for entry of the requestors name and the date of the request.
- Approved for Evaluation Name & Date – for the name, signature and date of the person(s) approving the change request for evaluation.

- Base Line Affected – for entry of the configuration information of the affected software, computer or computerized system or systems (e.g.- software release numbers and serial number ranges).
- Raw Data Affected – for specifying the databases, files or elements and/or schema affected.
- Effect On Logistics, Support, Interfaces – for specifying the effect the changes will have on logistics, support functions and processes, interfaces, processes, products
- Revalidation Required? Validation Team Approval & Date – for specifying validation and revalidation needs and for entry of project team approval or denial and the date the action was taken.
- Programmer(s) / Team Member(s) Assigned – for listing the names and functions of the people assigned tasks for the project.
- File(s) / Design(s) / Document(s) / Sops Affected – for listing SOPs, manuals, electronic files, and other documentation affected.
- Change Plan Approved By & Date – for entry of the name, signature and date of the approver(s) of the plan.
- SOP(s) / Documentation Updated & Date – for entry of the name of the SOPs, manuals and other documentation that was updated with the date of the activity.
- File(s) / Design(s) Checked In & Date – for use in control of software code where a check-out/check-in procedure is used to control access.
- Testing Completed On –Date – for entry of the date when testing was completed.
- Enter Documentation Reference – for entry of references to the documentation used for testing.
- Training Completed On – Date – for entry of the date(s) training was completed.
- Enter Documentation Reference – for entry of references to the documentation used for training.
- Final Approval By / Installed to Production – Signature & Date – for entry of the name(s), signature(s) and date(s) of final approval for use in the production environment.
- Additional Information

Document Management and Storage

A Document Management and Storage section is included in the Change Control SOP. It defines the Document Management and Storage methodology that is followed for issue, revision, replacement, control and management of this SOP. This methodology is in accordance with an established Change Control SOP (this SOP) and a Document Management and Storage SOP.

Revision History

A Revision History section is included in the Change Control SOP. The Change Control SOP Version Number and all Revisions made to the document are recorded in this section. These changes are in accordance with a Change Control Procedure SOP.

The Revision History section consists of a summary of the changes, a brief statement explaining the reasons the document was revised, and the name and affiliation of the person responsible for the entries. It includes the numbers of all items the Change Control SOP replaces and is written so that the original Change Control SOP is not needed to understand the revisions or the revised document being issued.

When the Change Control SOP is a new document, a "000" as the Revision Number is entered with a statement such as "Initial Issue" in the comments section.

Regulatory or audit commitments require preservation from being inadvertently lost or modified in subsequent revisions. All entries or modifications made as a result of regulatory or audit commitments require indication as such in the Revision History and should indicate the organization, date and applicable references. A complete and accurate audit trail must be developed and maintained throughout the project lifecycle.

Attachments

It should be required that any attachments to the Performance Qualification SOP be in accordance If needed, an Attachments section is included in the Change Control SOP in accordance with the SOP on SOPs. An Attachment Index and Attachment section follow the text of the Change Control SOP. The top of each attachment page indicates "Attachment # x," where 'x' is the sequential number of each attachment. The second line of each attachment page contains the attachment title. The number of pages and date of issue corresponding to each attachment are listed in the Attachment Index.

Attachments are placed directly after the Attachment Index and are sequentially numbered (e.g., Attachment # 1, Attachment # 2) with each page of an attachment containing the corresponding document number, page numbering, and total number of pages in the attachment (e.g., Page x of y). Each attachment is labeled consistent with the title of the attachment itself.

Attachments are treated as part of the Change Control SOP. If an attachment needs to be changed, the entire Change Control SOP requires a new revision number and is reissued. New attachments or revisions of existing attachments are in accordance with an established Change Control SOP and a Document Management and Storage SOP.

Chapter 28 – Document Management and Storage

In this chapter we will discuss the Standard Operating Procedure (SOP) for document management, storage, maintenance, and modification for a validation project. This SOP governs the preparation and implementation of Change Control Requests for documentation. It helps track and determine the compliance of a project with the business needs of the System Owner in support of a Good Clinical, Laboratory, Manufacturing or Distribution Practices (GxP) project.

Best industry practices and FDA Guidance documents indicate that in order to demonstrate that a computer or computerized system performs as it purports and is intended to perform, a document management and storage procedure is required to demonstrate that the system has been documented as compliant in accordance with the business needs of the project.

This SOP is used for all systems that create, modify, maintain, retrieve, transmit or archive records in electronic format. It is used in conjunction with an overlying corporate policy to provide sustainable compliance with the regulations.

Everyone involved in projects that modify the system documentation uses this SOP. It applies to software applications, computers, and computerized systems that control equipment and/or produce electronic data or records in support of all regulated laboratory, manufacturing, clinical or distribution activities. This includes processes, workflows, and procedures.

Title Block, Headers and Footers

The Document Management and Storage SOP should have a title block, title or cover page along with a header and/or footer on each page of the document that clearly identifies the document. The minimum information that should be displayed for the Document Management and Storage SOP is as follows:

- Document Title
- Document Number
- Document Revision Number
- Project Name and Number
- Site/Location/Department Identification
- Date Document was issued, last revised or draft status
- Effective Date
- Pagination: page of total pages (e.g., Page x of y)

The Document Title should be in large type font that enables clear identification. It should clearly and concisely describe the contents and intent of the document. Since this document will be part of a set or package of documents within a project, it should be consistent with other documents used for the project and specify that it is the Document Management and Storage SOP. This information should be on every page of the document.

The Document Number should be in large type font that enables clear identification. Since the Document Management and Storage SOP will be part of a set or package of documents within a project, it should be consistent with other documents used for the project. This information should appear on every page of the document.

The Document Revision Number, which identifies the revision of the document, should appear on each page of the document. Many documents will go through one or more changes over time, particularly during the draft stage and the review and approval cycle. It is important to maintain accurate revision control because even a

subtle change in a document can have an immense impact on the outcomes of activities. The SOP on SOPs should govern the numbering methodology.

See Chapter 6 – "The SOP on SOPs."

The Project Name and Number, if applicable, should be indicated on the cover or title page. This information identifies the overall project to which this document belongs.

The Site/Location/Department Identification should be indicated on the cover or title page. It identifies the Site/Location/Department that has responsibility for this document and its associated project.

The Date that the Document was issued, last revised or issued with draft status should be on each page of the document (e.g., 12-Dec-01). This will clearly identify the date on which the document was last modified.

The Effective Date of the Document should be on the title or cover page of the document (e.g., 12-Dec-01). This will clearly identify when to begin using the current version of the document.

Pagination, page of total pages (e.g., Page x of y), should be on each page of the document. This will help identify whether or not all of the pages of the document are present. Pages intentionally left blank should clearly be designated as such.

Approval Section

The Approval Section should contain a table with the Author and Approvers printed names, titles and a place to sign and date the SOP. The SOP on SOPs should govern who the approvers of the document should be. The functional department (e.g., Production / Quality Assurance / Engineering) of the signatory should be placed below each printed name and title along with a statement associated with each approver indicating the person's role or qualification in the approval process. Indicate the significance of the associated signature (e.g. This document meets the requirements of Corporate Policy 9055, Electronic Records; Electronic Signatures, dated 22-Mar-2000).

The completion (signing) of this section indicates that the contents have been reviewed and approved by the listed individuals.

Table of Contents

The SOP should contain a Table of Contents, which appears towards the beginning of the document. This Table of Contents provides a way to easily locate various topics contained in the document.

Purpose

The Purpose section should clearly state that the intention of the document is to provide a Standard Operating Procedure to be used to manage and store documents in support of compliance of a project with the business needs of the system owner in support of a Good Clinical, Laboratory, Manufacturing or Distribution Practices (GxP) project.

This introduction to the SOP concisely describes the document's purpose in sufficient detail so that a non-technical reader can understand its purpose.

Scope

The Scope section identifies the limitations of the Document Management and Storage SOP. It limits the scope to all GxP systems and is applicable to all computers, computerized systems, infrastructure, LANs and WANs including clinical, laboratory, manufacturing and distribution computer systems. It applies to all Sites, Locations and Departments within the company.

This SOP applies to the modification and storage of all documentation, except SOPs, for software, computer and computerized systems governed by GxP regulations.

Equipment / Materials / Tools

This section lists any required or recommended equipment, materials, tools, templates or guides. Provide the name and a brief description of the equipment, instruments, materials and/or tools to be used in the procedures or testing covered by the Document Management and Storage SOP. This is for reference purposes and requires the user to utilize the specific item or a designated equivalent. The purpose and use is explained elsewhere in the document. Substitution of comparable equipment may be made in accordance with procedures for Change Control.

If the information is included in a job aid or operating manual, a description of where the information may be found with a specific reference to the document number, title, date and revision number should be specified.

Warnings / Notes / General Information / Safety

Warnings, notes, general or safety information applicable to the Document Management and Storage SOP are listed in this section.

Policies, Guidelines, References

This section lists references to appropriate regulatory requirements, SOPs, policies, guidelines, or other relevant documents referred to in the Document Management and Storage SOP. The name and a brief description of applicable related SOPs, corporate policies, procedures and guidelines, Quality Control monographs, books, technical papers, publications, specific regulations, batch records, vendor or system manuals, and all other relevant documents that are required by the Document Management and Storage SOP should be included. This is for reference purposes and requires the user to use the specific item or a designated equivalent. The purpose and use are explained elsewhere in the document.

SOPs cite references and incorporate them in whole by reference rather than specifying details that duplicate these items. This is done so that changes in policy, guidelines or reference materials require no changes to SOPs themselves and the most current version of the referenced material is used. Examples are:

- Policy ABC-123, Administrative Policy for Electronic Records; Electronic Signatures
- Corporate, Plan for Compliance with 21 CFR Part 11, Electronic Records; Electronic Signatures
- SOP-456, Electronic Records; Electronic Signatures
- FDA, 21 CFR Part 11, Electronic Records; Electronic Signatures
- FDA, General Principals of Software Validation; Final Guidance for Industry and FDA Staff, January 11, 2002

Assumptions / Exclusions / Limitations

Any and all assumptions, exclusions or limitations within the context of the SOP should be discussed. Specific issues should be directly addressed in this section of the Document Management and Storage SOP.

Glossary of Terms

A Glossary of Terms should be included that defines all of the terms and acronyms that are used in this document or that are specific to the operation being described and may not be generally known to the reader. The reader may also be referred to the FDA document, "Glossary of Computerized System and Software Development Terminology."

Roles and Responsibilities

The System Owner / End-User is responsible for preparing and maintaining the change control documentation for each the validation project and ensure that it is reviewed and approved by the System Owner, CVG and QA(U) and others as appropriate.

The System Owner / End-User is responsible for the storage of documentation using a sustainable compliant method.

The QA(U) Administrator is responsible for document storage, archive and retrieval.

The Departmental Coordinator is responsible for controlled copies of documents.

Procedure

The SOP should contain instructions and procedures for the management and controlled storage of all documentation. It should enable the creation and processing of Change Control documentation that track and govern all Change Requests for procedures and documentation other than SOPs.

Documents, in general, describe requirements, specifications, procurement information, results of procedural events (reports), guidelines, and policies. They are needed for many purposes within GxP environments. Their management and storage is important in demonstrating sustainable compliance with the regulations.

Management

Change Control

Requests for documentation change are initiated by the System Owner / End-User, who communicates these requests with CVG, QA(U), and the users of the software application, computer, or computerized system. This is a formal process that follows a defined method.

The end result of a request to change a document must provide a positive impact. If there is no benefit to changing a document, the document should not be changed.

Triggers for Document Change or Decommission

This section specifies the triggers that enable control of a document during its life cycle. Here are some examples of circumstances that might trigger the need to revise or decommission a document:

- Change in Business Conditions or pertinent input from another location, such as that caused by regulatory activity
- Change in Management Philosophy
- Change in Organization or Responsibilities
- Change in Procedure
- Change to an Attachment or Form
- Changes to document content
- Equipment Implementation, Change or Obsolescence
- New or modified Corporate Policies
- Periodic Document Review
- Process Implementation, Change or Obsolescence
- Regulatory Change or a commitment to a regulatory agency
- Technology Implementation, Change or Obsolescence

Revised Document

When an existing document does not cover a process or procedure that is necessary, then a document needs to be created or revised to remedy such condition. The Document Management and Storage SOP specifies how to accomplish this remediation. Some examples of what might be done follow.

Initiating a Document Revision

When a document needs to be created or revised, the Originator will:

Prepare the draft of the new or changed document in concert with a Subject Matter Expert (SME), if necessary.

Prepare a Document Change Form (DCF) from a controlled system (e.g. – computer network drive, intranet) by entering the following information:

- Originator's name and title
- Originator's department number and name
- SOP number and title
- Reason for the new or revised SOP
- Department / Documents affected
- Training requirements
- Originator's signature and date

Create a list of Reviewers and alternates, where possible. These include, at a minimum, the Originating Department Manager, System Owner, CVG, and QA(U).

Send the draft with the updated history revisions page, the DCF, and the list of Reviewers to the Departmental Coordinator for document format review and completion.

Document Review

For each document, the Departmental Coordinator sets up and maintains an electronic folder that contains read-only copies of the following:

- Original document
- Revised documents

- Document Change Forms
- Revision History pages
- Reviewers' Comment Forms

The Departmental Coordinator distributes the documentation set to the Reviewers for their review and comment indicating that the documentation must be reviewed by the close of business of a specific day. Not more than 5 business days should be allowed for completion.

Once each reviewer has completed reviewing the documentation, comments and an indication of acceptance or rejection is entered on the Reviewer's Comment Form and returned to the Departmental Coordinator.

The Originator reviews the comments and incorporates them into the new or revised document. The Originator contacts a reviewer directly to clarify any points of concern.

Document Re-review

After the new or revised document has been modified to include the comments, it, along with the documents listed below, is distributed for a subsequent review as described in the Document Approval process

Document Approval

The Departmental Coordinator sends the completed, reviewed new/revised document to the Company Document Administrator who performs the following:

Log receipt of the document

Create a circulation folder that includes:

- Reviewer's Comments Form
- Signature Routing Form
- Original Document
- Revised Document
- Document Change Control Form
- History of Revisions

Route the circulation folder for approval

Track the circulation folder

When there is a required completion date, the process provides for sufficient lead-time so that all necessary Approval signatures are in place by that date.

The Approvers complete the approval of the documentation by the close of business of the specified date.

Approvers use the "Signature Routing Form" to route the circulation folder.

Note: Minor typographical or grammatical errors are corrected and should not delay the approval process.

The author of the revised document is not an Approver.

Representatives or designees of at least the following departments approve the document: Originating Department Manager, CVG and QA(U)

QA(U) is the last department to approve the revised document.

After all Approvers have signed the documentation, it is returned to the Company Document Administrator who proceeds to implement the document in accordance with the section on Document Implementation.

Decommissioning a Document

When a document no longer represents a process or procedure that is necessary or is currently performed, then that document is decommissioned and no longer made available for use. This might be achieved follows.

Initiating the Decommissioning Process

The Originator of the request to decommission a document prepares a Document Change Control Form (DCF) and a Decommission Request Form (DRF) from the controlled system (e.g. – computer network drive, intranet) by entering the following information:

- Originator's name and title
- Originator's department number and name
- Document number and title
- Reason for decommissioning the document
- Department / Documents affected
- Originator's signature and date

The Originator creates a list of Reviewers and alternates, where possible. These include, at a minimum, the manager of the department that originally issued the document, the System Owner, if applicable, CVG and QA(U).

The Originator sends the draft with the updated history revisions page, the DCF, the DRF and the list of Reviewers to the Departmental Coordinator.

Decommissioning Approval Process

The Departmental Coordinator forwards these to the Company Document Administrator who performs the following:

- Retrieve the original signed and dated document from the corporate records management file room
- Make a copy of document
- File 'original' document in file Documents Routing for Deletion
- The Document Database is updated to indicate that the document has been routed for decommission
- Log receipt of the document
- Create a circulation folder that includes:
- Document Signature Routing Form
- A copy of the original Document
- Revised document(s)
- Document Change Control Form
- Decommission Request Form

- History of Revisions
- Track the circulation folder

The Company Document Administrator distributes the documentation set to the Approvers for their review and approval indicating that the documentation must be reviewed by the close of business of a specific day. Not more than 5 business days should be allowed for completion.

When there is a required completion date, the process provides for sufficient lead-time so that all necessary Approval signatures are in place by that date.

The Approvers complete the approval of the documentation by the close of business of the specified date.

Representatives of at least the following departments approve decommissioning the document: the Originator, Cognizant Departmental Manager, Manager of any Department affected by the document and QA(U).

The Decommission Approvers approve the document decommissioning by signing the Deletion Request Form and routing the document package to the next approver using the "Signature Routing Form" to route the circulation folder.

QA(U) is the last department to approve the decommissioning.

After all Approvers have signed the documentation, it is returned to the Company Document Administrator who retrieves the original signed/dated (hard copy) from the file Documentation Routing for Deletion.

Decommissioning

The Company Document Administrator is responsible for the decommissioning of the document.

All original "wet ink signature" documents have the following procedure performed:

Stamp the top of the cover page with the following in red ink:

DECOMMISSIONED	
Replaced by:	____________
Document #:	____________
Version:	____________
Initials:	____________
Date:	____________

Figure 3 – Decommissioned Stamp

Fill in the stamp as appropriate.

Stamp all other pages at the top in red ink with the word DECOMMISSIONED.

The Company Document Administrator places the decommissioned document in a clear, side loading sheet protector and places it in a red interior file folder labeled with the deleted document number.

The Company Document Administrator places the file in the document History file in the corporate records management file room.

On the workday (evening) prior to the document's decommissioned date, the QA Administrator s moves the electronic record of the decommissioned document to the designated area for decommissioned documents on the network.

The Company Document Administrator sends a Notice of Decommissioning to each Departmental Coordinator notifying the department the date that the decommissioning is effective.

The Departmental Coordinator is responsible for retrieving the issued copies of the decommissioned document and destroying each copy on or about the decommissioned date. No copy is permitted to be in circulation or use after the decommissioned date.

Maintenance

All documents are required to follow specific maintenance procedures during their life cycle in order to assure that they are current and appropriate for their intended use. Examples of these procedures follow.

Periodic Review

Documents are reviewed periodically by the Originating Department to ensure that the content is accurate. The periodic review period is that stated in the document and must not exceed (2) years from the Effective Date of the document.

Approximately six months prior to the required review period completion date the Company Document Administrator begins the review process by sending a completed Document Review Form to the Departmental Coordinator.

Within 10 working days, the Departmental Coordinator of the Originating Department returns the completed Document Review Formto the Company Document Administrator indicating the status of the review as acceptable as written, in need of revision or to be decommissioned.

The Company Document Administrator files the completed Document Review Form in the applicable document file and proceeds in accordance with the sections that discuss No Revision Required or Decommission as appropriate.

No Revision Required

The Company Document Administrator, upon receipt of the completed Document Review Form, obtains agreement from QA(U). QA(U) reviews and approves/disapproves the Document Review Form.

If QA(U) approves the Document Review Form, then the cover page is stamped as shown below. The Company Document Administrator re-issues the document with the stamped cover page of the document.

Periodic Review Completed **Content Accurate as Written**
Date Completed: ____________________
Initials: ____________________
Date: ____________________

Figure 4 – Periodic Review Required Stamp

Revision Required

The Originating Department, upon returning the completed Document Review Form, initiates the revision process in accordance with the section that discusses New or Revised Documents.

Decommission Required

The Originating Department, upon returning the completed Document Review Form, initiates the decommissioning in accordance with the section that discusses Decommissioning of Document.

Document Implementation

All documents are required to follow specific implementation procedures in order to assure that they are current, appropriate for their intended use and properly controlled. Examples of these procedures follow.

Each document has an effective date assigned by the Company Document Administrator in accordance with the following procedure. The Company Document Administrator verifies that:

- All parties have approved the document by signing and dating their respective signature blocks on the Approval Page of the document
- The cover page has a header and footer printed with "Official Copy" in red for official copies
- Each official copy has been sequentially numbered

- There is an Effective Date that is at least 30 days later than the last approval date or following approval by a Regulatory Agency unless specified otherwise by the Originating Department. This is to allow time for training and distribution.

The Company Document Administrator updates the Master Documentation Index and distributes the document(s) in accordance with the section that discusses Distribution and Tracking of Controlled Copies of a document.

The document is used on or after the Effective Date and after training requirements have been satisfied.

Distribution and Tracking of Controlled Copies

Numbered copies are distributed based on the content of the document and the requirements of each department. The Company Document Administrator distributes documents based on requests from Departmental Coordinators that have been authorized by the requesting Department's Manager. The Departmental Coordinator tracks and distributes the documents.

The Departmental Coordinator is responsible for the distribution of documentation updates, attachments, decommission notices for use in the areas where they are required.

Official documents must not be copied, except as follows:

Documents may be copied for training purposes, provided that:

- The first copy is made from an official document and the first page of that first copy is dated and stamped in red ink with the words: "For Training Purposes Only." This copy is to remain in the hands of the trainer.
- Each succeeding copy made from the stamped copy, bears the statement "For Training Purposes Only."
- Distribution and destruction of all copies is under the supervision of the trainer. All copies are considered expired 30 days after the date-stamp.

Documents may be copied for requests made by the FDA or other regulatory agencies. These copies are marked "Provided to the <regulatory agency name> on <date>" before issuance and recorded in the QA(U) files.

Documents may be copied for documentation purposes such as, for inclusion in Validation Protocols, provided that they are identified for this purpose and marked "For Validation Use Only."

Documents may be copied for the purpose of revisioning and are considered to be drafts (unapproved) and are marked "Not for Production Use," "Draft," or "For ReviewPurposes Only," as appropriate.

Superseded copies and/or attachments are destroyed unless they are required to be archived for reference purposes, in which case they are marked OBSOLETE – FOR REFERENCE ONLY.

The Company Document Administrator issues a Destruction of Document Notice to each Departmental Coordinator for that document. The Destruction of Document Notice includes the copy numbers of each copy issued to that department.

Each Departmental Coordinator is responsible for the destruction of all copies in that department. Upon destroying the obsolete document, the Departmental Coordinator signs and returns the Destruction of Document Notice to the Company Document Coordinator.

The Company Document Administrator maintains a file of current, superseded and decommissioned documents.

Security

Documents contain information that may not be generally known or available to the public. Therefore, employees, vendors or contractors, without express written consent of an authorized individual of the company, except in the performance of contractual services, must not:

- Use or disclose this information outside the company
- Publish an article with respect thereto
- Remove or aid in the removal from the company any confidential information, property or material

Corrections and Dates

Manual changes or corrections to documentation are made in black or blue indelible ink by placing a single line through the entity to be corrected, entering any correction in a clear and concise manner, initialing and dating the correction. Do *not* obliterate any item.

Do *not* pre- or post-date anything. If a date is in error or missed, enter the current date with initials and an explanation as to what transpired.

Storage

The Company Document Administrator is responsible for storing validation documents in a secure, controlled access, document storage room or facility. This may be on-site or off-site depending on user needs and company policy.

Document Control Room

The Document Control Room is a centralized secure controlled-access facility operated by a Company Document Administrator. Original and "wet ink signature" copies of all GxP related physical documents, including all validation documents, are maintained there.

Electronic versions of the same documentation are similarly stored in a centralized secure controlled-access environment operated by a Company Document Administrator.

Indexing and Filing

All documents are assigned an index number, storage location, listed in a master index, and are filed or stored in their assigned location.

File documents by system, validation effort, and index number. File the most current version in front of the superceded versions of the same document.

Chapter 28 – Document Management and Storage

Documents involved in a current or ongoing validation effort are stored in an Work in Progress folder until the work is completed.

Log

A log is maintained that records all transactions surrounding documents. Entries are made indicating when document is received, issued, re-issued, modified, and decommissioned. For each document, the log tracks, at a minimum, the following information:

- Document Number
- Title
- Index Number
- Cognizant Manager
- Departmental Coordinator
- Company Document Administrator
- Commission Date
- List of Individuals Issued Authorized Copies with Date of Issue
- Dates of Modification
- Decommission Date
- Storage Location
- Requested / Viewed By
- Superceding Information
- Recall Information

Chain of Custody

A Chain of Custody form identifying the document(s) being transferred is signed and dated by the person relinquishing custody and by the person receiving custody of a controlled document. At a minimum, the following information is recorded on the Chain of Custody form:

- Document Number
- Document Title
- Indexing information
- Relinquished By, Date, Time
- Received By, Date, Time

The recipient of a controlled document is responsible for all documents issued to them. Persons issued controlled documents must notify the Departmental Coordinator and the Company Document Administrator of any change of status of the document in his or her possession. The Company Document Administrator and Departmental Coordinator are jointly responsible for the transfer of controlled documents.

Archival and Backup

Once a system has been validated or re-validated, the complete validation package is transferred to an Archive within a predefined period of time. Electronic packages are backed up or archived to a designated media and the media is treated in the same manner as a physical document.

Decommissioned Systems

Documents for decommissioned systems along with all relevant documentation not previously archived is placed in the validation archive. The document authorizing the decommissioning is the placed at the end of the archive as the last indexed item. The entire documentation set is then transferred to the Archive within a predefined period of time.

Retrieval / Access

The System Owner authorizes document access or issuance. All information surrounding each transaction is entered into the log and documented on a Chain of Custody form.

Retention

Documents for software applications, computer and/or computerized systems that are active are maintained for the duration of the active status.

Where software applications, computer and/or computerized systems have been decommissioned, the documentation is retained for the same time period as is required for the associated GxP data. The System Owner and QA(U) determine this based on regulatory requirements (Predicate Rule).

Document Management and Storage

A Document Management and Storage section is included in the Document Management and Storage SOP. It defines the Document Management and Storage methodology that is followed for issue, revision, replacement, control and management of this SOP. This methodology is in accordance with an established Change Control SOP and a Document Management and Storage SOP – this SOP.

Revision History

A Revision History section is included in the Document Management and Storage SOP. The Document Management and Storage SOP Version Number and all Revisions made to the document are recorded in this section. These changes are in accordance with a Change Control Procedure SOP.

The Revision History section consists of a summary of the changes, a brief statement explaining the reasons the document was revised, and the name and affiliation of the person responsible for the entries. It includes the numbers of all items the Document Management and Storage SOP replaces and is written so that the original Document Management and Storage SOP is not needed to understand the revisions or the revised document being issued.

When the Document Management and Storage SOP is a new document, a "000" as the Revision Number is entered with a statement such as "Initial Issue" in the comments section.

Regulatory or audit commitments require preservation from being inadvertently lost or modified in subsequent revisions. All entries or modifications made as a result of regulatory or audit commitments require indication as such in the Revision History and should indicate the organization, date and applicable references. A complete and accurate audit trail must be developed and maintained throughout the project lifecycle.

Attachments

If needed, an Attachments section is included in the Document Management and Storage SOP in accordance with the SOP on SOPs. An Attachment Index and Attachment section follow the text of the Document Management and Storage SOP. The top of each attachment page indicates "Attachment # x," where 'x' is the sequential number of each attachment. The second line of each attachment page contains the attachment title. The number of pages and date of issue corresponding to each attachment are listed in the Attachment Index.

Attachments are placed directly after the Attachment Index and are sequentially numbered (e.g., Attachment # 1, Attachment # 2) with each page of an attachment containing the corresponding document number, page numbering, and total number of pages in the attachment (e.g., Page x of y). Each attachment is labeled consistent with the title of the attachment itself.

Attachments are treated as part of the Document Management and Storage SOP. If an attachment needs to be changed, the entire Document Management and Storage SOP require a new revision number and is reissued. New attachments or revisions of existing attachments are in accordance with an established Change Control SOP and a Document Management and Storage SOP.

Chapter 29 – Configuration Management

In this chapter we will discuss the Standard Operating Procedure (SOP) for controlling and managing a validated system's configuration. It specifies the requirements for applying technical and administrative direction and surveillance to identify and document the functional and physical characteristics of a configuration item, control changes to those characteristics, record and report change processing and implementation status, and verifying compliance with specified requirements.[12] It provides a means to support sustainable compliance with cGxPs and 21 CFR Part 11 by controlling how a system is configured and managed.

Best industry practices and FDA Guidance documents indicate that in order to demonstrate that a comp uter or computerized system performs as it purports and is intended to perform, a configuration management system is required to manage, maintain and control the system configuration as sustainedly compliant in accordance with the business needs of the project. It is needed to control all configuration documentation, physical media and parts representing or comprising the system. For computer or computerized systems, hardware, software, firmware, operating systems, procedures and documentation, it addresses the various and changing configurations, operations, and support environments surrounding its use.

This SOP is used for all systems that create, modify, maintain, retrieve, transmit or archive records in electronic format. It is used in conjunction with an overlying corporate policy to provide sustainable compliance with the regulations in planning and implementing a configuration management program.

Everyone involved in projects for a software application, computer or computerized system that controls equipment and/or produces electronic data or records in support of all regulated laboratory, manufacturing, clinical or distribution activities where modification of the software, computer or computeri zed system, process, workflow or procedure is involved uses this SOP.

Title Block, Headers and Footers

The Configuration Management SOP should have a title block, title or cover page along with a header and/or footer on each page of the document that clearly identifies the document. The minimum information that should be displayed for the Configuration Management SOP is as follows:

- Document Title
- Document Number
- Document Revision Number
- Project Name and Number
- Site/Location/Department Identification
- Date Document was issued, last revised or draft status
- Effective Date
- Pagination: page of total pages (e.g., Page x of y)

The Document Title should be in large type font that enables clear identification. It should clearly and concisely describe the contents and intent of the document. Since this document will be part of a set or package of documents within a project, it should be consistent with other documents used for the project and specify that it is the Configuration Management SOP. This information should be on every page of the document.

---

[12] Taken from the IEEE definition for Configuration Management.

The Document Number should be in large type font that enables clear identification. Since the Configuration Management SOP will be part of a set or package of documents within a project, it should be consistent with other documents used for the project. This information should appear on every page of the document.

The Document Revision Number, which identifies the revision of the document, should appear on each page of the document. Many documents will go through one or more changes over time, particularly during the draft stage and the review and approval cycle. It is important to maintain accurate revision control because even a subtle change in a document can have an immense impact on the outcomes of activities. The SOP on SOPs should govern the numbering methodology.

See Chapter 6 – "The SOP on SOPs."

The Project Name and Number, if applicable, should be indicated on the cover or title page. This information identifies the overall project to which this document belongs.

The Site/Location/Department Identification should be indicated on the cover or title page. It identifies the Site/Location/Department that has responsibility for this document and its associated project.

The Date that the Document was issued, last revised or issued with draft status should be on each page of the document (e.g., 12-Dec-01). This will clearly identify the date on which the document was last modified.

The Effective Date of the Document should be on the title or cover page of the document (e.g., 12-Dec-01). This will clearly identify when to begin using the current version of the document.

Pagination, page of total pages (e.g., Page x of y), should be on each page of the document. This will help identify whether or not all of the pages of the document are present. Pages intentionally left blank should clearly be designated as such.

Approval Section

The Approval Section should contain a table with the Author and Approvers printed names, titles and a place to sign and date the SOP. The SOP on SOPs should govern who the approvers of the document should be. The functional department (e.g., Production / Quality Assurance / Engineering) of the signatory should be placed below each printed name and title along with a statement associated with each approver indicating the person's role or qualification in the approval process. Indicate the significance of the associated signature (e.g. This document meets the requirements of Corporate Policy 9055, Electronic Records; Electronic Signatures, dated 22-Mar-2000).

The completion (signing) of this section indicates that the contents have been reviewed and approved by the listed individuals.

Table of Contents

The SOP should contain a Table of Contents, which appears towards the beginning of the document. This Table of Contents provides a way to easily locate various topics contained in the document.

Purpose

The Purpose section should clearly state that the intention of the document is to provide a Standard Operating Procedure to be used to manage and control configuration management of new or upgraded computer or computerized systems. It provides one of the means to maintain compliance of a system with the business needs of the system owner in support of a Good Clinical, Laboratory, Manufacturing or Distribution Practices (GxP) and current regulation.

This introduction to the SOP concisely describes the document's purpose in sufficient detail so that a non-technical reader can understand its purpose. For example:

> "The purpose of this SOP is to specify the requirements for applying technical and administrative direction and surveillance to identify and document the functional and physical characteristics of a configuration item, control changes to those characteristics, record and report change processing and implementation status, and verifying compliance with specified requirements. It provides a means to support sustainable compliance with cGxPs and 21 CFR Part 11 by controlling how a system is configured and managed."

Scope

The Scope section identifies the limitations of the Configuration management SOP. It limits the scope to all GxP systems and is applicable to all computers, computerized systems, infrastructure, LANs and WANs including clinical, laboratory, manufacturing and distribution computer systems. It applies to all Sites, Locations and Departments within the company. For example:

> "This document applies to all computers, computerized systems, infrastructure, LANs and WANs including clinical, laboratory, manufacturing and distribution computer systems that are governed by cGxPs. This SOP applies to the control all configuration documentation, physical media and parts representing or comprising the computer or computerized system including hardware, software, firmware, operating systems, procedures and documentation. It applies to all departments and divisions at all locations of the Example Pharmaceuticals Corporation."

Equipment / Materials / Tools

This section lists any required or recommended equipment, materials, tools, templates or guides. Provide the name and a brief description of the equipment, instruments, materials and/or tools to be used in the procedures or testing covered by the Configuration Management SOP. This is for reference purposes and requires the user to utilize the specific item or a designated equivalent. The purpose and use is explained elsewhere in the document. Substitution of comparable equipment may be made in accordance with procedures for Change Control.

If the information is included in a job aid or operating manual, a description of where the information may be found with a specific reference to the document number, title, date and revision number should be specified.

Warnings / Notes / General Information / Safety

Warnings, notes, general or safety information applicable to the Configuration Management SOP are listed in this section.

Policies, Guidelines, References

This section lists references to appropriate regulatory requirements, SOPs, policies, guidelines, or other relevant documents referred to in the Configuration Management SOP. The name and a brief description of applicable related SOPs, corporate policies, procedures and guidelines, Quality Control monographs, books, technical papers, publications, specific regulations, batch records, vendor or system manuals, and all other relevant documents that are required by the Configuration Management SOP should be included. This is for reference purposes and requires the user to use the specific item or a designated equivalent. The purpose and use are explained elsewhere in the document.

SOPs cite references and incorporate them in whole by reference rather than specifying details that duplicate these items. This is done so that changes in policy, guidelines or reference materials require no changes to SOPs themselves and the most current version of the referenced material is used. Examples are:

- Policy ABC-123, Administrative Policy for Electronic Records; Electronic Signatures
- Corporate, Plan for Compliance with 21 CFR Part 11, Electronic Records; Electronic Signatures
- SOP-456, Electronic Records; Electronic Signatures
- FDA, 21 CFR Part 11, Electronic Records; Electronic Signatures
- FDA, General Principals of Software Validation; Final Guidance for Industry and FDA Staff, January 11, 2002

Assumptions / Exclusions / Limitations

Any and all assumptions, exclusions or limitations within the context of the SOP should be discussed. Specific issues should be directly addressed in this section of the Configuration Management SOP.

Glossary of Terms

The Glossary of Terms defines all of the terms and acronyms that are used in this document or that are specific to the operation being described and may not be generally known to the reader. The reader may also be referred to the FDA document, "Glossary of Computerized System and Software Development Terminology." For example:

> "Configuration. (IEEE) (1) The arrangement of a computer system or component as defined by the number, nature, and interconnections of its constituent parts. (2) In configuration management, the functional and physical characteristics of hardware or software as set forth in technical documentation or achieved in a product."

Roles and Responsibilities

The System Owner / End-User is responsible for preparing and maintaining the Configuration Management documentation for each validation project and to ensure that it is reviewed and approved by the System Owner, CVG and QA(U) and others as appropriate.

The System Owner / End-User is responsible for the storage of documentation using a sustainedly compliant method.

Configuration Manager, the System Owner designee, coordinates and oversees all aspects of configuration management.

The QA(U) Administrator is responsible for document storage, archive and retrieval.

The Departmental Coordinator is responsible for controlled copies of documents.

Procedure

The SOP should contain instructions and procedures that enable the management and control of each system's configuration and its associated documentation and for the controlled storage of that documentation.

A configuration management system is implemented for each software application, computer, and/ or computerized system. Configuration Management procedures are used to control all configuration documentation, physical media, and parts representing or comprising a computer or computerized system. This includes, but is not limited to, hardware, software, firmware, operating systems, procedures and documentation. It addresses the various and changing configurations, operations and support environments surrounding its use. The integrity of the system and the ability to trace changes as they occur throughout the lifetime of the system is provided by this procedure. It is needed in order to provide a means to reconstruct, by design, any version of the system at any point in time while inhibiting the implementation of unauthorized changes.

Risks

The consequences of failing to perform and maintain configuration management for a computer or computerized system are many. Changes may be made with or without malicious intent and tend to introduce errors and problems. They may conflict with existing configurations or other changes. Multiple changes made to the same item may negate previous changes or conflict with one another. They may cause irreconcilable operational problems or errors in data. Rollback to a previous configuration may not be possible and may cause problems with investigations. All of these could result in an inadequate or non-existent audit trail as required by 21 CFR Part 11. Under FDA inspection, such changes may cause the system to be considered non-compliant with GxP requirements.

Elements

The configuration management system is used to manage and control the following configuration elements:

Configuration Identification

Define the functional, performance, interface and physical attributes of the system. Establish configuration identification by selecting and identifying Configuration Items (CI). At a minimum, configuration management governs the following areas and items:

- Program Objectives
- Organization and Relationships
- Roles and Responsibilities
- Configuration Items
- Hardware
- Software
- Peripherals
- Interfaces
- Infrastructure
- Network and Communications
- Parameters and Settings
- Processes and procedures
- Specifications
- Qualification Tests

- System SOPs
- Manuals
- Documentation
- Resources (tools, techniques, and methodologies)
- Physical Media

PHYSICAL PARTS

Determine the types of configuration documentation required for each CI.

Issue numbers and other identifiers affixed to the CIs and to the technical documentation that defines the CIs configuration, including internal and external interfaces.

Establish the configuration baselines for CIs.[13]

Configuration Control

Establish configuration control for the systematic proposal, justification, evaluation, coordination, approval or disapproval of proposed changes, and the implementation of all approved changes, in the configuration of a CI after establishment of the configuration baseline(s) for the CI.[14] It ensures effective control of the configured system.

Document all proposed changes to the computing environment and ensure they are prioritized according to the needs of the business units.

See Chapter 27 – "Change Control."

Obtain the written approval of the relevant business unit and support management before scheduling a proposed change. Publish guidelines on who has the authority to propose a change and who can approve it. No single person can have dual authority.

Once a change has been approved and is scheduled for action, create a detailed project plan.

See Chapter 7 – "Project Plan."

Provide for a period of concurrent operation and/or a back-out plan.

Identify conflicting changes and ensure that they are not scheduled concurrently.

Formally notify the System Owner, End-Users, and other affected business areas of the implementation timetable and impact.

Ensure that End-Users review changes after they are tested and prior to migration to the production environment.

---

[13] Based on MIL-STD-973 definitions.

[14] Based on MIL-STD-973 definitions.

Once a component has passed acceptance testing, further updates are prohibited. This ensures that the production version is the one that was tested.

Assess the impact on vendor/contractor responsibilities and warranties.

Establish a process for the migration of tested changes into production that does not support a conflict of interest. The Configuration Manager must not have a role in approving, developing or implementing changes. Persons having access control authorities to perform the task must not have a vested interest in the process.

Formal management approval is required to migrate a change to production.

Monitor changes in progress. Ensure that implementation schedules are being met and, if not, review the reasons and ensure that a revised timetable is issued.

Document all changes and maintain the documentation in accordance with a Document Management and Storage SOP.

Configuration Status Accounting

Perform configuration status accounting (CSA) to document and report information needed in the management CIs. The CSA system accomplishes the following:

- Identification of the approved configuration
- Documenting and reporting of the status of proposed changes
- Documenting and reporting of the status and results of all configuration audits
- Documenting and reporting of requests for deviations and waivers regarding configuration
- Documenting and reporting of the status of changes to the configuration
- Document the traceability of all changes from baseline through current configuration
- Documenting and reporting of installation status and effectiveness of all configuration changes

Configuration Verification and Audit

Conduct a Functional Configuration Audit (FCA) and Physical Configuration Audit (PCA) to establish a baseline configuration for each CI and resolve all discrepancies identified during the conduct of the FCA/PCA.

Functional Configuration Audit

Prior to release for pre-production testing, functional configuration audits are conducted for each configuration item of a system. They are performed to verify that the performance of each configuration item and each system is in accordance with the configuration documentation. Where verification can only be ascertained after integration and testing in the production environment, the functional configuration audit is conducted using the results of these tests.

Physical Configuration Audit

The physical configuration audit verifies that each configuration item has been built or configured in accordance with its design specifications. These include, but are not limited to the items listed in Configuration Identification above. Physical configuration audits are generally preceded by functional configuration audits, but may be performed concurrently. When the physical configuration audit is complete, the system is baselined and all further changes are in accordance with a formal change control procedure.

See Chapter 27 – "Change Control."

Versioning

Version numbers are used to identify a unique construct of a system. In this way the life cycle of a system is tracked and recorded. Also, if a system needs to be reconstructed to a prior point in time, all of the configuration information will be available to perform that reconstruction.

For all system releases, the Configuration Manager assigns a unique version number to pre-production testing in accordance with the following format:

version.release.build{.ppatchnumber.patchbuild}

For example: 6.2.43.p1 = Version 6, Release 2, Build 43, patch 1

Where,

"version" is the version number for the product and is incremented when there is a major release.

"release" is incremented when there is a minor release. "0" is used for the first release of a new major version.

A build is created for a particular version. For example, 5.0.10 would be build 10 of version 5.0.

Each component of the version number is an integer. Letters are not used.

Do not reference components that are not included in the release (e.g. – If a release is not patched, no patch designation should be included in the version number).

Do not duplicate or re-use version numbers. Only one build of any version is ever distributed for production use. If a change is required to a build currently in production, the number of the release is incremented.

Numbers are not changed once assigned. The assigned number is the version identifier throughout the life cycle of the release.

A "patch" is a special build used to replace specific file(s) within a version.release.build. Not all releases have patches.

Patches are sequentially numbered starting with the number 1 and followed by a build identifier for the build of the patch.

For example, if the server code has a bug in version 10.2.6, a patch is created to correct the bug. The first build of the patch is identified as 10.2.6p1.1. If testing determines that the patch does not resolve the problem as expected, then the next patch build will be 10.2.6p1.2.

A patched version of a production system requiring a new patch would receive a new integer, followed by a build identifier. For example, if there were a subsequent problem to the patch, the new patch would be 10.2.6p2.1.

The version.release combination is the release number published for users and represents the particular version of a specific system.

In both version and patch numbers, the build ID is not a key component from a User or Functional Requirements perspective because the scope of the requirements does not change from build to build. However, the build number is a key component in distinguishing exactly which build of a version meets the requirements and is suitable for production and that the correct build is distributed and in use.

Exception to this policy may occur under the following conditions:

- To bring ancillary utilities in alignment with the versioning of other systems
- To address emergency conditions
- To include / change functionality between builds during pre-release testing of version.revision

Justification for the versioning to a system or ancillary application that is not in compliance with the governing SOP must be documented in the corresponding release notes.

Document Management and Storage

A Document Management and Storage section is included in the Configuration Management SOP. It defines the Document Management and Storage methodology that is followed for issue, revision, replacement, control and management of this SOP. This methodology is in accordance with an established Change Control SOP and a Document Management and Storage SOP.

Revision History

A Revision History section is included in the Configuration Management SOP. The Configuration Management SOP Version Number and all Revisions made to the document are recorded in this section. These changes are in accordance with a Change Control Procedure SOP.

The Revision History section consists of a summary of the changes, a brief statement explaining the reasons the document was revised, and the name and affiliation of the person responsible for the entries. It includes the numbers of all items the Configuration Management SOP replaces and is written so that the original Configuration Management SOP is not needed to understand the revisions or the revised document being issued.

When the Configuration Management SOP is a new document, a "000" as the Revision Number is entered with a statement such as "Initial Issue" in the comments section.

Regulatory or audit commitments require preservation from being inadvertently lost or modified in subsequent revisions. All entries or modifications made as a result of regulatory or audit commitments require indication as such in the Revision History and should indicate the organization, date and applicable references. A complete and accurate audit trail must be developed and maintained throughout the project lifecycle.

Attachments

If needed, an Attachments section is included in the Configuration Management SOP in accordance with the SOP on SOPs. An Attachment Index and Attachment section follow the text ofthe Configuration Management SOP. The top of each attachment page indicates "Attachment # x," where 'x' is the sequential number of each attachment. The second line of each attachment page contains the attachment title. The number of pages and date of issue corresponding to each attachment are listed in the Attachment Index.

Attachments are placed directly after the Attachment Index and are sequentially numbered (e.g., Attachment # 1, Attachment # 2) with each page of an attachment containing the corresponding document number, page numbering, and total number of pages in the attachment (e.g., Page x of y). Each attachment is labeled consistent with the title of the attachment itself.

Attachments are treated as part of the Configuration Management SOP. If an attachment needs to be changed, the entire Configuration Management SOP requires a new revision number and is reissued. New attachments or revisions of existing attachments are in accordance with an established Change Control SOP and a Document Management and Storage SOP.

Chapter 30 – Periodic Review

In this chapter we will discuss the Standard Operating Procedure (SOP) for performing a periodic review of a validated software application, computer or computerized system. It specifies the requirements for the performance of periodic reviews of validated systems.

Best industry practices and FDA Guidance documents indicate that in order to demonstrate that a software application, computer or computerized system performs as it purports and is intended to perform, a periodic review is required to ensure that the system is compliant with its business needs, requirements, and current regulation. It is needed to control all documentation, physical media and parts representing or comprising the software application, computer or computerized system including hardware, software, firmware, operating systems, procedures, and documentation.

This SOP is used for all systems that create, modify, maintain, retrieve, transmit or archive records in electronic format whether or not they have been previously validated. It is used in conjunction with an overlying corporate policy to provide sustainable compliance with the regulations by performing periodic reviews.

The System Owner, Computer Validation Group and Quality Assurance (Unit) use this SOP to evaluate a software application, computer or computerized system that controls equipment and/or produces electronic data or records in support of all regulated laboratory, manufacturing, clinical or distribution activities where modification of the software, computer or computerized system, process, workflow or procedure is involved.

Title Block, Headers and Footers

The Periodic ReviewSOP should have a title block, title or cover page along with a header and/or footer on each page of the document that clearly identifies the document. The minimum information that should be displayed for the Periodic ReviewSOP is as follows:

- Document Title
- Document Number
- Document Revision Number
- Project Name and Number
- Site/Location/Department Identification
- Date Document was issued, last revised or draft status
- Effective Date
- Pagination: page of total pages (e.g., Page x of y)

The Document Title should be in large type font that enables clear identification. It should clearly and concisely describe the contents and intent of the document. Since this document will be part of a set or package of documents within a project, it should be consistent with other documents used for the project and specify that it is the Periodic Review SOP. This information should be on every page of the document.

The Document Number should be in large type font that enables clear identification. Since the Periodic Review SOP will be part of a set or package of documents within a project, it should be consistent with other documents used for the project. This information should appear on every page of the document.

The Document Revision Number, which identifies the revision of the document, should appear on each page of the document. Many documents will go through one or more changes over time, particularly during the draft stage and the review and approval cycle. It is important to maintain accurate revision control because even a

subtle change in a document can have an immense impact on the outcomes of activities. The SOP on SOPs should govern the numbering methodology.

See Chapter 6 – "The SOP on SOPs."

The Project Name and Number, if applicable, should be indicated on the cover or title page. This information identifies the overall project to which this document belongs.

The Site/Location/Department Identification should be indicated on the cover or title page. It identifies the Site/Location/Department that has responsibility for this document and its associated project.

The Date that the Document was issued, last revised or issued with draft status should be on each page of the document (e.g., 12-Dec-01). This will clearly identify the date on which the document was last modified.

The Effective Date of the Document should be on the title or cover page of the document (e.g., 12-Dec-01). This will clearly identify when to begin using the current version of the document.

Pagination, page of total pages (e.g., Page x of y), should be on each page of the document. This will help identify whether or not all of the pages of the document are present. Pages intentionally left blank should clearly be designated as such.

Approval Section

The Approval Section should contain a table with the Author and Approvers printed names, titles and a place to sign and date the SOP. The SOP on SOPs should govern who the approvers of the document should be. The functional department (e.g. – Production / Quality Assurance / Engineering) of the signatory should be placed below each printed name and title along with a statement associated with each approver indicating the person's role or qualification in the approval process. Indicate the significance of the associated signature (e.g. This document meets the requirements of Corporate Policy 9055, Electronic Records; Electronic Signatures, dated 22-Mar-2000).

The completion (signing) of this section indicates that the contents have been reviewed and approved by the listed individuals.

Table of Contents

The SOP should contain a Table of Contents, which appears towards the beginning of the document. This Table of Contents provides a way to easily locate various topics contained in the document.

Purpose

The Purpose section should clearly state that the intention of the document is to provide a Standard Operating Procedure that is used to perform periodic reviews of validated systems to determine the compliance of a system with the business needs and requirements of the System Owner in support of a Good Clinical, Laboratory, Manufacturing or Distribution Practices (GxP) and current regulation.

This introduction to the SOP concisely describes the document's purpose in sufficient detail so that a non-technical reader can understand its purpose. For example:

"The purpose of this SOP is to specify the requirements for performing periodic reviews of validated systems to determine the compliance of a system with the business needs of the system owner in support of a Good Clinical, Laboratory, Manufacturing or Distribution Practices (GxP) and current regulation."

Scope

The Scope section identifies the limitations of the Periodic Review SOP. It limits the scope to all GxP systems and is applicable to all computers, computerized systems, infrastructure, LANs and WANs including clinical, laboratory, manufacturing and distribution computer systems. It applies to all Sites, Locations and Departments within the company. For example:

"This document applies to all computers, computerized systems, infrastructure, LANs, WANs and SANs including clinical, laboratory, manufacturing and distribution computer systems that are governed by cGxPs. This SOP applies to the periodic review of all documentation, physical media and parts representing or comprising the computer or computerized system including hardware, software, firmware, operating systems and procedures. It applies to all departments and divisions at all locations of the Example Pharmaceuticals Corporation."

Equipment / Materials / Tools

This section lists any required or recommended equipment, materials, tools, templates or guides. Provide the name and a brief description of the equipment, instruments, materials and/or tools to be used in the procedures or testing covered by the Periodic Review SOP. This is for reference purposes and requires the user to utilize the specific item or a designated equivalent. The purpose and use is explained elsewhere in the document. Substitution of comparable equipment may be made in accordance with procedures for Change Control.

If the information is included in a job aid or operating manual, a description of where the information may be found with a specific reference to the document number, title, date and revision number should be specified.

Warnings / Notes / General Information / Safety

Warnings, notes, general or safety information applicable to the Periodic Review SOP are listed in this section.

Policies, Guidelines, References

This section lists references to appropriate regulatory requirements, SOPs, policies, guidelines, or other relevant documents referred to in the Periodic Review SOP. The name and a brief description of applicable related SOPs, corporate policies, procedures and guidelines, Quality Control monographs, books, technical papers, publications, specific regulations, batch records, vendor or system manuals, and all other relevant documents that are required by the Periodic Review SOP should be included. This is for reference purposes and requires the user to use the specific item or a designated equivalent. The purpose and use are explained elsewhere in the document.

SOPs cite references and incorporate them in whole by reference rather than specifying details that duplicate these items. This is done so that changes in policy, guidelines or reference materials require no changes to SOPs themselves and the most current version of the referenced material is used. Examples are:

- Policy ABC-123, Administrative Policy for Electronic Records; Electronic Signatures
- Corporate, Plan for Compliance with 21 CFR Part 11, Electronic Records; Electronic Signatures
- SOP-456, Electronic Records; Electronic Signatures
- FDA, 21 CFR Part 11, Electronic Records; Electronic Signatures

- FDA, General Principals of Software Validation; Final Guidance for Industry and FDA Staff, January 11, 2002

Assumptions / Exclusions / Limitations

Any and all assumptions, exclusions or limitations within the context of the SOP should be discussed. Specific issues should be directly addressed in this section of the Periodic Review SOP.

Glossary of Terms

A Glossary of Terms should be included that defines all of the terms and acronyms that are used in this document or that are specific to the operation being described and may not be generally known to the reader. The reader may also be referred to the FDA document, "Glossary of Computerized System and Software Development Terminology."

Roles and Responsibilities

The System Owner is responsible for preparing and conducting the Periodic Review for each system that is required to be sustainedly compliant with cGxPs, 21 CFR Part 11and other regulations as well as to ensure that it is reviewed and approved by the System Owner, CVG and QA(U) and others as appropriate.

The System Owner / End-User is responsible for the storage of documentation using a sustainedly compliant method.

CVG and QA(U) are responsible for the review and approval of the periodic review.

The QA(U) Administrator is responsible for document storage, archive and retrieval.

The Departmental Coordinator is responsible for controlled copies of documents.

Procedure

The SOP should contain instructions and procedures that require and enable the periodic review of a system's validation status. It should cover:

- Compliance with 21 CFR Part 11 and other regulations
- System operation and use
- System documentation and use
- Controlled storage of associated documentation.

A periodic review of the validation status of validated systems is mandatory. This review assesses whether the system has changed and whether it continues to operate as it was previously validated. Reviewof the system and accompanying validation must be documented. Review frequency is discretionary, but should be less than a three-year review cycle and stated in the Validation Plan or other official document. However, if not formally stated, all systems should default to a two-year review cycle.

Identify the computer system to undergo periodic review.

Identify, locate, collect and inventory all validation and other documents that pertain to the system. This includes, but is not limited to:

- Project Plans
- Validation Plans
- Requirements and Specifications
- Procurement Documents
- Validation Protocols and Reports
- Change Control Requests
- Training Materials
- Vendor Supplied Manuals
- Manuals and SOPs.

Review the system including documentation, physical media and parts representing or comprising the computer or computerized system. Include hardware, software, firmware, operating systems and procedures. Compare these to the current best industry practice, expectations, and requirements for a sustainedly compliant system. In the absence of validation documentation, perform a retrospective validation.

Review the documentation for compliance with current best practices and compliance with current regulations. Document any deviations or deficiencies.

Compare the findings to the requirements for the computer system. Base the comparison on the GAMP five-tier categorization that defines the complexity of the computer systems to be validated. The computer system's complexity is used to determine the validation requirements.

See Chapter 3 – "Overview of Computer and Computerized System Validation."

Prepare the Periodic Review form, conduct the review and complete the form, and circulate the completed form for review and approval.

Create a Periodic Review Summary Report with a synopsis of the findings of the review. Indicate the key activities that are necessary to meet validation requirements. Where activities are not performed, explain the business risk and GxP implications of not completing these tasks. If necessary, detail problems and deficiencies identified during the review with recommended actions to bring the system into compliance.

Remediate indicated problems and perform the actions detailed in the Periodic Review Summary Report.

Triggers

When a computer or computerized system experiences any of the following, it must undergo periodic review:

- Has a failure
- Undergoes an upgrade or modification to its:
    - Hardware configuration
    - Software configuration
    - Functionality
    - Use
- Moved
- Repaired
- Replaced either physically or logically
- Changes to the preventive maintenance, or calibration program

Periodic review is also triggered by the elapse of time specified in the system validation documentation. In the absence of such specification, it should not exceed a two-year period. The System Owner with the approval of QA(U) may define additional triggers.

Continued Functionality and Applicability

Establish that the system as currently configured, installed, operated, and used meets the business requirements for which it is intended. Frequently the needs of the business change. Sometimes the system becomes obsolete, but its use continues. Sometimes, the system changes to meet those needs. Often the process of formal change control to maintain conformance and compliance is not completed.

Verify Changes

Verify that all adds, changes, deletes and modifications of any kind to the documentation, physical media, and parts representing or comprising the computer or comp uterized system have been performed in accordance with company policies and procedures as well as all applicable current regulations. This includes hardware, software, firmware, operating systems and procedures

Verify Periodic Evaluations

Verify that all previously required reviews have been performed; they are in proper order; and have been formally reviewed and approved. Missing or skipped reviews are cause for re-validation of the system.

Documentation Review

The system documentation is considered acceptable if the following conditions are met:

The system has been validated or had a periodic review in accordance with its specifications or the default review period in the absence of a required period.

The following documents or their equivalent, as applicable, are present and in compliance with current best industry practice and regulation:

- Letter of Notification / System Proposal
- Change Control Documents
- Project Team Charter
- Project Plan
- Risk Assessment
- Validation Plan
- User Requirements
- Functional Requirements
- Vendor Audit
- Vendor Audit Report
- Hardware Design Specification
- Software Design Specification
- Unit Testing Documentation
- Factory Acceptance Test Documents
- IQ Protocol
- IQ Summary Report
- OQ Protocol
- OQ Summary Report

- PQ Protocol
- PQ Summary Report
- 21 CFR Part 11 Assessment
- 21 CFR Part 11 Assessment Report
- Traceability Matrix Document
- Manuals
- Miscellaneous Documents
- Training Documents
- SOPs
- Project Validation Final Report
- Appropriate Period Reviews

Conformance of System to Documentation

Compare the documentation to the current system and how it is being used. Changes to the system or its use that are not reflected in the documentation require proper administration of configuration management, change control, regression testing and validation. The system and its use must be in accordance with the documentation and the documentation must reflect the system and its use.

Problem Resolution Review

Assess prior problem resolutions to ensure that they have been performed and are in accordance with the best practice and current regulation at the time of their resolve.

Performance Evaluation

Evaluate the overall performance of the system as a whole. Ensure that it meets or exceeds the business needs and requirements it is intended to serve. This includes operation of the software, hardware, processes and procedures in general along with the training and qualification of the people who use or operate the system.

21 CFR Part 11 Compliance

In the process of performing the periodic review, consider the compliance of each item or area examined with respect to its compliance with 21 CFR Part 11, Electronic Records; Electronic Signatures. Keep in mind that all GxP systems need to comply with this regulation, including legacy systems. If necessary, concurrently conduct a 21 CFR Part 11 Assessment.

See Chapter 24 – 21 CFR Part 11 Assessment.

Periodic Review Form

Prepare a Periodic Review Form and record the details of the findings of the review. Be sure to enter the sources of information provided. Examples of information to record are as follows:

System Name – the name by which the system is known should be consistent across all documentation. If the system has not been assigned an official name, then create one in accordance with formal change control procedures.

System Type – Instrumentation Control, Equipment Control, Global Application, Site Application or Single-User Application, Infrastructure, Server, Firewall

Software Name / Version / Vendor / Support – list the name of each software application including operating systems, firmware and ancillary programs with their version number and source of acquisition and support.

Peripheral Equipment – printers, plotters, instrument(s) or equipment interfaced to or being controlled by the computer as part of the system including the type of interface. This includes such operations as Distributed Control Systems (DCS) or Supervisory Control and Data Acquisition (SCADA) systems.

Location – list the location of each hardware item that is part of the system. Except for servers, it is not necessary to list infrastructure items unless an infrastructure system is being reviewed. Indicate the site, building, floor and room by its physical plant designation or description.

Equipment – list the make, model, serial number, asset number and any other designation of each piece of equipment that comprises the s ystem.

Network Information – indicate network connectivity and information, if networked, including addresses, computer name, ID, workgroup and domain.

System SOPs – list all SOPs applicable the system's operation and use.

Preventative Maintenance – list all preventive maintenance specifications and requirements applicable to the system. Associated equipment or instrumentation not under review is excluded.

Validation Documentation – list the documents that comprise the current baseline of the system's validation. These include the last retrospective validation, gap analysis or periodic review along with all change control documents since the last periodic review for all hardware including interfaced equipment and instrumentation.

Logs – list all equipment and software, use, back-up, archive, calibration and maintenance logs. Indicate any unresolved issues or missed calibration items.

Documentation Assessment – Enter the assessment of the accuracy of representation of the documentation of the system, its operation and use.

System SOPs Assessment – Enter the assessment of the accuracy of representation of the system SOPs of the system, its operation and use.

Training Assessment – Indicate the assessment of training and qualification of the End-Users, support, maintenance, contractors and vendors. List the location of training and qualification records.

Attachments – list the title of each attachment and the total number of items attached. Attach all supporting documents.

Summary Report

Prepare a Periodic Review Summary Report and have it reviewed and approved by the System Owner, CVG and QA(U).

See Chapter 25 – "Summary Reports."

Document Management and Storage

A Document Management and Storage section is included in the Periodic Review SOP. It defines the Document Management and Storage methodology that is followed for issue, revision, replacement, control and management of this SOP. This methodology is in accordance with an established Change Control SOP and a Document Management and Storage SOP.

Revision History

A Revision History section is included in the Periodic Review SOP. The Periodic Review SOP Version Number and all Revisions made to the document are recorded in this section. These changes are in accordance with a Change Control Procedure SOP.

The Revision History section consists of a summary of the changes, a brief statement explaining the reasons the document was revised, and the name and affiliation of the person responsible for the entries. It includes the numbers of all items the Periodic Review SOP replaces and is written so that the original Periodic Review SOP is not needed to understand the revisions or the revised document being issued.

When the Periodic Review SOP is a new document, a "000" as the Revision Number is entered with a statement such as "Initial Issue" in the comments section.

Regulatory or audit commitments require preservation from being inadvertently lost or modified in subsequent revisions. All entries or modifications made as a result of regulatory or audit commitments require indication as such in the Revision History and should indicate the organization, date and applicable references. A complete and accurate audit trail must be developed and maintained throughout the project lifecycle.

Attachments

If needed, an Attachments section is included in the Periodic Review SOP in accordance with the SOP on SOPs. An Attachment Index and Attachment section follow the text of the Periodic Review SOP. The top of each attachment page indicates "Attachment # x," where 'x' is the sequential number of each attachment. The second line of each attachment page contains the attachment title. The number of pages and date of issue corresponding to each attachment are listed in the Attachment Index.

Attachments are placed directly after the Attachment Index and are sequentially numbered (e.g., Attachment # 1, Attachment # 2) with each page of an attachment containing the corresponding document number, page numbering, and total number of pages in the attachment (e.g., Page x of y). Each attachment is labeled consistent with the title of the attachment itself.

Attachments are treated as part of the Periodic Review SOP. If an attachment needs to be changed, the entire Periodic Review SOP requires a new revision number and is reissued. New attachments or revisions of existing attachments are in accordance with an established Change Control SOP and a Document Management and Storage SOP.

Chapter 31 – Manuals and SOPs

In this chapter we will discuss the Standard Operating Procedure (SOP) for managing system specific documentation, manuals, and SOPs for software applications, computer or computerized systems that are validated systems. It specifies the requirements for obtaining or creating, using, managing and storing system specific manuals and SOPs.

Best industry practices and FDA Guidance documents indicate that in order to demonstrate that a computer or computerized system performs as it purports and is intended to perform, management of documentation, manuals, and SOPs is required to ensure that the system, its operation, and use are compliant with the business needs and requirements it is intended to fulfill. It is needed to control *all* system specific documentation comprising the computer or computerized system especially manuals and SOPs.

This SOP is used for all systems that create, modify, maintain, retrieve, transmit or archive records in electronic format whether or not they have been previously validated. It is used in conjunction with an overlying corporate policy to provide sustainable compliance with the regulations by obtaining or creating, using, managing and storing system specific documentation, manuals, and SOPs.

The System Owner and End-Users use this SOP to control the proper use and operation a software application, computer or computerized system that controls equipment and/or produces electronic data or records in support of all regulated laboratory, manufacturing, clinical or distribution activities.

Title Block, Headers and Footers

The SOP for Manuals and SOPs should have a title block, title or cover page along with a header and/or footer on each page of the document that clearly identifies the document. The minimum information that should be displayed for the SOP for Manuals and SOPs is as follows:

- Document Title
- Document Number
- Document Revision Number
- Project Name and Number
- Site/Location/Department Identification
- Date Document was issued, last revised or draft status
- Effective Date
- Pagination: page of total pages (e.g., Page x of y)

The Document Title should be in large type font that enables clear identification. It should clearly and concisely describe the contents and intent of the document. Since this document will be part of a set or package of documents within a project, it should be consistent with other documents used for the project and specify that it is the SOP for Manuals and SOPs. This information should be on every page of the document.

The Document Number should be in large type font that enables clear identification. Since the SOP for Manuals and SOPs will be part of a set or package of documents within a project, it should be consistent with other documents used for the project. This information should appear on every page of the document.

The Document Revision Number, which identifies the revision of the document, should appear on each page of the document. Many documents will go through one or more changes over time, particularly during the draft stage and the review and approval cycle. It is important to maintain accurate revision control because even a subtle change in a document can have an immense impact on the outcomes of activities. The SOP on SOPs should govern the numbering methodology.

See Chapter 6 – "The SOP on SOPs."

The Project Name and Number, if applicable, should be indicated on the cover or title page. This information identifies the overall project to which this document belongs.

The Site/Location/Department Identification should be indicated on the cover or title page. It identifies the Site/Location/Department that has responsibility for this document and its associated project.

The Date that the Document was issued, last revised or issued with draft status should be on each page of the document (e.g., 12-Dec-01). This will clearly identify the date on which the document was last modified.

The Effective Date of the Document should be on the title or cover page of the document (e.g., 12-Dec-01). This will clearly identify when to begin using the current version of the document.

Pagination, page of total pages (e.g., Page x of y), should be on each page of the document. This will help identify whether or not all of the pages of the document are present. Pages intentionally left blank should clearly be designated as such.

Approval Section

The Approval Section contains a table with the Author and Approvers printed names, titles and a place to sign and date the SOP. The SOP on SOPs governs the approvers of the document. The functional department (e.g., Production / Quality Assurance / Engineering) of the signatory is placed below each printed name and title along with a statement associated with each approver indicating the person's role or qualification in the approval process. Indicate the significance of the associated signature (e.g. This document meets the requirements of Corporate Policy 9055, Electronic Records; Electronic Signatures, dated 22-Mar-2000).

The completion (signing) of this section indicates that the contents have been reviewed and approved by the listed individuals.

Table of Contents

The SOP should contain a Table of Contents, which appears towards the beginning of the document. This Table of Contents provides a way to easily locate various topics contained in the document.

Purpose

The Purpose section clearly states that the intention of the document is to provide a Standard Operating Procedure used to provide procedures for managing system specific documentation, manuals, and SOPs of validated systems to maintain sustainable compliance of a system with the business needs and requirements of the System Owner in support of a Good Clinical, Laboratory, Manufacturing or Distribution Practices (GxP) and current regulation. This introduction to the SOP concisely describes the document's purpose in sufficient detail so that a non-technical reader can understand its purpose. For example:

> "The purpose of this SOP is to specify the requirements for obtaining or creating, using, managing and storing system specific manuals and SOPs of validated systems to maintain sustainable compliance of a system with the business needs of the system owner in support of a Good Clinical, Laboratory, Manufacturing or Distribution Practices (GxP) and current regulation."

Scope

The Scope section identifies the limitations of the SOP for Manuals and SOPs. It limits the scope to all GxP systems and is applicable to all computers, computerized systems, infrastructure, LANs and WANs including clinical, laboratory, manufacturing and distribution computer systems. It applies to all Sites, Locations and Departments within the company.

> "This document applies to all computers, computerized systems, infrastructure, LANs and WANs including clinical, laboratory, manufacturing and distribution computer systems that are governed by cGxPs. This SOP applies to system specific manuals and SOPs of computer or computerized systems. It applies to all departments and divisions at all locations of the Example Pharmaceuticals Corporation."

Equipment / Materials / Tools

This section lists any required or recommended equipment, materials, tools, templates or guides. Provide the name and a brief description of the equipment, instruments, materials and/or tools to be used in the procedures or testing covered by the SOP for Manuals and SOPs. This is for reference purposes and requires the user to utilize the specific item or a designated equivalent. The purpose and use is explained elsewhere in the document. Substitution of comparable equipment may be made in accordance with procedures for Change Control.

If the information is included in a job aid or operating manual, a description of where the information may be found with a specific reference to the document number, title, date and revision number should be specified.

Warnings / Notes / General Information / Safety

Warnings, notes, general or safety information applicable to the SOP for Manuals and SOPs are listed in this section.

Policies, Guidelines, References

This section lists references to appropriate regulatory requirements, SOPs, policies, guidelines, or other relevant documents referred to in the SOP for Manuals and SOPs. The name and a brief description of applicable related SOPs, corporate policies, procedures and guidelines, Quality Control monographs, books, technical papers, publications, specific regulations, batch records, vendor or system manuals, and all other relevant documents that are required by the SOP for Manuals and SOPs should be included. This is for reference purposes and requires the user to use the specific item or a designated equivalent. The purpose and use are explained elsewhere in the document.

SOPs cite references and incorporate them in whole by reference rather than specifying details that duplicate these items. This is done so that changes in policy, guidelines or reference materials require no changes to SOPs themselves and the most current version of the referenced material is used. Examp les are:

- Policy ABC-123, Administrative Policy for Electronic Records; Electronic Signatures
- Corporate, Plan for Compliance with 21 CFR Part 11, Electronic Records; Electronic Signatures
- SOP-456, Electronic Records; Electronic Signatures
- FDA, 21 CFR Part 11, Electronic Records; Electronic Signatures
- FDA, General Principals of Software Validation; Final Guidance for Industry and FDA Staff, January 11, 2002

Assumptions / Exclusions / Limitations

Any and all assumptions, exclusions or limitations within the context of the SOP should be discussed. Specific issues should be directly addressed in this section of the SOP for Manuals and SOPs.

Glossary of Terms

A Glossary of Terms should be included that defines all of the terms and acronyms that are used in this document or that are specific to the operation being described and may not be generally known to the reader. The reader may also be referred to the FDA document, "Glossary of Computerized System and Software Development Terminology."

Roles and Responsibilities

The System Owner is responsible for obtaining or creating, using, training, managing and storing system specific manuals and SOPs for each system that is required to be sustainedly compliant with cGxPs, 21 CFR Part 11and other regulations as well as to ensure that they are reviewed and approved by the System Owner, CVG and QA(U) and others as appropriate.

The System Owner / End-User is responsible for the storage of documentation using a sustainedly compliant method.

CVG and QA(U) are responsible for the review and approval of the system specific manuals and SOPs.

The QA(U) Administrator is responsible for document storage, archive and retrieval.

The Departmental Coordinator is responsible for controlled copies of documents.

End-Users are responsible for knowledge and proper use of system specific manuals and SOPs for each they use or operate that is required to be sustainedly compliant with cGxPs, 21 CFR Part 11and other regulations.

Procedure

The SOP should contain instructions and procedures that specify the requirements for obtaining or creating, using, managing and storing system specific manuals and SOPs of validated systems

Obtain or create documentation, manuals, and SOPs that are specific to the use and operation of systems that create, modify, maintain, retrieve, transmit or archive records in electronic format. Manuals should be obtained from the software or equipment vendor, where possible. Prepare an inventory the documents. After training the End-Users, ensure that the documents, manuals, and SOPs are used or followed in accordance with the procedures specified for the GxP system. It is imperative that the use and operation of the system reflect the documentation and vice versa. Failure to demonstrate conformance between specified procedures and actual practice might be cause for an FDA warning letter during a facility inspection.

The management, control, and storage of the system specific documentation, manuals, and SOPs are performed in accordance with the Roles and Responsibilities section above and a Document Management and Storage SOP.

Also see Chapter 28 – "Document Management and Storage."

Chapter 31 – Manuals and SOPs

Manuals

The vendor or developer of the specific software application, computer or computerized system generally supplies manuals. However, the company prepares some manuals. The types of manuals that are part of a computer or computerized system are as follows:

Hardware Manuals

Hardware manuals comprise vendor manuals, configuration diagrams and tables, wiring diagrams and tables, piping and instrumentation diagrams for computers, peripherals such as printers and plotters, equipment such as packaging and labels machines, instruments such as High Performance Liquid Chromatography (HPLC) or Mass Spectrometers.

Software Manuals

Software manuals contain programs, procedures, rules, and any associated documentation pertaining to the operation of a system.[15] They may also contain technical data or information, including computer listings and printouts, in human readable form, that describe or specify the design or details, explain the capabilities, or provide operating instructions for using the software to obtain desired results from a software system.[16]

End-User / Operator Manuals

End-User or Operator manuals are documentation that describe how to use a functional unit, and that may include description of the rights and responsibilities of the user, the owner, and the supplier of the unit.[17]

Administrator Manuals

Administrator manuals provide information for the System Administrator, the person(s) responsible for overall administration and operation of a system.

Training Manuals / Curricula

Training manuals provide information that is used to train the End-User or Operator of the system, the person(s) responsible for using and operating of a system.

See Chapter 33 – "Training."

Safety Manuals

Safety manuals point out possible hazards of unsafe operation and preventive measures to take in order to prevent personal injury, damage to the instrument, or compromising work output.

Job Aids

Job aids are controlled documents that are used by the people performing the work on the job and can indicate actions to be taken or be a tool that is used in the performance of the work. They generally provide assistance

---

[15] ANSI
[16] NIST
[17] ISO

or directions to the worker in performing a function or activity. Job aids can be in the form of templates, charts, tables, synopsized instructions sheets, checklists, algorithms, flowcharts, and pictures

Provide job aids to assist workers in the performance of routine tasks, lengthy or complex operations. Job aids supplement system documentation, manuals, and SOPs, not replace them.

Other Manuals

Some other manuals and documents that make up the documentation set for a system are: Theory of Operation, Overview, Specifications, Installation, Configuration or Set-up, Brochures, Guides and procurement documentation.

Standard Operating Procedures (SOPs)

The company generally creates customized Standard Operating Procedures for each computer or computerized system. They cover the day-to-day use and operation of the system. The types of SOPs that are part of a computer or computerized system are as follows:

End-User Operating Procedures

End-User Operating Procedures describe how the End-Users are to use and or operate the system on a routine basis. These procedures are specific to a given company, the department, and the task at hand. Reference End-User Manuals (documentation) to avoid duplicate description of standard items. Multiple procedures for use within the same or different departments or for various tasks may exist for the same system.

System Operating Procedures

System Operating Procedures specify the requirements for routine system operation and maintenance activities that are performed by System Administrator(s). These include, but are not limited to, system start-up, shutdown, task scheduling, and the use, review, and maintenance of logs.

Security Administration Procedures

Security Administration Procedures specify security features and functions, how they are authorized, administered, and maintained. They assign specific roles and responsibilities for the system. Areas addressed are logical and physical security as well as tiered or functional access and audit trails.

Database Administration Procedures

Database Administration Procedures specify the methods used to administer the database(s) associated with the system. They provide for a secure method for modifications of those database(s) including audit trails and logging.

Output Controls Procedures

Output Controls Procedures specify how outputs from the system that are controlled in accordance with company or regulatory requirements are managed. These procedural controls address physical and logical security.

Backup Procedures

Back-up Procedures specify routine periodic backup of the system and data. These procedures include, but are not limited to, what is to be backed-up, frequency or schedules, media, and procedures.

Purge and Archive Procedures

Purge and Archive Procedures specify how data is archived and purged. This includes, but is not limited to, storage and archival methods, meta-data creation and management, storage locations and schemes, retention periods, procedure validation, and media types.

Recovery Procedures

Recovery Procedures specify the method used in restoring a partial or complete data set or the entire system in the event of a failure or disaster. These procedures include, but are not limited to, backup and archive identification, recovery procedures, and procedure validation.

Business Continuity

The Business Continuity Plan provides a means to allow the business to continue operation in the event of interruption. This might be in the event of hardware or system failure or a disaster such as a fire or flood. The plan includes contingency procedures for manual operation of the business or business unit, disaster recovery procedures, system reconstruction, and procedure validation.

Disaster Recovery Procedures

Disaster recovery procedures specify alternate resources in the event of the primary system being unavailable. These procedures are used when manual procedures to operate the business for long periods of time are not realistic. Disaster recovery procedures include, but are not limited to, alternate hardware, software, equipment, peripherals, and network communication.

Periodic ReviewProcedures

Periodic ReviewProcedures specify the requirements for performing periodic reviews of validated systems to determine the compliance of a system with the business needs of the System Owner in support of a Good Clinical, Laboratory, Manufacturing or Distribution Practices (GxP) and current regulation.

See Chapter 30 – "Periodic Review."

Safety Procedures

Safety procedures specify the requirements for performing work related functions in a safe and hazard free manner. These are generally in accordance with an overall safety policy with procedures specific to the system or job related work provided in detail.

Document Management and Storage

A Document Management and Storage section is included in the SOP for Manuals and SOPs . It defines the Document Management and Storage methodology that is followed for issue, revision, replacement, control and management of this SOP. This methodology is in accordance with an established Change Control SOP and a Document Management and Storage SOP.

Revision History

A Revision History section is included in the SOP for Manuals and SOPs. The SOP for Manuals and SOPs Version Number and all Revisions made to the document are recorded in this section. These changes are in accordance with a Change Control Procedure SOP.

The Revision History section consists of a summary of the changes, a brief statement explaining the reasons the document was revised, and the name and affiliation of the person responsible for the entries. It includes the numbers of all items the SOP for Manuals and SOPs replaces and is written so that the original SOP for Manuals and SOPs is not needed to understand the revisions or the revised document being issued.

When the SOP for Manuals and SOPs is a new document, a "000" as the Revision Number is entered with a statement such as "Initial Issue" in the comments section.

Regulatory or audit commitments require preservation from being inadvertently lost or modified in subsequent revisions. All entries or modifications made as a result of regulatory or audit commitments require indication as such in the Revision History and should indicate the organization, date and applicable references. A complete and accurate audit trail must be developed and maintained throughout the project lifecycle.

Attachments

If needed, an Attachments section is included in the SOP for Manuals and SOPs in accordance with the SOP on SOPs. An Attachment Index and Attachment section follow the text of the SOP for Manuals and SOPs. The top of each attachment page indicates "Attachment # x," where 'x' is the sequential number of each attachment. The second line of each attachment page contains the attachment title. The number of pages and date of issue corresponding to each attachment are listed in the Attachment Index.

Attachments are placed directly after the Attachment Index and are sequentially numbered (e.g., Attachment # 1, Attachment # 2) with each page of an attachment containing the corresponding document number, page numbering, and total number of pages in the attachment (e.g., Page x of y). Each attachment is labeled consistent with the title of the attachment itself.

Attachments are treated as part of the SOP for Manuals and SOPs. If an attachment needs to be changed, the entire SOP for Manuals and SOPs requires a new revision number and is reissued. New attachments or revisions of existing attachments are in accordance with an established Change Control SOP and a Document Management and Storage SOP.

Chapter 32 – Logs

In this chapter we will discuss the Standard Operating Procedure (SOP) for creating, maintaining and reviewing logs of software applications, computer or computerized system that are validated systems. It specifies the requirements for creating, using, managing, and storing system specific logs.

Best industry practices and FDA Guidance documents indicate that in order to demonstrate that a software application, computer or computerized system performs as it purports and is intended to perform, management of logs is required to ensure that the system, its operation, and use are compliant with the business needs and requirements of the System Owner and current regulation. It is needed to document all system, server, contents, hardware, software, change control, and incident report information for validated GxP compliant systems.

This SOP is used for all systems that create, modify, maintain, retrieve, transmit or archive records in electronic format whether or not they have been previously validated. It is used in conjunction with an overlying corporate policy to provide sustainable compliance with the regulations by obtaining or creating, using, managing and storing computer and computerized system specific logs.

The System Owner, End-Users, and System Administrators use this SOP to control the proper use and operation of a software application, computer or computerized system that controls equipment and/or produces electronic data or records in support of all regulated laboratory, manufacturing, clinical or distribution activities.

Title Block, Headers and Footers

The Logs SOP should have a title block, title or cover page along with a header and/or footer on each page of the document that clearly identifies the document. The minimum information that should be displayed for the Logs SOP is as follows:

- Document Title
- Document Number
- Document Revision Number
- Project Name and Number
- Site/Location/Department Identification
- Date Document was issued, last revised or draft status
- Effective Date
- Pagination: page of total pages (e.g., Page x of y)

The Document Title should be in large type font that enables clear identification. It should clearly and concisely describe the contents and intent of the document. Since this document will be part of a set or package of documents within a project, it should be consistent with other documents used for the project and specify that it is the Logs SOP. This information should be on every page of the document.

The Document Number should be in large type font that enables clear identification. Since the Logs SOP will be part of a set or package of documents within a project, it should be consistent with other documents used for the project. This information should appear on every page of the document.

The Document Revision Number, which identifies the revision of the document, should appear on each page of the document. Many documents will go through one or more changes over time, particularly during the draft stage and the review and approval cycle. It is important to maintain accurate revision control because even a

subtle change in a document can have an immense impact on the outcomes of activities. The SOP on SOPs should govern the numbering methodology.

See Chapter 6 – "The SOP on SOPs."

The Project Name and Number, if applicable, should be indicated on the cover or title page. This information identifies the overall project to which this document belongs.

The Site/Location/Department Identification should be indicated on the cover or title page. It identifies the Site/Location/Department that has responsibility for this document and its associated project.

The Date that the Document was issued, last revised or issued with draft status should be on each page of the document (e.g., 12-Dec-01). This will clearly identify the date on which the document was last modified.

The Effective Date of the Document should be on the title or cover page of the document (e.g., 12-Dec-01). This will clearly identify when to begin using the current version of the document.

Pagination, page of total pages (e.g., Page x of y), should be on each page of the document. This will help identify whether or not all of the pages of the document are present. Pages intentionally left blank should clearly be designated as such.

Approval Section

The Approval Section contains a table with the Author and Approvers printed names, titles and a place to sign and date the SOP. The SOP on SOPs governs the approvers of the document. The functional department (e.g., Production / Quality Assurance / Engineering) of the signatory is placed below each printed name and title along with a statement associated with each approver indicating the person's role or qualification in the approval process. Indicate the significance of the associated signature (e.g. This document meets the requirements of Corporate Policy 9055, Electronic Records; Electronic Signatures, dated 22-Mar-2000).

The completion (signing) of this section indicates that the contents have been reviewed and approved by the listed individuals.

Table of Contents

The SOP should contain a Table of Contents, which appears towards the beginning of the document. This Table of Contents provides a way to easily locate various topics contained in the document.

Purpose

The Purpose section clearly states that the intention of the document is to provide a Standard Operating Procedure used for creating, using, managing and storing system specific logs of validated systems to maintain sustainable compliance of a system with the business needs of the system owner in support of a Good Clinical, Laboratory, Manufacturing or Distribution Practices (GxP) and current regulation. This introduction to the SOP concisely describes the document's purpose in sufficient detail so that a non-technical reader can understand its purpose. For example:

> "The purpose of this SOP is to specify the requirements for creating, using, managing and storing system specific logs of validated systems to maintain sustainable compliance of a system with the business needs of the system owner in support of a Good Clinical, Laboratory, Manufacturing or Distribution Practices (GxP) and current regulation."

Scope

The Scope section identifies the limitations of the Logs. It limits the scope to all GxP systems and is applicable to all computers, computerized systems, infrastructure, LANs and WANs including clinical, laboratory, manufacturing and distribution computer systems. It applies to all Sites, Locations and Departments within the company. For example:

> "This document applies to all computers, computerized systems, infrastructure, LANs and WANs including clinical, laboratory, manufacturing and distribution computer systems that are governed by cGxPs. This SOP applies to system specific logs of computer or computerized systems. It applies to all departments and divisions at all locations of the Example Pharmaceuticals Corporation."

Equipment / Materials / Tools

This section lists any required or recommended equipment, materials, tools, templates or guides. Provide the name and a brief description of the equipment, instruments, materials and/or tools to be used in the procedures or testing covered by the Logs SOP. This is for reference purposes and requires the user to utilize the specific item or a designated equivalent. The purpose and use is explained elsewhere in the document. Substitution of comparable equipment may be made in accordance with procedures for Change Control.

If the information is included in a job aid or operating manual, a description of where the information may be found with a specific reference to the document number, title, date and revision number should be specified.

Warnings / Notes / General Information / Safety

Warnings, notes, general or safety information applicable to the Logs SOP are listed in this section.

Policies, Guidelines, References

This section lists references to appropriate regulatory requirements, SOPs, policies, guidelines, or other relevant documents referred to in the Logs SOP. The name and a brief description of applicable related SOPs, corporate policies, procedures and guidelines, Quality Control monographs, books, technical papers, publications, specific regulations, batch records, vendor or system manuals, and all other relevant documents that are required by the Logs SOP should be included. This is for reference purposes and requires the user to use the specific item or a designated equivalent. The purpose and use are explained elsewhere in the document.

SOPs cite references and incorporate them in whole by reference rather than specifying details that duplicate these items. This is done so that changes in policy, guidelines or reference materials require no changes to SOPs themselves and the most current version of the referenced material is used. Examples are:

- Policy ABC-123, Administrative Policy for Electronic Records; Electronic Signatures
- Corporate, Plan for Compliance with 21 CFR Part 11, Electronic Records; Electronic Signatures
- SOP-456, Electronic Records; Electronic Signatures
- FDA, 21 CFR Part 11, Electronic Records; Electronic Signatures
- FDA, General Principals of Software Validation; Final Guidance for Industry and FDA Staff, January 11, 2002

Assumptions / Exclusions / Limitations

Any and all assumptions, exclusions or limitations within the context of the SOP should be discussed. Specific issues should be directly addressed in this section of the Logs SOP.

Glossary of Terms

A Glossary of Terms should be included that defines all of the terms and acronyms that are used in this document or that are specific to the operation being described and may not be generally known to the reader. The reader may also be referred to the FDA document, "Glossary of Computerized System and Software Development Terminology."

Roles and Responsibilities

The System Owner is responsible for obtaining or creating, using, training, managing and storing system specific logs for each system that is required to be sustainedly compliant with cGxPs, 21 CFR Part 1,1and other regulations as well as to ensure that they are reviewed and approved by the System Owner, CVG and QA(U) and others as appropriate.

QA(U) is responsible for the review and approval of the system specific logs.

The Document Administrator is responsible for document storage, archive and retrieval.

The Departmental Coordinator is responsible for controlled copies of documents.

End-Users and System Administrators are responsible for knowledge and proper use of system specific logs for each system they use or operate that is required to be sustainedly compliant with cGxPs, 21 CFR Part 11and other regulations.

Procedure

The SOP should contain instructions and procedures that specify the requirements for creating, using, managing and storing logs for validated systems.

Create logs that are specific to the use and operation of systems that create, modify, maintain, retrieve, transmit or archive records in electronic format. After training, the End-Users and System Administrators ensure that the logs are used and maintained in accordance with the procedures specified for the GxP system. It is imperative that the use and operation of the system reflect the documentation and vice versa. Failure to demonstrate conformance between specified procedures and actual practice might be cause for an FDA warning letter during a facility inspection.

The management, control, and storage of system specific logs are performed in accordance with the Roles and Responsibilities section above, a Configuration Management SOP, Change Control SOP, and Document Management and Storage SOP.

See Chapter 27 – "Change Control," Chapter 28 – "Document Management and Storage," and Chapter 29 – "Configuration Management."

A log for each computer system maintained in compliance with GxP requirements and for each category listed below within each system of the servers, applications, and systems detailing areas of responsibility, service level agreements, points of contact and other pertinent information is created and maintained. The IT/IS Unit prepares a database with associated SOPs for Computer Operations with this and other information maintained in it. Use Logs are maintained for equipment and instruments to record their use along with other pertinent information. The log (database) can be a validated software application that is compliant with 21 CFR Part 11 (highly recommended) or simply a paper system such as pages in a binder that constitute a logbook. Paper records must conform to GxP requirements.

System Log

The indexed System Log contains the following:

- System Name
- Regulated Classification (GxP)
- Node ID / Machine Type
- Location(s)
- System Overview
- Operating System
- Maintenance Records
- Main Application Software
- System Owner
- System Administrator
- Application Manager(s)
- Additional Points of Contact
- Backup Media Storage
- Disaster Recovery Plan
- Licenses
- System Configuration Report
- Change Control Documentation
- Incident Report Documentation
- Inventory of Manuals and SOPs
- Service Level Agreement(s)

The log is updated, signed and dated each time a change to the system is made. Copies of all change control documentation and incident reports, preceded by an index, are kept in the system log. Changes to the system are performed in accordance with Configuration Management, Document Management and Storage, and Change Control SOPs. This includes changes made as a result of routine, non-routine, preventative maintenance, or repair activities. Log entries are dated and signed by the person making the change and must include the reason for the change.

Server Log

The indexed Server Log contains the following:

- Server Name
- Server Identification
- Node ID / Machine Type
- Applications / System(s) Installed
- Regulated Classification (GxP)
- Location
- Overview
- Operating System

- System Owner
- Server Managers
- System Administrator
- Application Manager(s)
- Additional Points of Contact
- Maintenance Records
- Backup Media Storage
- Disaster Recovery Plan
- Licenses
- System Configuration Report
- Change Control Documentation
- Incident Report Documentation
- Inventory of Manuals and SOPs

The log is updated, signed and dated each time a change to the server is made. Copies of all change control documentation and incident reports, preceded by an index, are kept in the system log. Changes to the server are performed in accordance with Configuration Management, Document Management and Storage, and Change Control SOPs. This includes changes made as a result of routine, non-routine, preventative maintenance, or repair activities. Log entries are dated and signed by the person making the change and must include the reason for the change.

Use Logs

Use Logs are created and maintained in accordance with SOPs specific to the equipment or instrumentation to which they apply. Use Logs are important as they provide a crucial link in the cradle to grave record keeping of information. They record the "who, what, when, where, why, and how" surrounding the use of equipment and instruments.

Corrections and Dates

Manual changes or corrections to documentation are made in black or blue indelible ink by placing a single line through the entity to be corrected, entering any correction in a clear and concise manner, initialing and dating the correction. Do NOT obliterate any item.

Do *not* pre- or post-date anything. If a date is in error or missed, enter the current date with initials and an explanation as to what transpired.

If an electronic records and signature system is used, it must have an audit trail compliant with 21 CFR Part 11.

Document Management and Storage

A Document Management and Storage section is included in the Logs SOP. It defines the Document Management and Storage methodology that is followed for issue, revision, replacement, control and management of this SOP. This methodology is in accordance with an established Change Control SOP and a Document Management and Storage SOP.

Revision History

A Revision History section is included in the Logs SOP. The Logs SOP Version Number and all Revisions made to the document are recorded in this section. These changes are in accordance with a Change Control Procedure SOP.

The Revision History section consists of a summary of the changes, a brief statement explaining the reasons the document was revised, and the name and affiliation of the person responsible for the entries. It includes the numbers of all items the Logs SOP replaces and is written so that the original Logs SOP is not needed to understand the revisions or the revised document being issued.

When the Logs SOP is a new document, a "000" as the Revision Number is entered with a statement such as "Initial Issue" in the comments section.

Regulatory or audit commitments require preservation from being inadvertently lost or modified in subsequent revisions. All entries or modifications made as a result of regulatory or audit commitments require indication as such in the Revision History and should indicate the organization, date and applicable references. A complete and accurate audit trail must be developed and maintained throughout the project lifecycle.

Attachments

If needed, an Attachments section is included in the Logs SOP in accordance with the SOP on SOPs. An Attachment Index and Attachment section follow the text of the Logs SOP. The top of each attachment page indicates "Attachment # x," where 'x' is the sequential number of each attachment. The second line of each attachment page contains the attachment title. The number of pages and date of issue corresponding to each attachment are listed in the Attachment Index.

Attachments are placed directly after the Attachment Index and are sequentially numbered (e.g., Attachment # 1, Attachment # 2) with each page of an attachment containing the corresponding document number, page numbering, and total number of pages in the attachment (e.g., Page x of y). Each attachment is labeled consistent with the title of the attachment itself.

Attachments are treated as part of the Logs SOP. If an attachment needs to be changed, the entire Logs SOP requires a new revision number and is reissued. New attachments or revisions of existing attachments are in accordance with an established Change Control SOP and a Document Management and Storage SOP.

Chapter 33 – Training

In this chapter we will discuss the Standard Operating Procedure (SOP) for creating, maintaining, and reviewing training for software applications, computer or computerized systems that are validated systems. It specifies the requirements for the development, curricula, assessment, records, management, and control of the documentation and procedures.

Best industry practices and FDA Guidance documents indicate that in order to demonstrate that a software application, computer or computerized system performs as it purports and is intended to perform, training and assessment is necessary to ensure that the system, its operation, and use are compliant with the business needs of the project and current regulation. It is needed to perform and document all employee, consultant, and contractor training and assessment for validated GxP systems.

This SOP is used for all systems that create, modify, maintain, retrieve, transmit, or archive records in electronic format whether or not they have been previously validated. It is used in conjunction with an overlying corporate policy to provide sustainable compliance with the regulations by preparing, implementing, managing and maintaining, recording and storing computer and computerized system training curricula.

The System Owner and Training Department use this SOP to prepare training curricula and train users of a software application, computer or computerized system that controls equipment and/or produces electronic data or records in support of all regulated laboratory, manufacturing, clinical or distribution activities.

Title Block, Headers and Footers

The Training SOP should have a title block, title or cover page along with a header and/or footer on each page of the document that clearly identifies the document. The minimum information that should be displayed for the Training SOP is as follows:

- Document Title
- Document Number
- Document Revision Number
- Project Name and Number
- Site/Location/Department Identification
- Date Document was issued, last revised or draft status
- Effective Date
- Pagination: page of total pages (e.g., Page x of y)

The Document Title should be in large type font that enables clear identification. It should clearly and concisely describe the contents and intent of the document. Since this document will be part of a set or package of documents within a project, it should be consistent with other documents used for the project and specify that it is the Training SOP. This information should be on every page of the document.

The Document Number should be in large type font that enables clear identification. Since the Training SOP will be part of a set or package of documents within a project, it should be consistent with other documents used for the project. This information should appear on every page of the document.

The Document Revision Number, which identifies the revision of the document, should appear on each page of the document. Many documents will go through one or more changes over time, particularly during the draft stage and the review and approval cycle. It is important to maintain accurate revision control because even a subtle change in a document can have an immense impact on the outcomes of activities. The SOP on SOPs should govern the numbering methodology.

See Chapter 6 – "The SOP on SOPs."

The Project Name and Number, if applicable, should be indicated on the cover or title page. This information identifies the overall project to which this document belongs.

The Site/Location/Department Identification should be indicated on the cover or title page. It identifies the Site/Location/Department that has responsibility for this document and its associated project.

The Date that the Document was issued, last revised or issued with draft status should be on each page of the document (e.g., 12-Dec-01). This will clearly identify the date on which the document was last modified.

The Effective Date of the Document should be on the title or cover page of the document (e.g., 12-Dec-01). This will clearly identify when to begin using the current version of the document.

Pagination, page of total pages (e.g., Page x of y), should be on each page of the document. This will help identify whether or not all of the pages of the document are present. Pages intentionally left blank should clearly be designated as such.

Approval Section

The Approval Section contains a table with the Author and Approvers printed names, titles and a place to sign and date the SOP. The SOP on SOPs governs the approvers of the document. The functional department (e.g., Production / Quality Assurance / Engineering) of the signatory is placed below each printed name and title along with a statement associated with each approver indicating the person's role or qualification in the approval process. Indicate the significance of the associated signature (e.g. This document meets the requirements of Corporate Policy 9055, Electronic Records; Electronic Signatures, dated 22-Mar-2000).

The completion (signing) of this section indicates that the contents have been reviewed and approved by the listed individuals.

Table of Contents

The SOP should contain a Table of Contents, which appears towards the beginning of the document. This Table of Contents provides a way to easily locate various topics contained in the document.

Purpose

The Purpose section clearly states that the intention of the document is to provide a Standard Operating Procedure used for preparing, implementing, managing and maintaining, recording and storing computer and computerized system training curricula, which involves development, curricula, assessment, job aids, records, management and control for training in the use of validated systems to maintain sustainable compliance of a system with the business needs of the system owner in support of a Good Clinical, Laboratory, Manufacturing or Distribution Practices (GxP) and current regulation. This introduction to the SOP concisely describes the document's purpose in sufficient detail so that a non-technical reader can understand its purpose. For example:

> "The purpose of this SOP is to specify the requirements for preparing, implementing, managing and maintaining, recording and storing computer and computerized system training curricula, which involves development, curricula, assessment, job aids, records, management and control for training in the use of validated systems to maintain sustainable compliance of a system with the business needs of the system owner in support of a Good Clinical, Laboratory, Manufacturing or Distribution Practices (GxP) and current regulation."

Scope

The Scope section identifies the limitations of the Training SOP. It limits the scope to all GxP systems and is applicable to all computers, computerized systems, infrastructure, LANs and WANs including clinical, laboratory, manufacturing and distribution computer systems. It applies to all Sites, Locations and Departments within the company. For example:

> "This document applies to all computers, computerized systems, infrastructure, LANs and WANs including clinical, laboratory, manufacturing and distribution computer systems that are governed by cGxPs. This SOP applies to training for computer or computerized systems. It establishes the general requirements of developing, updating and maintaining employee, contractor and vendor training curriculum. It applies to all departments and divisions at all locations of the Example Pharmaceuticals Corporation."

Equipment / Materials / Tools

This section lists any required or recommended equipment, materials, tools, templates or guides. Provide the name and a brief description of the equipment, instruments, materials and/or tools to be used in the procedures or testing covered by the Training SOP. This is for reference purposes and requires the user to utilize the specific item or a designated equivalent. The purpose and use is explained elsewhere in the document. Substitution of comparable equipment may be made in accordance with procedures for Change Control.

If the information is included in a job aid or operating manual, a description of where the information may be found with a specific reference to the document number, title, date and revision number should be specified.

Warnings / Notes / General Information / Safety

Warnings, notes, general or safety information applicable to the Training SOP are listed in this section.

Policies, Guidelines, References

This section lists references to appropriate regulatory requirements, SOPs, policies, guidelines, or other relevant documents referred to in the Training SOP. The name and a brief description of applicable related SOPs, corporate policies, procedures and guidelines, Quality Control monographs, books, technical papers, publications, specific regulations, batch records, vendor or system manuals, and all other relevant documents that are required by the Training SOP should be included. This is for reference purposes and requires the user to use the specific item or a designated equivalent. The purpose and use are explained elsewhere in the document.

SOPs cite references and incorporate them in whole by reference rather than specifying details that duplicate these items. This is done so that changes in policy, guidelines or reference materials require no changes to SOPs themselves and the most current version of the referenced material is used. Examples are:

- Policy ABC-123, Administrative Policy for Electronic Records; Electronic Signatures
- Corporate, Plan for Compliance with 21 CFR Part 11, Electronic Records; Electronic Signatures
- SOP-456, Electronic Records; Electronic Signatures
- FDA, 21 CFR Part 11, Electronic Records; Electronic Signatures
- FDA, General Principals of Software Validation; Final Guidance for Industry and FDA Staff, January 11, 2002

Assumptions / Exclusions / Limitations

Any and all assumptions, exclusions or limitations within the context of the SOP should be discussed. Specific issues should be directly addressed in this section of the Training SOP.

Glossary of Terms

The Glossary of Terms defines all of the terms and acronyms that are used in this document or that are specific to the operation being described and may not be generally known to the reader. The reader may also be referred to the FDA document, "Glossary of Computerized System and Software Development Terminology."

Roles and Responsibilities

The System Owner and Training Department are responsible for obtaining or preparing, implementing, managing, maintaining, recording, and storing computer and computerized system training and curricula. This involves development, curricula, training, assessment, job aids, records, management, and control required for sustainable compliance with cGxPs, 21 CFR Part 11and other regulations. The System Owner ensures that training documentation are reviewed and approved by the System Owner, CVG, QA(U), and others as appropriate.

Each System Owner and first line supervisor is responsible to assure that all persons using or working on a GxP system have been trained in accordance with the job-specific requirements.

The Training Department is responsible for the preparation of training curricula, manuals, and materials based on current best industry practice for the teaching profession. They perform the training and assessment of each End-User, Vendor, Consultant or Contractor who is to use or perform work on the computer or computerized system along with the required record keeping.

QA(U) is responsible for the review and approval of the training materials, records, and training management.

The Document Administrator is responsible for document storage, archive, and retrieval.

The Departmental Coordinator is responsible for controlled copies of documents.

End-Users and System Administrators are responsible to be trained and knowledgeable in the proper use of each system they use or operate, which is required to be sustainedly compliant with cGxPs, 21 CFR Part 11,and other regulations.

The Human Resources is responsible for providing timely, accurate updates of employee data to support training.

The Procurement Department is responsible for maintaining and providing timely, accurate updates of contractor and vendor data to support training.

Everyone must be trained prior to performing work involving a validated system.

Procedure

The SOP should contain instructions and procedures that specify the requirements for preparing, implementing, managing and maintaining, recording, and storing computer and computerized system training curricula, which involves development, curricula, assessment, job aids, records, management, and control for training in the use of validated systems.

All polices, procedures and guidelines encompassing the use, maintenance and operation of a validated computer or computerized system have associated job aids, training curricula, training manuals and assessment documents generated and implemented.

Training and an assessment of that training indicating knowledge and understanding provides assurance that individuals are qualified to perform the work required.

Provide curricula and training with assessment that are specific to the use and operation of systems that create, modify, maintain, retrieve, transmit, or archive records in electronic format. After training the End-Users, System Administrators, and others who use or work on validated systems, ensure that the materials are used and maintained in accordance with the procedures specified for the GxP system. It is imperative that the use and operation of the system reflect the documentation and vice versa. Failure to demonstrate conformance between specified procedures and actual practice might be cause for an FDA warning letter during a facility inspection.

Departmental or job specific training is based on the position within a given area (e.g. – instruction on the standard operating procedures, job activities, operation of equipment, and processes). Assessments for critical positions are included as part of the training requirements.

Curriculum

Current Good Practices (cGxP) Training

Annual Current Good Practices (cGxP) Training is given as prerequisite to all other training in order to acquire and maintain an understanding of the regulations, examine their history, and emphasize their importance in carrying out job functions. This is normally provided in accordance with company policy and procedures for GxP requirements and is also part of new employee, vendor, or contractor orientation.

Initial Training

Initial training is given prior to using or performing work on a software application, computer or computerized system. Initial training applies to End-Users using the system in the performance of job functions as well as to System Administrators, Contractors, and Vendors performing any work on the system or its components. Initial training is specific to the function performed and includes an overview, detailed operation, and use. It includes familiarization with the following list of SOPs: Training, Security, Change Control, Periodic Review, Backup/Restore/Archive, Disaster Recovery, System Use, Preventative Maintenance and Safety.

See Chapter 31 – "Manuals and SOPs."

Continuing Training

Continuing training is generally performed on an annual basis as a refresher course or on modifications and upgrades to the system. All End-Users, System Administrators, Vendors and Contractors must be given Continuing Training to maintain their qualification to use or work on the system.

Remedial Training

Remedial training is given when the necessary knowledge or skills to perform a job or task are deficient. Knowledge and skills must be maintained at an acceptable level in order for the people doing the work to be deemed qualified.

Triggers for Training

Training or re-training, as determined by the System Owner or End-User's Supervisor and approved by QA(U), is performed when one or more of the following conditions occur:

- System Change
- Procedural Change
- Process Change
- Continuing Training Due
- New Employee, Contractor, or Vendor
- Skill or Knowledge Deficiency
- Safety Incident, Violation or Deficiency

Training Curricula / Manuals / Materials

Development

The System Owner and Department Manager work with the Training Department to develop curricula, manuals, job aids, and materials for each GxP job function and process requirement. The curricula are based on system manuals, job and task descriptions, and the SOPs that have been developed for the system, which include assessments, quality assurance methods and procedures, investigations, and change control. The training curricula are reviewed and approved by the System Owner, CVG, and QA(U).

See Chapter 31 – "Manuals and SOPs."

Job Aids, which are created in accordance with an SOP on Manuals and SOPs to provide assistance or directions to a worker in performing a function or activity, require training and instruction for use. These Job Aids are used to indicate actions to take or are a tool used in the performance of the work. They can be in the form of templates, charts, tables, synopsized instructions sheets, checklists, algorithms, flowcharts, or pictures[18]

Format

The format of training curricula, manuals, and materials is determined by the Training Department with the approval of QA(U). It is generally governed by best industry practice for the teaching profession and is beyond the scope of this handbook.

[18] See *How to Write a Job Aid*, Tony Moore, Moore Performance Improvement, Inc.: www.expertojt.com

Updating Training Curricula / Manuals / Materials

Training curricula, manuals, and materials, as determined by the System Owner, CVG, and QA(U), is updated when one or more of the following conditions occur:

- System Change
- Procedural Change
- Process Change
- Safety Deficiency
- Change in Responsibility
- Change in Regulation

The nature of updates is based on what has changed or occurred; changes to manuals and SOPs. Modifications to the training curricula, manuals, and materials are put through a review and approval process in accordance with change control procedures.

See Chapter 27 – "Change Control."

Management and Control

The Training Department, in conjunction with the System Owner and Departmental Coordinator, manages and controls all training documentation and materials. They are reviewed prior to use and periodically thereafter for accuracy, completeness, and applicability of use. All training documentation, manuals, and materials are controlled documents and must be tracked for possession and use.

See Chapter 31 – "Manuals and SOPs"

Assessment

Formal assessment following training is performed for each person trained. This includes a determination as to whether each individual has acquired sufficient knowledge and/or skill to perform the tasks required by the job associated with the training session. Training must have taken place before an assessment is performed. Only individuals passing an assessment or meeting assessment criteria are considered qualified.

The content of the assessment is based on the training curricula, manuals, and materials presented in the training session being assessed. The assessment can be skills based, whereby the individual is observed performing the requisite job tasks and a determination of the outcome of the performance is the qualifier.

The person performing the assessment must be qualified to perform the assessment by being familiar with the training materials and the job requirements. He or she must be qualified as a "Trainer."

Each trainee's functional supervisor signs the training and assessment records to provide documented evidence of their awareness of the level of knowledge and skill attained.

Records

A record must be maintained for each employee, contractor, or vendor's employee that indicates what training has been given along with when, where, and by whom that training took place. The results of the associated assessment or test are part of the record. These records are necessary to provide proof that an individual is qualified to perform the assigned work. Each department normally has access to these records for reference purposes with no ability to alter a record directly. Additionally, any other qualifications, including items such as a resume, outside certifications and training, are part of their training record.

For each function or job performed by an individual, there must be documentation on file that training and assessment of that training has been performed. Qualification of the people performing the work must be documented prior to the performance of work with or on the system.

Training records are retained for a minimum period of 10 years unless otherwise required by regulation or Predicate Rules. Records in support of marketed product, for example, are required to be retained for the same retention period as records for that product.

Document Management and Storage (Training Documentation)

A Document Management and Storage section is included in each Training document. It defines the Document Management and Storage methodology that is followed for issue, revision, replacement, control and management of each Training document. This methodology is in accordance with an established Change Control SOP and a Document Management and Storage SOP.

Revision History (Training Documentation)

A Revision History section is included in each Training document. The Training document Version Number and all Revisions made to the document are recorded in this section. These changes are in accordance with a Change Control Procedure SOP.

The Revision History section consists of a summary of the changes, a brief statement explaining the reasons the document was revised, and the name and affiliation of the person responsible for the entries. It includes the numbers of all items the Training document replaces and is written so that the original Training document is not needed to understand the revisions or the revised document being issued.

When the Training document is a new document, a "000" as the Revision Number is entered with a statement such as "Initial Issue" in the comments section.

Regulatory or audit commitments require preservation from being inadvertently lost or modified in subsequent revisions. All entries or modifications made as a result of regulatory or audit commitments require indication as such in the Revision History and should indicate the organization, date and applicable references. A complete and accurate audit trail must be developed and maintained throughout the project lifecycle.

Attachments (Training Documentation)

If needed, an Attachments section is included in each Training document in accordance with the SOP on SOPs. An Attachment Index and Attachment section follow the text of a Training document. The top of each attachment page indicates "Attachment # x," where 'x' is the sequential number of each attachment. The second line of each attachment page contains the attachment title. The number of pages and date of issue corresponding to each attachment are listed in the Attachment Index.

Attachments are placed directly after the Attachment Index and are sequentially numbered (e.g., Attachment # 1, Attachment # 2) with each page of an attachment containing the corresponding document number, page numbering, and total number of pages in the attachment (e.g., Page x of y). Each attachment is labeled consistent with the title of the attachment itself.

Attachments are treated as part of the Training document. If an attachment needs to be changed, the entire Training document requires a new revision number and is reissued. New attachments or revisions of existing attachments are in accordance with an established Change Control SOP and a Document Management and Storage SOP.

Document Management and Storage *(SOP)*

A Document Management and Storage section is included in the Training SOP. It defines the Document Management and Storage methodology that is followed for issue, revision, replacement, control and management of this SOP. This methodology is in accordance with an established Change Control SOP and a Document Management and Storage SOP.

Revision History *(SOP)*

A Revision History section is included in the Training SOP. The Training SOP Version Number and all Revisions made to the document are recorded in this section. These changes are in accordance with a Change Control Procedure SOP.

The Revision History section consists of a summary of the changes, a brief statement explaining the reasons the document was revised, and the name and affiliation of the person responsible for the entries. It includes the numbers of all items the Training SOP replaces and is written so that the original Training SOP is not needed to understand the revisions or the revised document being issued.

When the Training SOP is a new document, a "000" as the Revision Number is entered with a statement such as "Initial Issue" in the comments section.

Regulatory or audit commitments require preservation from being inadvertently lost or modified in subsequent revisions. All entries or modifications made as a result of regulatory or audit commitments require indication as such in the Revision History and should indicate the organization, date and applicable references. A complete and accurate audit trail must be developed and maintained throughout the project lifecycle.

Attachments *(SOP)*

If needed, an Attachments section is included in the Training SOP in accordance with the SOP on SOPs. An Attachment Index and Attachment section follow the text of the Training SOP. The top of each attachment page indicates "Attachment # x," where 'x' is the sequential number of each attachment. The second line of each attachment page contains the attachment title. The number of pages and date of issue corresponding to each attachment are listed in the Attachment Index.

Attachments are placed directly after the Attachment Index and are sequentially numbered (e.g., Attachment # 1, Attachment # 2) with each page of an attachment containing the corresponding document number, page numbering, and total number of pages in the attachment (e.g., Page x of y). Each attachment is labeled consistent with the title of the attachment itself.

Attachments are treated as part of the Training SOP. If an attachment needs to be changed, the entire Training SOP requires a new revision number and is reissued. New attachments or revisions of existing attachments are in accordance with an established Change Control SOP and a Document Management and Storage SOP.

Chapter 34 – Project Validation Final Report

In this chapter we will discuss the Standard Operating Procedure (SOP) for creating a Project Validation Final Report for a validation project. It specifies the requirements for preparing a final report for a Good Clinical, Laboratory, Manufacturing or Distribution Practices (GxP) project.

Best industry practices and FDA Guidance documents indicate that in order to demonstrate that a software application, computer, or computerized system performs as it purports and is intended to perform, a Project Validation Final Report is required to demonstrate, in a summarized format, the outcomes of a validation project.

This SOP is used for all systems that create, modify, maintain, retrieve, transmit or archive records in electronic format. It is used in conjunction with an overlying corporate policy to provide sustainable compliance with the regulation. All GxP and 21 CFR Part 11 validation projects must have their outcomes summarized in accordance with this SOP.

The System Owner uses this SOP to summarize the outcome of a validation project for projects for a software application, computer or computerized system that controls equipment and/or produces electronic data or records in support of all regulated laboratory, manufacturing, clinical or distribution activities.

Title Block, Headers and Footers

The Project Validation Plan Final Report SOP should have a title block, title or cover page along with a header and/or footer on each page of the document that clearly identifies the document. The minimum information that should be displayed for the Project Validation Plan Final Report SOP is as follows:

- Document Title
- Document Number
- Document Revision Number
- Project Name and Number
- Site/Location/Department Identification
- Date Document was issued, last revised or draft status
- Effective Date
- Pagination: page of total pages (e.g., Page x of y)

The Document Title should be in large type font that enables clear identification. It should clearly and concisely describe the contents and intent of the document. Since this document will be part of a set or package of documents within a project, it should be consistent with other documents used for the project and specify that it is the Project Validation Plan Final Report SOP. This information should be on every page of the document.

The Document Number should be in large type font that enables clear identification. Since the Project Validation Plan Final Report SOP will be part of a set or package of documents within a project, it should be consistent with other documents used for the project. This information should appear on every page of the document.

The Document Revision Number, which identifies the revision of the document, should appear on each page of the document. Many documents will go through one or more changes over time, particularly during the draft stage and the review and approval cycle. It is important to maintain accurate revision control because even a subtle change in a document can have an immense impact on the outcomes of activities. The SOP on SOPs should govern the numbering methodology.

See Chapter 6 – "The SOP on SOPs."

The Project Name and Number, if applicable, should be indicated on the cover or title page. This information identifies the overall project to which this document belongs.

The Site/Location/Department Identification should be indicated on the cover or title page. It identifies the Site/Location/Department that has responsibility for this document and its associated project.

The Date that the Document was issued, last revised or issued with draft status should be on each page of the document (e.g., 12-Dec-01). This will clearly identify the date on which the document was last modified.

The Effective Date of the Document should be on the title or cover page of the document (e.g., 12-Dec-01). This will clearly identify when to begin using the current version of the document.

Pagination, page of total pages (e.g., Page x of y), should be on each page of the document. This will help identify whether or not all of the pages of the document are present. Pages intentionally left blank should clearly be designated as such.

Approval Section *(SOP)*

The Approval Section should contain a table with the Author and Approvers printed names, titles and a place to sign and date the SOP. The SOP on SOPs should govern who the approvers of the document should be. The functional department (e.g., Production / Quality Assurance / Engineering) of the signatory should be placed below each printed name and title along with a statement associated with each approver indicating the person's role or qualification in the approval process. Indicate the significance of the associated signature (e.g. This document meets the requirements of Corporate Policy 9055, Electronic Records; Electronic Signatures, dated 22-Mar-2000).

The completion (signing) of this section indicates that the contents have been reviewed and approved by the listed individuals.

Table of Contents *(SOP)*

The SOP should contain a Table of Contents, which appears towards the beginning of the document. This Table of Contents provides a way to easily locate various topics contained in the document.

Purpose *(SOP)*

The Purpose section should clearly state that the intention of the document is to provide a Standard Operating Procedure to be used to create a Project Validation Final Report for a validation project of new or upgraded computer or computerized systems.

This introduction to the SOP concisely describes the document's purpose in sufficient detail so that a non-technical reader can understand its purpose.

Scope *(SOP)*

The Scope section identifies the limitations of the Project Validation Final Report SOP. It limits the scope to all GxP systems and is applicable to all computers, computerized systems, infrastructure, LANs and WANs including clinical, laboratory, manufacturing and distribution computer systems. It applies to all Sites, Locations and Departments within the company.

All validation projects must have their outcomes summarized in accordance with this SOP.

Equipment / Materials / Tools

This section lists any required or recommended equipment, materials, tools, templates or guides. Provide the name and a brief description of the equipment, instruments, materials and/or tools to be used in the procedures or testing covered by the Project Validation Final Report SOP. This is for reference purposes and requires the user to utilize the specific item or a designated equivalent. The purpose and use is explained elsewhere in the document. Substitution of comparable equipment may be made in accordance with procedures for Change Control.

If the information is included in a job aid or operating manual, a description of where the information may be found with a specific reference to the document number, title, date and revision number should be specified.

Warnings / Notes / General Information / Safety

Warnings, notes, general or safety information applicable to the Project Validation Final Report SOP are listed in this section.

Policies, Guidelines, References

This section lists references to appropriate regulatory requirements, SOPs, policies, guidelines, or other relevant documents referred to in the Project Validation Final Report SOP. The name and a brief description of applicable related SOPs, corporate policies, procedures and guidelines, Quality Control monographs, books, technical papers, publications, specific regulations, batch records, vendor or system manuals, and all other relevant documents that are required by the Project Validation Final Report SOP should be included. This is for reference purposes and requires the user to use the specific item or a designated equivalent. The purpose and use are expla ined elsewhere in the document.

SOPs cite references and incorporate them in whole by reference rather than specifying details that duplicate these items. This is done so that changes in policy, guidelines or reference materials require no changes to SOPs themselves and the most current version of the referenced material is used. Examples are:

- Policy ABC-123, Administrative Policy for Electronic Records; Electronic Signatures
- Corporate, Plan for Compliance with 21 CFR Part 11, Electronic Records; Electronic Signatures
- SOP-456, Electronic Records; Electronic Signatures
- FDA, 21 CFR Part 11, Electronic Records; Electronic Signatures
- FDA, General Principals of Software Validation; Final Guidance for Industry and FDA Staff, January 11, 2002

Assumptions / Exclusions / Limitations (SOP)

Any and all assumptions, exclusions or limitations within the context of the SOP should be discussed. Specific issues should be directly addressed in this section of the Project Validation Final Report SOP.

Glossary of Terms

A Glossary of Terms should be included that defines all of the terms and acronyms that are used in this document or that are specific to the operation being described and may not be generally known to the reader. The reader may also be referred to the FDA document, "Glossary of Computerized System and Software Development Terminologys."

Roles and Responsibilities

The System Owner is responsible for the preparation of the Project Validation Final Report for each validation project performed and to ensure that it is reviewed and approved by the System Owner, CVG and QA(U).

Procedure

The SOP should contain instructions and procedures that enable the creation of a Project Validation Final Report for a validation project.

The final report is prepared as stand-alone document for all projects. It is not good practice to include it as part of other documents in an attempt to reduce the amount of workload or paperwork. This only causes more work over the length of the project resulting from misunderstandings and confusion. Clear, concise documents that address specific parts of any given project are desirable.

Purpose (report)

This section should clearly state that the purpose of the Project Validation Final Report is to summarize the outcome of a specific GxP validation project. This short introduction should be concise and present details so that an unfamiliar reader can understand the purpose of the document with regard to its use.

Scope (report)

The Scope section identifies the limitations of a specific project validation final report in relationship to validating a specific software application, computer or computerized system. The objective of a final report is to summarize the results of the entire project.

Assumptions / Exclusions / Limitations (report)

Discuss any assumptions, exclusions or limitations within the context of the report. Specific issues should be directly addressed in this section of the Project Validation Final Report.

Overview of the Process

Provide an overview of the business operations for the system including workflow, material flow and the control philosophy of what functions, calculations and processes are to be performed by the system, as well as interfaces that the system or application must be provided. Outline the activities required for acceptance of the entire system.

Participants

List the participants in the procedure by name, title, department or company affiliation and responsibility in a table.

Accomplishments and Impact

Discuss the accomplishments of the project and its overall impact.

Key Learning

List the key items learned from the experience with recommendations, as appropriate.

Rework & Delay Occurrences

Explain any rework that was required. What caused it. Discuss any delays that it may have caused and their impact on the project.

Summary

Summarize the overall work performed and the outcome of the project.

Conclusions

Provide the conclusions drawn from the results of the project. Specify whether the results of the project are acceptable. Does the outcome of the work warrant approval and/or qualification for use of the system? Is it permissible to proceed with the commissioning of the system?

Document Management and Storage

A Document Management and Storage section is included in the Project Validation Final Report SOP. It defines the Document Management and Storage methodology that is followed for issue, revision, replacement, control and management of this SOP and the Project Validation Final Report. This methodology is in accordance with an established Change Control SOP and a Document Management and Storage SOP.

Revision History

A Revision History section is included in the Project Validation Final Report SOP. The Project Validation Final Report SOP Version Number and all Revisions made to the document are recorded in this section. These changes are in accordance with a Change Control Procedure SOP.

The Revision History section consists of a summary of the changes, a brief statement explaining the reasons the document was revised, and the name and affiliation of the person responsible for the entries. It includes the numbers of all items the Project Validation Final Report SOP replaces and is written so that the original Project Validation Final Report SOP is not needed to understand the revisions or the revised document being issued.

When the Project Validation Final Report SOP is a new document, a "000" as the Revision Number is entered with a statement such as "Initial Issue" in the comments section.

Regulatory or audit commitments require preservation from being inadvertently lost or modified in subsequent revisions. All entries or modifications made as a result of regulatory or audit commitments require indication as such in the Revision History and should indicate the organization, date and applicable references. A complete and accurate audit trail must be developed and maintained throughout the project lifecycle.

Attachments

If needed, an Attachments section is included in the Project Validation Final Report SOP in accordance with the SOP on SOPs. An Attachment Index and Attachment section follow the text of the Project Validation Final Report SOP. The top of each attachment page indicates "Attachment # x," where 'x' is the sequential number of each attachment. The second line of each attachment page contains the attachment title. The number of pages and date of issue corresponding to each attachment are listed in the Attachment Index.

Attachments are placed directly after the Attachment Index and are sequentially numbered (e.g., Attachment # 1, Attachment # 2) with each page of an attachment containing the corresponding document number, page numbering, and total number of pages in the attachment (e.g., Page x of y). Each attachment is labeled consistent with the title of the attachment itself.

Attachments are treated as part of the Project Validation Final Report SOP. If an attachment needs to be changed, the entire Project Validation Final Report SOP requires a new revision number and is reissued. New attachments or revisions of existing attachments are in accordance with an established Change Control SOP and a Document Management and Storage SOP.

Chapter 35 – Ongoing Operation

In this chapter we will discuss the Standard Operating Procedure (SOP) for establishing protocols that provide for compliant operation of validated software applications, computer and computerized systems. The Ongoing Operation of each system has specific methods and procedures to govern its day-to-day operation, use, maintenance, and control.

The FDA has determined that in order to maintain the integrity, accuracy, reliability and consistency of operation of computer and computerized systems, the subsequent electronic data and provide a vehicle for the use of electronic signatures that are considered the equivalent of hand written signatures, an Ongoing Operation protocol is needed to govern the day-to-day operation, use, maintenance, and control of software and systems. Systems identified as requiring validation must be validated and maintained in a validated, compliant state in the interest of health and safety.

This document affirms the overall company policy and commitment to validation in support of sustained regulatory compliance. It further defines commitments to equipment, method, software, process, and system sustainable compliance with regulations.

This SOP is used for all systems that create, modify, maintain, retrieve, transmit or archive records in electronic format where validation and compliance with the regulations is required. It is used to produce SOPs that govern the day-to-day operation, use, maintenance, and control of software applications, computer and computerized systems. It applies to all legacy systems, as no system is grandfathered. It is used in conjunction with an overlying corporate policy to provide sustainable compliance with the regulations.

Everyone involved in the operation of a software application, computer or computerized system that controls equipment and/or produces electronic data or records in support of all regulated laboratory, manufacturing, clinical or distribution activities uses this SOP. The System Owner, CVG and QA(U) use it to create and enforce procedures that govern the day-to-day operation, use, maintenance and control of software applications, computer and computerized systems. An auditor or team uses it in evaluating the state of validation of a system and its compliance with current regulation.

Title Block, Headers and Footers

The Ongoing Operation SOP should have a title block, title or cover page along with a header and/or footer on each page of the document that clearly identifies the document. The minimum information that should be displayed for the Ongoing Operation SOP is as follows:

- Document Title
- Document Number
- Document Revision Number
- Project Name and Number
- Site/Location/Department Identification
- Date Document was issued, last revised or draft status
- Effective Date
- Pagination: page of total pages (e.g., Page x of y)

The Document Title should be in large type font that enables clear identification. It should clearly and concisely describe the contents and intent of the document. Since this document will be part of a set or package of documents within a project, it should be consistent with other documents used for the project and specify that it is the Ongoing Operation SOP. This information should be on every page of the document.

The Document Number should be in large type font that enables clear identification. Since the Ongoing Operation SOP will be part of a set or package of documents within a project, it should be consistent with other documents used for the project. This information should appear on the every page of the document.

The Document Revision Number, which identifies the revision of the document, should appear on each page of the document. Many documents will go through one or more changes over time, particularly during the draft stage and the review and approval cycle. It is important to maintain accurate revision control because even a subtle change in a document can have an immense impact on the outcomes of activities. The SOP on SOPs should govern the numbering methodology.

See Chapter 6 – The SOP on SOPs.

The Project Name and Number, if applicable, should be indicated on the cover or title page. This information identifies the overall project to which this document belongs.

The Site/Location/Department Identification should be indicated on the cover or title page. It identifies the Site/Location/Department that has responsibility for this document and its associated project.

The Date that the Document was issued, last revised or issued with draft status should be on each page of the document (e.g., 12-Dec-01). This will clearly identify the date on which the document was last modified.

The Effective Date of the Document should be on the title or cover page of the document (e.g., 12-Dec-01). This will clearly identify when to begin using the current version of the document.

Pagination, page of total pages (e.g., Page x of y), should be on each page of the document. This will help identify whether or not all of the pages of the document are present. Pages intentionally left blank should clearly be designated as such.

Approval Section

The Approval Section should contain a table with the Author and Approvers printed names and titles and a place to sign and date the SOP. This SOP on SOPs governs who the approvers of the document should be. The functional department (e.g., Production / Quality Assurance / Engineering) of the signatory should be placed below each printed name and title along with a statement associated with each approver indicating the person's role or qualification in the approval process. Indicate the significance of the associated signature (e.g. This document meets the requirements of Corporate Policy 9055, Electronic Records; Electronic Signatures, dated 22-Mar-2000).

The completion (signing) of this section indicates that the contents have been reviewed and approved by the listed individuals.

Table of Contents

The SOP should contain a Table of Contents, which appears towards the beginning of the document. This Table of Contents provides a way to easily locate various topics contained in the document.

Purpose

The Purpose section should clearly state that the intention of the document is to provide a Standard Operating Procedure to be used to create an Ongoing Operation protocol for software applications, computer and computerized system or infrastructure maintenance and use. Its purpose is to provide a means to ensure sustainable compliance with company policies, procedures and current regulation. This introduction to the SOP concisely describes the document's purpose in sufficient detail so that a non-technical reader can understand its purpose. For example:

"The purpose of this SOP is to establish a requirement for protocols for compliant operation of validated software applications, computer and computerized systems. The Ongoing Operation of each system must specify the method and procedures used in its day-to-day operation, use, maintenance, and control."

Scope

The Scope section identifies the limitations of the Ongoing Operation SOP. It limits the scope to all GxP systems and is applicable to all computers, computerized systems, infrastructure, LANs and WANs including clinical, laboratory, manufacturing and distribution computer systems. It applies to all Sites, Locations and Departments within the company.

The Ongoing Operation SOP includes electronic records, electronic signatures and handwritten signatures executed to electronic records. It applies to all records that are created, modified, maintained, archived, retrieved or transmitted in electronic form under any requirements set forth in FDA regulations. Provide the scope or range of control and authority of the SOP. State who, what, when, where, why, and how the SOP is to control the processes or procedures specified in the Purpose section. For example:

"This document applies to all computers, computerized systems, infrastructure, LANs, WANs, and SANs including clinical, laboratory, manufacturing, and distribution computer systems that are governed by cGxPs. It establishes the general requirements of developing, updating and maintaining software application, computer and computerized system ongoing operation. It applies to all departments and divisions at all locations of the Example Pharmaceuticals Corporation."

Equipment / Materials / Tools

This section lists any required or recommended equipment, materials, tools, templates or guides. Provide the name and a brief description of the equipment, instruments, materials and/or tools to be used in the procedures or testing covered by the Ongoing Operation SOP. This is for reference purposes and requires the user to utilize the specific item or a designated equivalent. The purpose and use is explained elsewhere in the document. Substitution of comparable equipment may be made in accordance with procedures for Change Control.

If the information is included in a job aid or operating manual, a description of where the information may be found with a specific reference to the document number, title, date and revision number should be specified.

Warnings / Notes / General Information / Safety

Warnings, notes, general or safety information applicable to the Ongoing Operation SOP are listed in this section.

Policies, Guidelines, References

This section lists references to appropriate regulatory requirements, SOPs, policies, guidelines, or other relevant documents referred to in the Ongoing Operation SOP. The name and a brief description of applicable related SOPs, corporate policies, procedures and guidelines, Quality Control monographs, books, technical papers, publications, specific regulations, batch records, vendor or system manuals, and all other relevant documents that are required by the Ongoing Operation SOP should be included. This is for reference purposes and requires the user to use the specific item or a designated equivalent. The purpose and use are explained elsewhere in the document.

SOPs cite references and incorporate them in whole by reference rather than specifying details that duplicate these items. This is done so that changes in policy, guidelines or reference materials require no changes to SOPs themselves and the most current version of the referenced material is used. Examples are:

- Policy ABC-123, Administrative Policy for Electronic Records; Electronic Signatures
- Corporate, Plan for Compliance with 21 CFR Part 11, Electronic Records; Electronic Signatures
- SOP-456, Electronic Records; Electronic Signatures
- FDA, 21 CFR Part 11, Electronic Records; Electronic Signatures
- FDA, General Principals of Software Validation; Final Guidance for Industry and FDA Staff, January 11, 2002

Assumptions / Exclusions / Limitations

Any and all assumptions, exclusions or limitations within the context of the SOP should be discussed. Specific issues should be directly addressed Ongoing Operation SOP. Examples are:

"Legacy systems are not grandfathered under 21 CFR Part 11, there is no limitation with respect to legacy systems."

Glossary of Terms

A Glossary of Terms should be included that defines all of the terms and acronyms that are used in this document or that are specific to the operation being described and may not be generally known to the reader. The reader may also be referred to the FDA document, "Glossary of Computerized System and Software Development Terminology."

Roles and Responsibilities

Validation activities are the result of a cooperative effort involving many people, including the System Owners, End Users, the Computer Validation Group, Subject Matter Experts (SME) and Quality Assurance (Unit).

This section should clearly specify the participants by name, title, department or company affiliation and role in a table.

The Ongoing Operation protocol is the responsibility of the System Owner, CVG and QA(U).

The QA(U) is responsible to verify that the details in the protocol are compliant with the applicable corporate practices and policies. They also assure that the approach meets appropriate cGxP regulations, good business practices and compliance with 21 CFR Part 11.

The qualifications of the individuals involved in the project are kept on file and made available as needed in accordance with cGxP. The qualifications are kept with the project documentation or are available from the Human Resources Department. The location of the qualification documentation is stated in this section.

Procedure *(SOP)*

A key responsibility of the System Owner is the implementation of standard operating procedures that specify and require that all validated systems be maintained in a sustainedly compliant, validated state. All modifications and changes to a system must be documented in accordance with the SDLC methodology. This includes hardware, software, training, and all documentation. Specify the frequency of periodic review and training.

The protocol contains instructions and procedures that specify requirements for Ongoing Operation of a software application, computer or computerized system.

Purpose (protocol)

This section requires clearly stating the Ongoing Operation protocol's purpose with information as to why it is considered necessary to create it. This short introduction should be concise and present details so that an unfamiliar reader can understand the purpose of the document. For example:

> "The purpose of this protocol is to define the requirements for compliant operation of the XYZ computerized system. The Ongoing Operation of the XYZ system must be in accordance with the method and procedures specified herein for its day-to-day operation, use, maintenance, and control."

Scope (protocol)

The Scope section identifies the limitations of a specific Ongoing Operation protocol in relationship to the maintenance and use of the system. For example:

"This document applies to the XYZ computer system used by the tableting manufacturing department of the Example Pharmaceuticals Corporation located in Anytown, US. It establishes the requirements for developing, updating, and maintaining software application, computer and computerized system ongoing operation."

Equipment / Materials / Tools (protocol)

This section lists any required or recommended equipment, materials, tools, templates or guides. Provide the name and a brief description of the equipment, instruments, materials and/or tools to be used in the procedures or testing covered by the Ongoing Operation protocol. This is for reference purposes and requires the user to utilize the specific item or a designated equivalent. The purpose and use is explained elsewhere in the document. Substitution of comparable equipment may be made in accordance with procedures for Change Control.

If the information is included in a job aid or operating manual, a description of where the information may be found with a specific reference to the document number, title, date and revision number should be specified.

Warnings / Notes / General Information / Safety (protocol)

Warnings, notes, general or safety information applicable to the Ongoing Operation protocol are listed in this section.

Policies / Guidelines / References (protocol)

The sustainable compliance, maintenance and operation of the system must comply with specific requirements. This section lists references to appropriate regulatory requirements, SOPs, policies, guidelines, or other relevant documents referred to in the Ongoing Operation protocol. The name and a brief description of applicable related SOPs, corporate policies, procedures and guidelines, Quality Control monographs, books, technical papers, publications, specific regulations, batch records, vendor or system manuals, and all other relevant documents that are required by the Ongoing Operation protocol should be included. This is for reference purposes and requires the user to use the specific item or a designated equivalent. The purpose and use are explained elsewhere in the document.

Assumptions / Exclusions / Limitations (protocol)

Discuss any assumptions, exclusions or limitations within the context of the Software Design Specification Specific issues should be directly addressed in this section of the Ongoing Operation protocol.

Glossary of Terms (protocol)

The Glossary of Terms should define all of the terms and acronyms that are used in the document or that are specific to the operation being described and may not be generally known to the reader. The reader may also be referred to the FDA document, "Glossary of Computerized System and Software Development Terminology" or other specific glossaries that may be available.

Titles, terms, and definitions that are used in the corporate policies, procedures and guidelines should be defined.

Roles and Responsibilities (protocol)

Ongoing Operation activities are the result of a cooperative effort involving many people, including the System Owners, End Users, CVG, Subject Matter Experts (SME) and Quality Assurance (Unit).

This section should clearly specify the participants by name, title, department or company affiliation and role in a table.

The Ongoing Operation protocol is the responsibility of System Owner, CVG and QA(U).

The System Owner is the responsible for ensuring that the system is maintained and operates in accordance with the Ongoing Operation protocol.

The qualifications of the individuals involved are kept on file and made available as needed in accordance with cGxP. The qualifications are kept with the system documentation or are available from the Human Resources Department. The location of the qualification documentation is stated in this section.

Creation and Maintenance

The Ongoing Operation protocol creation and maintenance is the joint responsibility of System Owner, CVG and QA(U).

The Ongoing Operation protocol will go through constant revisions, as it is a living document intended to maintain compliance with cGxP and 21 CFR Part 11. As a routine normal process, it ensures that the activities are in compliance with cGxPs, FDA regulations, company policies, and industry standards and guidelines.

Chapter 35 – Ongoing Operation

The roles and responsibilities for preparing the Ongoing Operation protocol are addressed in accordance with established Standard Operating Procedures.

After initial creation, the System Owners is responsible for putting all changes or modifications through the review and approval process prior to inclusion and implementation. Any modification or addition is considered a trigger for this requirement.

Use

The people using and maintaining the system are required to be familiar with the Ongoing Operation protocol in order to fulfill the regulatory requirements. The QA(U) is required to use the Ongoing Operation protocol in concert with other project documentation as a basis for periodic review and audit of the system.

The Ongoing Operation protocol is the governing document for use and maintenance of software applications, computer and computerized systems, processes, and procedures within the company, site or division.

Procedure (protocol)

Background Information

Provide an overview or outline of the business operations and activities that are performed by the specific site or department that uses the software application, computer or computerized system.

Maintenance and Operation

The system must be maintained and operated in a validated state that is compliant with company policies and procedures, cGxP and 21 CFR Part 11 requirements. This applies to the software, hardware, people, processes and procedures. It means that the system as a whole must work, be maintained and operated as defined and purported, including the people following the written procedures.

Documentation

The System Owner is responsible for the distribution, access and use of documentation for system operation and maintenance. He or she is also responsible for revision and change control procedures to maintain an audit trail that documents time-sequenced development and modification of systems documentation.

Systems documentation explains how a software application, computer or computerized system functions, is used, and maintained. This includes SOPs, administration and user manuals, SDLC documentation, maintenance manuals and logs, periodic validation and qualification reports, documentation storage, and change control procedures. The may be in paper or electronic form. Electronic versions must comply with 21 CFR Part 11 requirements.

Only the current version of any given document is allowed to be in use. This must correspond to the version of the associated system in use. There must be a method to restrict, track, and control who is issued documentation along with the version issued.

All documentation must be retained for the same period of time as the data its system collects based on Predicate Rules.

See Chapter 28 – "Document Management and Storage."

Training

Establish the general requirements of developing, updating, and maintaining employee, Contractor and Vendor training curriculum.

Create, maintain and review training for software applications, computer or computerized systems that are validated/qualified systems. Specify the requirements for the development, curricula, assessment, records, management and control of the documentation and procedures. Document the assessment of training and qualification of the End-Users, support, maintenance, contractors and vendors. Identify the location of training and qualification records.

Training and an assessment of that training indicating knowledge and understanding provides assurance that individuals are qualified to perform the work required.

This applies to all computers, computerized systems, infrastructure, LANs, WANs, and SANs including clinical, laboratory, manufacturing and distribution computer systems that are governed by cGxPs.

See Chapter 33 – "Training."

Resources

Assign Roles and Responsibilities to ensure that the Ongoing Operation of the system is the responsibility of specific individuals who have been trained in their assigned tasks and are qualified to perform their respective roles.

See Chapter 4 – "Roles and Responsibilities."

Periodic Review

Specify the requirements for performing periodic reviews of validated systems to determine the compliance of a system with the business needs of the System Owner in support of a Good Clinical, Laboratory, Manufacturing or Distribution Practices (GxP) and current regulation. This applies to all computers, computerized systems, infrastructure, LANs, WANs and SANs including clinical, laboratory, manufacturing, and distribution computer systems that are governed by cGxPs.

A periodic review of the validation status of validated systems is mandatory. This review assesses whether the system has changed and whether it continues to operate as it was previously validated or qualified. Review of the system and its validation must be documented. Review frequency is discretionary, should be less than a three-year review cycle and should be stated in the validation plan or other official document. However, if not formally stated, all systems should default to a two-year review cycle.

When a computer or computerized system has a failure, undergoes an upgrade or modification to its hardware, software configuration, functionality, use, or it is moved, repaired or replaced either physically or logically, has had changes to the preventive maintenance or calibration program, it must undergo periodic review. Periodic review is also triggered by the elapse of time specified in the system validation/qualification documentation. In the absence of such specification, it should not exceed a two-year period. The System Owner, with the approval of QA(U), may define additional triggers.

In the process of performing the periodic review, consider the compliance of each item or area examined with respect to its compliance with 21 CFR Part 11. Keep in mind that all GxP systems need to comply with this regulation, including legacy systems. If necessary, concurrently conduct a 21 CFR Part 11 Assessment.

Chapter 35 – Ongoing Operation

See Chapter 30 – "Periodic Review."

Document Management and Storage (protocol)

A Document Management and Storage section is included in each Ongoing Operation protocol. It defines the Document Management and Storage methodology that is followed for issue, revision, replacement, control and management of each Ongoing Operation protocol. This methodology is in accordance with an established Change Control SOP and a Document Management and Storage SOP.

Revision History (protocol)

A Revision History section is included in each Ongoing Operation protocol. The Ongoing Operation protocol Version Number and all Revisions made to the document are recorded in this section. These changes are in accordance with a Change Control Procedure SOP.

The Revision History section consists of a summary of the changes, a brief statement explaining the reasons the document was revised, and the name and affiliation of the person responsible for the entries. It includes the numbers of all items the Ongoing Operation protocol replaces and is written so that the original Ongoing Operation protocol is not needed to understand the revisions or the revised document being issued.

When the Ongoing Operation protocol is a new document, a "000" as the Revision Number is entered with a statement such as "Initial Issue" in the comments section.

Regulatory or audit commitments require preservation from being inadvertently lost or modified in subsequent revisions. All entries or modifications made as a result of regulatory or audit commitments require indication as such in the Revision History and should indicate the organization, date and applicable references. A complete and accurate audit trail must be developed and maintained throughout the project lifecycle.

Attachments (protocol)

If needed, an Attachments section is included in each Ongoing Operations protocol in accordance with the SOP on SOPs. An Attachment Index and Attachment section follow the text of an Ongoing Operations protocol. The top of each attachment page indicates "Attachment # x," where 'x' is the sequential number of each attachment. The second line of each attachment page contains the attachment title. The number of pages and date of issue corresponding to each attachment are listed in the Attachment Index.

Attachments are placed directly after the Attachment Index and are sequentially numbered (e.g., Attachment # 1, Attachment # 2) with each page of an attachment containing the corresponding document number, page numbering, and total number of pages in the attachment (e.g., Page x of y). Each attachment is labeled consistent with the title of the attachment itself.

Attachments are treated as part of the Ongoing Operations protocol. If an attachment needs to be changed, the entire Ongoing Operations protocol requires a new revision number and is reissued. New attachments or revisions of existing attachments are in accordance with an established Change Control SOP and a Document Management and Storage SOP.

Document Management and Storage *(SOP)*

A Document Management and Storage section is included in the Ongoing Operation SOP. It defines the Document Management and Storage methodology that is followed for issue, revision, replacement, control and management of this SOP. This methodology is in accordance with an established Change Control SOP and a Document Management and Storage SOP.

Revision History (SOP)

A Revision History section is included in the Ongoing Operation SOP. The Ongoing Operation SOP Version Number and all Revisions made to the document are recorded in this section. These changes are in accordance with a Change Control Procedure SOP.

The Revision History section consists of a summary of the changes, a brief statement explaining the reasons the document was revised, and the name and affiliation of the person responsible for the entries. It includes the numbers of all items the Ongoing Operation SOP replaces and is written so that the original Ongoing Operation SOP is not needed to understand the revisions or the revised document being issued.

When the Ongoing Operation SOP is a new document, a "000" as the Revision Number is entered with a statement such as "Initial Issue" in the comments section.

Regulatory or audit commitments require preservation from being inadvertently lost or modified in subsequent revisions. All entries or modifications made as a result of regulatory or audit commitments require indication as such in the Revision History and should indicate the organization, date and applicable references. A complete and accurate audit trail must be developed and maintained throughout the project lifecycle.

Attachments (SOP)

If needed, an Attachments section is included in the Ongoing Operations SOP in accordance with the SOP on SOPs. An Attachment Index and Attachment section follow the text of the Ongoing Operations SOP. The top of each attachment page indicates "Attachment # x," where 'x' is the sequential number of each attachment. The second line of each attachment page contains the attachment title. The number of pages and date of issue corresponding to each attachment are listed in the Attachment Index.

Attachments are placed directly after the Attachment Index and are sequentially numbered (e.g., Attachment # 1, Attachment # 2) with each page of an attachment containing the corresponding document number, page numbering, and total number of pages in the attachment (e.g., Page x of y). Each attachment is labeled consistent with the title of the attachment itself.

Attachments are treated as part of the Ongoing Operations SOP. If an attachment needs to be changed, the entire Ongoing Operations SOP requires a new revision number and is reissued. New attachments or revisions of existing attachments are in accordance with an established Change Control SOP and a Document Management and Storage SOP.

Chapter 36 – Decommissioning and Retirement

In this chapter we will discuss the Standard Operating Procedure (SOP) for establishing protocols for compliant decommissioning and retirement of validated software applications, computer and computerized systems. The Decommissioning and Retirement of each system has specified the method and procedures used in the termination of its useful life.

The FDA has determined that in order to maintain the integrity, accuracy, reliability and consistency of operation of computer and computerized systems, the subsequent electronic data and provide a vehicle for the use of electronic signatures that are considered the equivalent of hand written signatures, a Decommissioning and Retirement Change Control Document and Project Plan are needed to govern the retirement from operation and use of software and systems. Systems identified as requiring validation must be decommissioned and retired in accordance with a formal procedure in order to maintain a sustainedly compliant state in the interest of health and safety.

This document affirms the overall company policy and commitment to SDLC process in support of sustained regulatory compliance and further defines commitments to equipment, method, software, process, and system sustainable compliance with regulations.

This SOP is used for all systems that create, modify, maintain, retrieve, transmit or archive records in electronic format where validation and compliance with the regulations is required. It is used to produce SOPs that govern the decommissioning and retirement of software applications, computer and computerized systems. It applies to all legacy systems, as no system is grandfathered. It is used in conjunction with an overlying corporate policy to provide sustainable compliance with the regulations.

Everyone involved in the operation of a software application, computer or computerized system that controls equipment and/or produces electronic data or records in support of all regulated laboratory, manufacturing, clinical or distribution activities uses this SOP. The System Owner, CVG and QA(U) use it to create and enforce procedures that govern the decommissioning and retirement of software applications, computer and computerized systems. An auditor or team uses it in evaluating the state of compliance with current regulation prior to decommissioning and retirement.

Title Block, Headers and Footers

The Decommissioning and Retirement SOP should have a title block, title or cover page along with a header and/or footer on each page of the document that clearly identifies the document. The minimum information that should be displayed for the Decommissioning and Retirement SOP is as follows:

- Document Title
- Document Number
- Document Revision Number
- Project Name and Number
- Site/Location/Department Identification
- Date Document was issued, last revised or draft status
- Effective Date
- Pagination: page of total pages (e.g., Page x of y)

The Document Title should be in large type font that enables clear identification. It should clearly and concisely describe the contents and intent of the document. Since this document will be part of a set or package of documents within a project, it should be consistent with other documents used for the project and specify that it is the Decommissioning and Retirement SOP. This information should be on every page of the document.

The Document Number should be in large type font that enables clear identification. Since the Decommissioning and Retirement SOP will be part of a set or package of documents within a project, it should be consistent with other documents used for the project. This information should appear on every page of the document.

The Document Revision Number, which identifies the revision of the document, should appear on each page of the document. Many documents will go through one or more changes over time, particularly during the draft stage and the review and approval cycle. It is important to maintain accurate revision control because even a subtle change in a document can have an immense impact on the outcomes of activities. The SOP on SOPs should govern the numbering methodology.

See Chapter 6 – "The SOP on SOPs."

The Project Name and Number, if applicable, should be indicated on the cover or title page. This information identifies the overall project to which this document belongs.

The Site/Location/Department Identification should be indicated on the cover or title page. It identifies the Site/Location/Department that has responsibility for this document and its associated project.

The Date that the Document was issued, last revised or issued with draft status should be on each page of the document (e.g., 12-Dec-01). This will clearly identify the date on which the document was last modified.

The Effective Date of the Document should be on the title or cover page of the document (e.g., 12-Dec-01). This will clearly identify when to begin using the current version of the document.

Pagination, page of total pages (e.g., Page x of y), should be on each page of the document. This will help identify whether or not all of the pages of the document are present. Pages intentionally left blank should clearly be designated as such.

Approval Section

The Approval Section should contain a table with the Author and Approvers printed names and titles and a place to sign and date the SOP. This SOP on SOPs governs who the approvers of the document should be. The functional department (e.g., Production / Quality Assurance / Engineering) of the signatory should be placed below each printed name and title along with a statement associated with each approver indicating the person's role or qualification in the approval process. Indicate the significance of the associated signature (e.g. This document meets the requirements of Corporate Policy 9055, Electronic Records; Electronic Signatures, dated 22-Mar-2000).

The completion (signing) of this section indicates that the contents have been reviewed and approved by the listed individuals.

Table of Contents

The SOP should contain a Table of Contents, which appears towards the beginning of the document. This Table of Contents provides a way to easily locate various topics contained in the document.

Purpose

The Purpose section should clearly state that the intention of the document is to provide a Standard Operating Procedure used to create a Decommissioning and Retirement protocol for software applications, computer and computerized system or infrastructure maintenance and use. Its purpose is to provide a means to ensure sustainable compliance with company policies, procedures and current regulation. This introduction to the SOP concisely describes the document's purpose in sufficient detail so that a non-technical reader can understand its purpose. For example:

> "The purpose of this SOP is to establish a requirement for compliant decommissioning and retirement of validated software applications, computer and computerized systems. The Decommissioning and Retirement of each system must specify the method and procedures used to decommission a software application or system."

Scope

The Scope section identifies the limitations of the Decommissioning and Retirement SOP. It limits the scope to all GxP systems and is applicable to all computer and computerized systems, infrastructure, LANs, WANs, and SANs including clinical, laboratory, manufacturing and distribution computer systems. It applies to all Sites, Locations and Departments within the company.

The Decommissioning and Retirement SOP includes electronic records, electronic signatures, and handwritten signatures executed to electronic records. It applies to all records that are created, modified, maintained, archived, retrieved, or transmitted in electronic form under any requirements set forth in FDA regulations. Provide the scope or range of control and authority of the SOP. State who, what, when, where, why, and how the SOP is to control the processes or procedures specified in the Purpose section. For example:

> "This document applies to all computer and computerized systems, infrastructure, LANs, WANs, and SANs including clinical, laboratory, manufacturing, and distribution computer systems that are governed by cGxPs. It establishes the general requirements of decommissioning and retirement of a software application, computer or computerized system. It applies to all departments and divisions at all locations of the Example Pharmaceuticals Corporation."

Equipment / Materials / Tools

This section lists any required or recommended equipment, materials, tools, templates or guides. Provide the name and a brief description of the equipment, instruments, materials and/or tools to be used in the procedures or testing covered by the Decommissioning and Retirement SOP. This is for reference purposes and requires the user to utilize the specific item or a designated equivalent. The purpose and use is explained elsewhere in the document. Substitution of comparable equipment may be made in accordance with procedures for Change Control.

If the information is included in a job aid or operating manual, a description of where the information may be found with a specific reference to the document number, title, date and revision number should be specified.

Warnings / Notes / General Information / Safety

Warnings, notes, general or safety information applicable to the Decommissioning and Retirement SOP are listed in this section.

Policies, Guidelines, References

This section lists references to appropriate regulatory requirements, SOPs, policies, guidelines, or other relevant documents referred to in the Decommissioning and Retirement SOP. The name and a brief description of applicable related SOPs, corporate policies, procedures and guidelines, Quality Control monographs, books, technical papers, publications, specific regulations, batch records, vendor or system manuals, and all other relevant documents that are required by the Decommissioning and Retirement SOP should be included. This is for reference purposes and requires the user to use the specific item or a designated equivalent. The purpose and use are explained elsewhere in the document.

SOPs cite references and incorporate them in whole by reference rather than specifying details that duplicate these items. This is done so that changes in policy, guidelines or reference materials require no changes to SOPs themselves and the most current version of the referenced material is used. Examples are:

- Policy ABC-123, Administrative Policy for Electronic Records; Electronic Signatures
- Corporate, Plan for Compliance with 21 CFR Part 11, Electronic Records; Electronic Signatures
- SOP-456, Electronic Records; Electronic Signatures
- FDA, 21 CFR Part 11, Electronic Records; Electronic Signatures
- FDA, General Principals of Software Validation; Final Guidance for Industry and FDA Staff, January 11, 2002

Assumptions / Exclusions / Limitations

Any and all assumptions, exclusions or limitations within the context of the SOP should be discussed. Specific issues should be directly addressed in this section of the Decommissioning and Retirement SOP. Examples are:

"Legacy systems are not grandfathered under 21 CFR Part 11, there is no limitation with respect to legacy systems."

Glossary of Terms

A Glossary of Terms should be included that defines all of the terms and acronyms that are used in this document or that are specific to the operation being described and may not be generally known to the reader. The reader may also be referred to the FDA document, "Glossary of Computerized System and Software Development Terminology."

Roles and Responsibilities

Validation activities are the result of a cooperative effort involving many people, including the System Owners, End Users, the Computer Validation Group, Subject Matter Experts (SME) and Quality Assurance (Unit).

This section should clearly specify the participants by name, title, department or company affiliation and role in a table.

The Decommissioning and Retirement protocol is the responsibility of the System Owner, CVG and QA(U).

The QA(U) is responsible to verify that the details in the Documentation produced are compliant with the applicable corporate practices and policies. They also assure that the approach meets appropriate cGxP regulations, good business practices, and compliance with 21 CFR Part 11 and Predicate Rules.

The qualifications of the individuals involved in the project are kept on file and made available as needed in accordance with cGxP. The qualifications are kept with the project documentation or are available from the Human Resources Department. The location of the qualification documentation is stated in this section.

Procedure *(SOP)*

A key responsibility of the System Owner is the implementation of SOPs that specify and require that all validated systems be maintained in a sustainedly compliant, validated state. All modifications and changes to a system must be documented in accordance with the SDLC methodology. This includes hardware, software, training, and all documentation.

ISO/IEC 122207 defines retirement as " Withdrawal of active support by the operation and maintenance organization, partial or total replacement by a new system, or installation of an upgraded system."

The documentation contains instructions and procedures that specify requirements for Decommissioning and Retirement of a software application, computer or computerized system.

When a computer system is no longer functionally useful, the system is decommissioned and retired. This process follows an approved plan prior to initiating the shutdown process. Validation documentation and a backup copy of the computer system must be retained when a computer system or any software version is decommissioned. These items are kept for an indefinite period of time, unless otherwise required by Predicate Rules or determined on advice from counsel, but under no circumstances, less than 5 years. In addition, data files must be archived and retained in accordance with appropriate procedures based on the data and records that were generated and Predicate Rules for records retention.

In both the Change Control document and Project Plan, provide an overview or outline of the business operations and activities that are performed by the specific software application, computer or computerized system being decommissioned and retired along with information about the site or department that uses it. Indicate why the system is no longer needed.

The steps required for the decommissioning process depend on the composition and complexity of the system being retired.

Change Control

The software application, computer or computerized system is decommissioned and retired at the request of the System Owner. The process is initiated in accordance with formal Change Control procedures.

See Chapter 27 – "Change Control."

Project Plan

When a computer system is to be decommissioned, a project plan is created to document the requirements for decommissioning the system. The plan includes how the data is to be archived and/or transferred to another system.

It is important to carefully plan the decommissioning of a system so that the data are preserved in a useable fashion for future use. If product is involved, the data and the system may need to be reconstructed to its original final state for FDA audit purposes.

A decommissioning and retirement plan in both narrative and chart format is created to remove a system from active operation. It contains the work breakdown structures to be followed in the decommissioning and retirement process.

See Chapter 7 – "Project Plan."

Steps to include in the project plan are:

- Advance Notice

Provide advance notice of decommissioning and retirement by means of a Letter of Notification or System Proposal to everyone involved in the operation, use, and maintenance of the software application or system. Notifications should include the following:

- Statement of why the software application or system is no longer needed
- Description of the replacement or upgrade, if any, and its date of availability
- Options for alternate functionality, operation, or use once the application or system has been retired

- Validation Status Review

Review the validation documentation to ensure that it is complete before the system is shutdown. Perform a gap analysis to help identify and remediate deficiencies.

See Chapter 38 – "Gap Analysis and Retrospective Validation."

- Parallel Operation

Plan the decommissioning process to ensure overlap with any replacement system, which must already be operating in its production environment. When a system is being superseded or upgraded, parallel operations of the retiring and the new system are conducted for smooth transition to the new system. During this period, user training is provided.

- Shutdown Procedure

Provide detailed procedures for gracefully shutting down the software application and/or system. Include the actual steps to be performed as a checklist. Where possible, use existing procedures from manuals, SOPs, and protocols specific to the system. It is important to have and follow a specific procedure to show compliance with validation activities to ensure data integrity.

- Equipment

Specify the disposition of any equipment (e.g.- labeling equipment, HPLC, Mass Spectrometer) associated with the system. If it is to be stored for the duration of the retention period, indicate how to store it, where to store it, and who is responsible. Provide a method to restore it to operational use in the event the system needs to be reconstructed.

- Hardware

Specify the disposition of any hardware (e.g.- scanners, printers, computers, cabling) associated with the system. If it is to be stored for the duration of the retention period, indicate how to store it, where to store it, and who is responsible. Provide a method to restore it to operational use in the event the system needs to be reconstructed.

- Software

Specify the disposition of any software applications, operating systems, support programs, and drivers associated with the system. If it is to be stored for the duration of the retention period, indicate how to store it, where to store it, and who is responsible. Provide a method to restore it to operational use in the event the system needs to be reconstructed.

- Electronic Records

Electronic records, including such items as meta data, security, audit trails, and associated electronic signatures, must be archived and retained in accordance with appropriate procedures based the original system requirements, 21 CFR Part 11, and Predicate Rules for records retention. All data must be archived electronically.

A means must be facilitated that enables future access and use of the archived data based on the original rules of use. It must be available to the FDA in both electronic and human readable format for the entire retention period. All security and integrity issues surrounding the original method of storage and access must be maintained. This includes data archived as a result of an upgrade or migration.

Continued on-line access to data may be achieved by data migration to a replacement system. This should be treated as a development project in its own right using full SDLC methodology.

Archived data must not be able to be modified.

See Chapter 28 – "Document Management and Storage."

- Documentation

All associated documentation such as manuals, logs, code, protocols, and SOPs are archived in accordance with a project plan that has been prepared in accordance with a Document Management and Storage SOP, 21 CFR Part 11, and Predicate Rule requirements.

All documentation must be available to the FDA for the entire retention period. If the documentation is in electronic form, then all of the rules applicable to Electronic Records apply and the documentation must be available in both electronic and human readable format. All security and integrity issues surrounding the original method of storage and access must be maintained.

See Chapter 28 – "Document Management and Storage."

- Support and Maintenance Termination

Issue formal notices to terminate the following:

- Licenses
- Service Level Agreements
- Support Services
- Contracts
- Updates
- Scheduled Services
- Preventive Maintenance
- Archival Processes
- Equipment Calibration
- Training
- Periodic Reviews
- Audits

Notice of Implementation

All concerned parties are notified when it is time to implement the process. Provide a summary of the scheduled activities contained in the project plan or the project plan itself. Allow sufficient time for problematic responses to be addressed.

Implementation

Following a reasonable waiting period after the Notice of Implementation has been issued, proceed with the decommissioning process. Follow all procedures as specified. Deviations must be handled in accordance with a Change Control SOP.

See Chapter 27 – "Change Control."

Final Report

A Decommission Report is issued upon the decommissioning and retirement of the software application or system.

See Chapter 25 – "Summary Reports."

Document Management and Storage

A Document Management and Storage section is included in the Decommissioning and Retirement SOP. It defines the Document Management and Storage methodology that is followed for issue, revision, replacement, control and management of this SOP and the documents it governs. This methodology is in accordance with an established Change Control SOP and a Document Management and Storage SOP.

Revision History

A Revision History section is included in the Decommissioning and Retirement SOP. The Decommissioning and Retirement SOP Version Number and all Revisions made to the document are recorded in this section. These changes are in accordance with a Change Control Procedure SOP.

The Revision History section consists of a summary of the changes, a brief statement explaining the reasons the document was revised, and the name and affiliation of the person responsible for the entries. It includes the numbers of all items the Decommissioning and Retirement SOP replaces and is written so that the original Decommissioning and Retirement SOP is not needed to understand the revisions or the revised document being issued.

When the Decommissioning and Retirement SOP is a new document, a "000" as the Revision Number is entered with a statement such as "Initial Issue" in the comments section.

Regulatory or audit commitments require preservation from being inadvertently lost or modified in subsequent revisions. All entries or modifications made as a result of regulatory or audit commitments require indication as such in the Revision History and should indicate the organization, date and applicable references. A complete and accurate audit trail must be developed and maintained throughout the project lifecycle.

Attachments

If needed, an Attachments section is included in the Decommissioning and Retirement SOP in accordance with the SOP on SOPs. An Attachment Index and Attachment section follow the text of the Decommissioning and Retirement SOP. The top of each attachment page indicates "Attachment # x," where 'x' is the sequential number of each attachment. The second line of each attachment page contains the attachment title. The number of pages and date of issue corresponding to each attachment are listed in the Attachment Index.

Attachments are placed directly after the Attachment Index and are sequentially numbered (e.g., Attachment # 1, Attachment # 2) with each page of an attachment containing the corresponding document number, page numbering, and total number of pages in the attachment (e.g., Page x of y). Each attachment is labeled consistent with the title of the attachment itself.

Attachments are treated as part of the Decommissioning and Retirement SOP. If an attachment needs to be changed, the entire Decommissioning and Retirement SOP requires a new revision number and is reissued. New attachments or revisions of existing attachments are in accordance with an established Change Control SOP and a Document Management and Storage SOP.

Chapter 37 – Validation Master Plan

In this chapter we will discuss the Standard Operating Procedure (SOP) for cataloging and controlling hardware, software and systems, especially those required to be maintained in a validated state. It identifies what is to be validated, where it is located, who is responsible for its validation and when it is scheduled for validation, re-validation, or periodic review independent of periodic review triggers. The Validation Master Plan may be separated into multiple plans based on areas of responsibility. For example, the IT/IS Unit may be responsible for software, patches and drivers whereas Computer Operations may be responsible for hardware and environmental items in a data center. A support group would cover desktops and laptops. However, one person or group is responsible for the overall coordination of the master plans so that nothing is overlooked.

The FDA has determined that in order to maintain the integrity, accuracy, reliability and consistency of operation of computer and computerized systems, the subsequent electronic data and provide a vehicle for the use of electronic signatures that are considered the equivalent of hand written signatures, a Validation Master Plan is needed to identify and control hardware, software, and systems. Systems identified as requiring validation must be validated and maintained in a validated and sustainedly compliant state in the interest of health and safety.

This document affirms the overall company policy and commitment to validation in support of sustained regulatory compliance and further defines commitments to equipment, method, software, process, and system validation.

This SOP is used for all systems that create, modify, maintain, retrieve, transmit, or archive records in electronic format where validation and compliance with the regulations is required. It applies to all legacy systems, as no system is grandfathered. It is used in conjunction with an overlying corporate policy to provide sustainable compliance with the regulations.

Everyone involved in the operation of a software application, computer or computerized system that controls equipment and/or produces electronic data or records in support of all regulated laboratory, manufacturing, clinical or distribution activities uses this SOP. An auditor or team uses it in evaluating the state of validation of a system and its compliance with current regulation.

Title Block, Headers and Footers

The Validation Master Plan SOP should have a title block, title or cover page along with a header and/or footer on each page of the document that clearly identifies the document. The minimum information that should be displayed for the Validation Master Plan SOP is as follows:

- Document Title
- Document Number
- Document Revision Number
- Project Name and Number
- Site/Location/Department Identification
- Date Document was issued, last revised or draft status
- Effective Date
- Pagination: page of total pages (e.g., Page x of y)

The Document Title should be in large type font that enables clear identification. It should clearly and concisely describe the contents and intent of the document. Since this document will be part of a set or package of documents within a project, it should be consistent with other documents used for the project and specify that it is the Validation Master Plan SOP. This information should be on every page of the document.

The Document Number should be in large type font that enables clear identification. Since the Validation Master Plan SOP will be part of a set or package of documents within a project, it should be consistent with other documents used for the project. This information should appear on every page of the document.

The Document Revision Number, which identifies the revision of the document, should appear on each page of the document. Many documents will go through one or more changes over time, particularly during the draft stage and the review and approval cycle. It is important to maintain accurate revision control because even a subtle change in a document can have an immense impact on the outcomes of activities. The SOP on SOPs should govern the numbering methodology.

See Chapter 6 – "The SOP on SOPs."

The Project Name and Number, if applicable, should be indicated on the cover or title page. This information identifies the overall project to which this document belongs.

The Site/Location/Department Identification should be indicated on the cover or title page. It identifies the Site/Location/Department that has responsibility for this document and its associated project.

The Date that the Document was issued, last revised or issued with draft status should be on each page of the document (e.g., 12-Dec-01). This will clearly identify the date on which the document was last modified.

The Effective Date of the Document should be on the title or cover page of the document (e.g., 12-Dec-01). This will clearly identify when to begin using the current version of the document.

Pagination, page of total pages (e.g., Page x of y), should be on each page of the document. This will help identify whether or not all of the pages of the document are present. Pages intentionally left blank should clearly be designated as such.

Approval Section

The Approval Section should contain a table with the Author and Approvers printed names and titles and a place to sign and date the SOP. This SOP on SOPs should govern who the approvers of the document should be. The functional department (e.g., Production / Quality Assurance / Engineering) of the signatory should be placed below each printed name and title along with a statement associated with each approver indicating the person's role or qualification in the approval process. Indicate the significance of the associated signature (e.g. This document meets the requirements of Corporate Policy 9055, Electronic Records; Electronic Signatures, dated 22-Mar-2000).

The completion (signing) of this section indicates that the contents have been reviewed and approved by the listed individuals.

Table of Contents

The SOP should contain a Table of Contents, which appears towards the beginning of the document. This Table of Contents provides a way to easily locate various topics contained in the document.

Purpose

The Purpose section should clearly state that the intention of the document is to provide a Standard Operating Procedure to be used to assess the state of validation and compliance with current regulation of all software applications, computer or computerized systems, or infrastructure and to bring them into compliance with current regulation. This introduction to the SOP concisely describes the document's purpose in sufficient detail so that a non-technical reader can understand its purpose. For example:

> "The purpose of this SOP is to catalog and control all hardware, software, and systems, especially those required to be maintained in a validated state. It is used to identify what is to be validated, where it is located, who is responsible for its validation and when it is scheduled for validation, re-validation or periodic review independent of periodic review triggers. It lists all systems so that documented proof is provided that no system is overlooked.
>
> This document affirms the overall company policy and commitment to validation in support of sustained regulatory compliance and further defines commitments to equipment, method, software, process and system validation."

Scope

The Scope section identifies the limitations of the Validation Master Plan SOP. It limits the scope to all GxP systems and is applicable to all computers, computerized systems, infrastructure, LANs, WANs, and SANs including clinical, laboratory, manufacturing and distribution computer systems. It applies to all Sites, Locations and Departments within the company.

The Validation Master Plan SOP includes electronic records, electronic signatures, and handwritten signatures executed to electronic records. It applies to all records that are created, modified, maintained, archived, retrieved or transmitted in electronic form under any requirements set forth in FDA regulations. Provide the scope or range of control and authority of the SOP. State who, what, where, when, why and how the SOP is to control the processes or procedures specified in the Purpose section.

Equipment / Materials / Tools

This section lists any required or recommended equipment, materials, tools, templates or guides. Provide the name and a brief description of the equipment, instruments, materials and/or tools to be used in the procedures or testing covered by the Validation Master Plan SOP. This is for reference purposes and requires the user to utilize the specific item or a designated equivalent. The purpose and use is explained elsewhere in the document. Substitution of comparable equipment may be made in accordance with procedures for Change Control.

If the information is included in a job aid or operating manual, a description of where the information may be found with a specific reference to the document number, title, date and revision number should be specified.

Warnings / Notes / General Information / Safety

Warnings, notes, general or safety information applicable to the Validation Master Plan SOP are listed in this section.

Policies, Guidelines, References

This section lists references to appropriate regulatory requirements, SOPs, policies, guidelines, or other relevant documents referred to in the Validation Master Plan SOP. The name and a brief description of applicable related SOPs, corporate policies, procedures and guidelines, Quality Control monographs, books, technical papers, publications, specific regulations, batch records, vendor or system manuals, and all other relevant documents that are required by the Validation Master Plan SOP should be included. This is for reference purposes and requires the user to use the specific item or a designated equivalent. The purpose and use are explained elsewhere in the document.

SOPs cite references and incorporate them in whole by reference rather than specifying details that duplicate these items. This is done so that changes in policy, guidelines or reference materials require no changes to SOPs themselves and the most current version of the referenced material is used. Examp les are:

- Policy ABC-123, Administrative Policy for Electronic Records; Electronic Signatures
- Corporate, Plan for Compliance with 21 CFR Part 11, Electronic Records; Electronic Signatures
- SOP-456, Electronic Records; Electronic Signatures
- FDA, 21 CFR Part 11, Electronic Records; Electronic Signatures
- FDA, General Principals of Software Validation; Final Guidance for Industry and FDA Staff, January 11, 2002

Assumptions / Exclusions / Limitations

Any and all assumptions, exclusions or limitations within the context of the SOP should be discussed. Specific issues should be directly addressed in this section of the Validation Master Plan SOP. For example:

> "Legacy systems are not grandfathered under 21 CFR Part 11, there is no limitation with respect to legacy systems."

Glossary of Terms

A Glossary of Terms should be included that defines all of the terms and acronyms that are used in this document or that are specific to the operation being described and may not be generally known to the reader. The reader may also be referred to the FDA document, "Glossary of Computerized System and Software Development Terminology."

Roles and Responsibilities

Validation activities are the result of a cooperative effort involving many people, including the System Owners, End Users, the Computer Validation Group, Subject Matter Experts (SME) and Quality Assurance (Unit).

This section should clearly specify the participants by name, title, department or company affiliation and role in a table. The information in this section should conform to the corresponding information in the Project Plan, if applicable.

The Validation Master Plan (VMP) protocol is the responsibility of VMP Manager, CVG, and QA(U).

The cataloging and assessment of the hardware, software, and systems in the entire company to produce a VMP is, at a minimum, the joint effort of the VMP Manager, each System Owner, and QA(U), who is the final signatory of both the protocol and the plan.

The QA(U) is responsible to verify that the details in the protocol and plan are compliant with the applicable corporate practices and policies. They also assure that the approach meets appropriate cGxP regulations, good business practices, and compliance with 21 CFR Part 11.

The qualifications of the individuals involved in the project are kept on file and made available as needed in accordance with cGxP. The qualifications are kept with the project documentation or are available from the Human Resources Department. The location of the qualification documentation is stated in this section.

Procedure *(SOP)*

The protocol contains instructions and procedures that require the cataloging and control of all hardware, software, and systems.

Create a separate project plan in both narrative and chart format that contains work breakdown structures for collecting information about each application, computer or computerized system. Specify what is to be done and by whom. Include a timetable and a list of resources that are needed for the fulfillment of the project. This information will become part of the Validation Master Plan with the details of the information collected.

Establish a quality plan that defines the quality practices, resources, and activities relevant to software applications, computer or computerized systems. Establish how the requirements for quality will be met and that may be encountered as part of their job functions.

This plan establishes performance criteria and identification of items that are subject to validation as well as those that are not. Every validation activity, such as routine validation or validation of a new process or procedure, are included in the Validation Master Plan.

- Clearly define the validation method for GxP systems
- Identify the background, facility, process, boundaries and critical systems
- Specify the minimum requirements for assuring a sustainable validated state
- Define support systems and procedures to assure sustained compliance throughout the systems validation lifecycle.

See Chapter 7 – “Project Plan.”

Purpose (protocol)

This section requires clearly stating the Validation Master Plan purpose with information as to why it is considered necessary. This short introduction should be concise and present details so that an unfamiliar reader can understand the purpose of the document. For example:

> “The purpose of the Validation Master Plan SOP is to specify the method for cataloging and controlling all hardware, software and systems, especially those required to be maintained in a validated state and be GxP compliant. The plan identifies what is to be validated, where it is located, who is responsible for its validation, and when it is scheduled for validation, re-validation or periodic review independent of periodic review triggers.”

Scope (protocol)

The Scope section identifies the limitations of the Validation Master Plan in relationship to the items being cataloged. For example:

> “The scope of this document is limited to software applications, computer and computerized systems and any hardware and equipment associated with them at the Example Pharmaceuticals Corporation site located at Anywhere, US. This document describes the methods used to catalog and control hardware, software and systems, especially those required being maintained in a validated state.”

Equipment / Materials / Tools (protocol)

This section lists any required or recommended equipment, materials, tools, templates or guides. Provide the name and a brief description of the equipment, instruments, materials and/or tools to be used in the procedures or testing covered by the Validation Master Plan. This is for reference purposes and requires the user to utilize the specific item or a designated equivalent. The purpose and use is explained elsewhere in the document. Substitution of comparable equipment may be made in accordance with procedures for Change Control.

If the information is included in a job aid or operating manual, a description of where the information may be found with a specific reference to the document number, title, date and revision number should be specified.

Warnings / Notes / General Information / Safety (protocol)

Warnings, notes, general or safety information applicable to the Validation Master Plan are listed in this section.

Policies / Guidelines / References (protocol)

This section lists references to appropriate regulatory requirements, SOPs, policies, guidelines, or other relevant documents referred to in the Validation Master Plan. The name and a brief description of applicable related SOPs, corporate policies, procedures and guidelines, Quality Control monographs, books, technical papers, publications, specific regulations, batch records, vendor or system manuals, and all other relevant documents that are required by the Validation Master Plan should be included. This is for reference purposes and requires the user to use the specific item or a designated equivalent. The purpose and use are explained elsewhere in the document.

Assumptions / Exclusions / Limitations (protocol)

Discuss any assumptions, exclusions or limitations within the context of the Software Design Specification Specific issues should be directly addressed in this section of the Validation Master Plan.

If part of a system being cataloged consists of a Commercial Off The Shelf software application and another part consists of manufacturing or laboratory equipment, then the part of the computerized system that is equipment specific may be excluded by stating that it is being addressed elsewhere. For example,

> "Equipment covered by other inventorying and cataloging plans is specifically excluded providing the location of the information is shown."

Glossary of Terms (protocol)

The Glossary of Terms should define all of the terms and acronyms that are used in the document or that are specific to the operation being described and may not be generally known to the reader. The reader may also be referred to the FDA document, "Glossary of Computerized System and Software Development Terminology" or other specific glossaries that may be available.

Titles, terms, and definitions that are used in the corporate policies, procedures and guidelines should be defined.

Roles and Responsibilities (protocol)

Validation activities are the result of a cooperative effort involving many people, including the System Owners, End Users, the Computer Validation Group, Subject Matter Experts (SME) and Quality Assurance (Unit).

This section should clearly specify the participants by name, title, department or company affiliation and role in a table. The information in this section should conform to the corresponding information in the Project Plan, if applicable.

The Validation Master Plan (VMP) protocol is the responsibility of VMP Manager, CVG, and QA(U) along with a designated Project Team.

The cataloging and assessment of the hardware, software, equipment, and systems in the entire company to produce a VMP is, at a minimum, the joint effort of the VMP Manager, each System Owner, and QA(U), who is the final signatory of both the protocol and the plan.

The QA(U) is responsible to verify that the details in the protocol and plan are compliant with the applicable corporate practices and policies. They also assure that the approach meets appropriate cGxP regulations, good business practices and compliance with 21 CFR Part 11.

The qualifications of the individuals involved in the project are kept on file and made available as needed in accordance with cGxP. The qualifications are kept with the project documentation or are available from the Human Resources Department. The location of the qualification documentation is stated in this section.

Creation and Maintenance

The Validation Master Plan is the joint responsibility of VMP Manager, CVG and QA(U).

The Validation Master Plan will go through constant revisions, as it is a living document intended to catalog, track and control systems requiring compliance with cGxP and 21 CFR Part 11 along with those are not. As a routine process, it ensures that the validation activities are in compliance with cGxPs, FDA regulations, company policies and industry standards and guidelines.

The roles and responsibilities for preparing the Validation Master Plan protocol and catalog documentation are addressed in accordance with established Standard Operating Procedures.

After the initial cataloging has been complied, the VMP Manager and System Owners are responsible for recording system modifications as well as new computer systems prior to acquisition and/or development in the Validation Master Plan catalog. Any modification or addition is considered a trigger for this requirement.

Use

The members of the Project Team must be familiar with the Validation Master Plan in order to fulfill the project requirements. The QA(U) and Project Team use the Validation Master Plan in concert with other project documentation as a contract that details the project and expectations.

The Validation Master Plan is the governing document for identification and tracking of applications, computer and computerized systems within the company, site or division.

Procedure (protocol)

Background Information

This section of the VMP provides an overview of the business operations for the company or division, including organizational structure, and the policies regarding GxP and 21 CFR Part 11 compliance. It outlines the activities that are performed at the company overall or the specific site preparing the VMP.

Cataloger Identification

This section provides identification of each person participating in the cataloging process. The following information is entered in a table for identification purposes:

- Full Name – printed
- Signature
- Initials
- Affiliation

Facilities

Specify the location of and describe the physical facilities to be surveyed.

Catalog

Collect and catalog the following information:

- System / Application
- Description
- Plant / Process / Product Description
- Corporate System
- GxP
- Criticality
- System Owner
- IS Responsibility
- IS/IT Manager Contact
- System Administrator
- Validation Status
- Validation Date
- Periodic Review Frequency
- Next Periodic Review Date
- Server
- Client Computers
- Specific Considerations
- Run Frequency
- Backup Type
- Backup Frequency
- Recovery Priority
- Recovery Procedures Documented
- Platform
- OS
- Input Dependencies
- Output Dependencies
- Acceptance Criteria
- Procedural Controls
- 21 CFR Part 11 Assessment

Indicate whether the following is required with a summary of the outcomes, if performed:

- 21 CFR Part 11 assessment
- Gap Analysis

If performed, were the responses used to create an upgrade plan with specified actions and timelines to bring each system into compliance? Is there a validation plan? Identify the Summary Report documents.

System Controls

In accordance with 21 CFR Part 11 requirements, identify and indicate the state of the following, which are intertwined:

- Accurate Records
- Protection of the records and system
- Limited Access to the records and system
- Audit Trails of all activities where records are involved
- Operational, Authority & Device Checks
- Education, Training, & Experience to qualify the people using the system
- Accountability
- Documentation Controls
- Revision & Change Control
- Authenticity, Integrity, Confidentiality in Open Systems
- Comments

Deliverable

The Validation Master Plan and its associated catalog containing the above collected information is the deliverable. It is the outcome of the work from the Project Plan and the Validation Master Plan protocol. From this, the various projects and their priorities are identified, created, tracked, controlled, and carried out.

Document Management and Storage (protocol)

A Document Management and Storage section is included in each Validation Master Plan. It defines the Document Management and Storage methodology that is followed for issue, revision, replacement, control and management of each Validation Master Plan. This methodology is in accordance with an established Change Control SOP and a Document Management and Storage SOP.

Revision History (protocol)

A Revision History section is included in each Validation Master Plan. The Validation Master Plan Version Number and all Revisions made to the document are recorded in this section. These changes are in accordance with a Change Control Procedure SOP.

The Revision History section consists of a summary of the changes, a brief statement explaining the reasons the document was revised, and the name and affiliation of the person responsible for the entries. It includes the numbers of all items the Validation Master Plan replaces and is written so that the original Validation Master Plan is not needed to understand the revisions or the revised document being issued.

When the Validation Master Plan is a new document, a "000" as the Revision Number is entered with a statement such as "Initial Issue" in the comments section.

Regulatory or audit commitments require preservation from being inadvertently lost or modified in subsequent revisions. All entries or modifications made as a result of regulatory or audit commitments require indication as such in the Revision History and should indicate the organization, date and applicable references. A complete and accurate audit trail must be developed and maintained throughout the project lifecycle.

Attachments (protocol)

If needed, an Attachments section is included in each Validation Master Plan in accordance with the SOP on SOPs. An Attachment Index and Attachment section follow the text of a Validation Master Plan. The top of each attachment page indicates "Attachment # x," where 'x' is the sequential number of each attachment. The second line of each attachment page contains the attachment title. The number of pages and date of issue corresponding to each attachment are listed in the Attachment Index.

Attachments are placed directly after the Attachment Index and are sequentially numbered (e.g., Attachment # 1, Attachment # 2) with each page of an attachment containing the corresponding document number, page numbering, and total number of pages in the attachment (e.g., Page x of y). Each attachment is labeled consistent with the title of the attachment itself.

Attachments are treated as part of the Validation Master Plan. If an attachment needs to be changed, the entire Validation Master Plan requires a new revision number and is reissued. New attachments or revisions of existing attachments are in accordance with an established Change Control SOP and a Document Management and Storage SOP.

Document Management and Storage *(SOP)*

A Document Management and Storage section is included in the Validation Master Plan SOP. It defines the Document Management and Storage methodology that is followed for issue, revision, replacement, control and management of this SOP. This methodology is in accordance with an established Change Control SOP and a Document Management and Storage SOP.

Revision History *(SOP)*

A Revision History section is included in the Validation Master Plan SOP. The Validation Master Plan SOP Version Number and all Revisions made to the document are recorded in this section. These changes are in accordance with a Change Control Procedure SOP.

The Revision History section consists of a summary of the changes, a brief statement explaining the reasons the document was revised, and the name and affiliation of the person responsible for the entries. It includes the numbers of all items the Validation Master Plan SOP replaces and is written so that the original Validation Master Plan SOP is not needed to understand the revisions or the revised document being issued.

When the Validation Master Plan SOP is a new document, a "000" as the Revision Number is entered with a statement such as "Initial Issue" in the comments section.

Regulatory or audit commitments require preservation from being inadvertently lost or modified in subsequent revisions. All entries or modifications made as a result of regulatory or audit commitments require indication as such in the Revision History and should indicate the organization, date and applicable references. A complete and accurate audit trail must be developed and maintained throughout the project lifecycle.

Attachments *(SOP)*

If needed, an Attachments section is included in the Validation Master Plan SOP in accordance with the SOP on SOPs. An Attachment Index and Attachment section follow the text of the Validation Master Plan SOP. The top of each attachment page indicates "Attachment # x," where 'x' is the sequential number of each attachment. The second line of each attachment page contains the attachment title. The number of pages and date of issue corresponding to each attachment are listed in the Attachment Index.

Attachments are placed directly after the Attachment Index and are sequentially numbered (e.g., Attachment # 1, Attachment # 2) with each page of an attachment containing the corresponding document number, page numbering, and total number of pages in the attachment (e.g., Page x of y). Each attachment is labeled consistent with the title of the attachment itself.

Attachments are treated as part of the Validation Master Plan SOP. If an attachment needs to be changed, the entire Validation Master Plan SOP requires a new revision number and is reissued. New attachments or revisions of existing attachments are in accordance with an established Change Control SOP and a Document Management and Storage SOP.

Chapter 38 – Gap Analysis and Retrospective Validation

In this chapter we will discuss the Standard Operating Procedure (SOP) for determining the state of validation of infrastructure, a software application, computer or computerized system, and for specifying the requirements to bring it into a validated state when an audit indicates deficiencies.

The FDA has determined that in order to maintain the integrity, accuracy, reliability and consistency of operation of computer and computerized systems, the subsequent electronic data, and provide a vehicle for the use of electronic signatures that are considered the equivalent of hand written signatures, all systems must be validated and maintained in a validated and sustainedly compliant state in the interest of health and safety.

This SOP is used for all systems that create, modify, maintain, retrieve, transmit, or archive records in electronic format where validation and compliance with the regulations is uncertain. It applies to all legacy systems, as no system is grandfathered. It is used in conjunction with an overlying corporate policy to provide sustainable compliance with the regulations.

Everyone involved in the operation of a software application, computer or computerized system that controls equipment and/or produces electronic data or records in support of all regulated laboratory, manufacturing, clinical or distribution activities uses this SOP. An auditor or team uses it in evaluating the state of validation of a system and its compliance with current regulation.

Title Block, Headers and Footers

The Gap Analysis and Retrospective Validation SOP should have a title block, title or cover page along with a header and/or footer on each page of the document that clearly identifies the document. The minimum information that should be displayed for the Gap Analysis and Retrospective Validation SOP is as follows:

- Document Title
- Document Number
- Document Revision Number
- Project Name and Number
- Site/Location/Department Identification
- Date Document was issued, last revised or draft status
- Effective Date
- Pagination: page of total pages (e.g., Page x of y)

The Document Title should be in large type font that enables clear identification. It should clearly and concisely describe the contents and intent of the document. Since this document will be part of a set or package of documents within a project, it should be consistent with other documents used for the project and specify that it is the Gap Analysis and Retrospective Validation SOP. This information should be on every page of the document.

The Document Number should be in large type font that enables clear identification. Since the Gap Analysis and Retrospective Validation SOP will be part of a set or package of documents within a project, it should be consistent with other documents used for the project. This information should appear on every page of the document.

The Document Revision Number, which identifies the revision of the document, should appear on each page of the document. Many documents will go through one or more changes over time, particularly during the draft stage and the review and approval cycle. It is important to maintain accurate revision control because even a

subtle change in a document can have an immense impact on the outcomes of activities. The SOP on SOPs should govern the numbering methodology.

See Chapter 6 – "The SOP on SOPs."

The Project Name and Number, if applicable, should be indicated on the cover or title page. This information identifies the overall project to which this document belongs.

The Site/Location/Department Identification should be indicated on the cover or title page. It identifies the Site/Location/Department that has responsibility for this document and its associated project.

The Date that the Document was issued, last revised or issued with draft status should be on each page of the document (e.g., 12-Dec-01). This will clearly identify the date on which the document was last modified.

The Effective Date of the Document should be on the title or cover page of the document (e.g., 12-Dec-01). This will clearly identify when to begin using the current version of the document.

Pagination, page of total pages (e.g., Page x of y), should be on each page of the document. This will help identify whether or not all of the pages of the document are present. Pages intentionally left blank should clearly be designated as such.

Approval Section

The Approval Section should contain a table with the Author and Approvers printed names and titles and a place to sign and date the SOP. This SOP on SOPs should govern who the approvers of the document should be. The functional department (e.g., Production / Quality Assurance / Engineering) of the signatory should be placed below each printed name and title along with a statement associated with each approver indicating the person's role or qualification in the approval process. Indicate the significance of the associated signature (e.g. This document meets the requirements of Corporate Policy 9055, Electronic Records; Electronic Signatures, dated 22-Mar-2000).

The completion (signing) of this section indicates that the contents have been reviewed and approved by the listed individuals.

Table of Contents

The SOP should contain a Table of Contents, which appears towards the beginning of the document. This Table of Contents provides a way to easily locate various topics contained in the document.

Purpose *(SOP)*

The Purpose section should clearly state that the intention of the document is to provide a Standard Operating Procedure to be used to assess the state of validation and compliance with current regulation of a software application, computer or computerized system, or infrastructure and to bring it into compliance with current regulation. This introduction to the SOP concisely describes the document's purpose in sufficient detail so that a non-technical reader can understand its purpose.

Scope (SOP)

The Scope section identifies the limitations of the Gap Analysis and Retrospective Validation SOP. It limits the scope to all GxP systems and is applicable to all computers, computerized systems, infrastructure, LANs, WANs, and SANs including clinical, laboratory, manufacturing and distribution computer systems. It applies to all Sites, Locations and Departments within the company.

The Gap Analysis and Retrospective Validation SOP includes electronic records, electronic signatures, and handwritten signatures executed to electronic records. It applies to all records that are created, modified, maintained, archived, retrieved or transmitted in electronic form under any requirements set forth in FDA regulations. Provide the scope or range of control and authority of the SOP. State who, what, where, when, why and how the SOP is to control the processes or procedures specified in the Purpose section.

Equipment / Materials / Tools

This section lists any required or recommended equipment, materials, tools, templates or guides. Provide the name and a brief description of the equipment, instruments, materials and/or tools to be used in the procedures or testing covered by the Gap Analysis and Retrospective Validation SOP. This is for reference purposes and requires the user to utilize the specific item or a designated equivalent. The purpose and use is explained elsewhere in the document. Substitution of comparable equipment may be made in accordance with procedures for Change Control.

If the information is included in a job aid or operating manual, a description of where the information may be found with a specific reference to the document number, title, date and revision number should be specified.

Warnings / Notes / General Information / Safety

Warnings, notes, general or safety information applicable to the Gap Analysis and Retrospective Validation SOP are listed in this section.

Policies, Guidelines, References

This section lists references to appropriate regulatory requirements, SOPs, policies, guidelines, or other relevant documents referred to in the Gap Analysis and Retrospective Validation SOP. The name and a brief description of applicable related SOPs, corporate policies, procedures and guidelines, Quality Control monographs, books, technical papers, publications, specific regulations, batch records, vendor or system manuals, and all other relevant documents that are required by the Gap Analysis and Retrospective Validation SOP should be included. This is for reference purposes and requires the user to use the specific item or a designated equivalent. The purpose and use are explained elsewhere in the document.

SOPs cite references and incorporate them in whole by reference rather than specifying details that duplicate these items. This is done so that changes in policy, guidelines or reference materials require no changes to SOPs themselves and the most current version of the referenced material is used. Examples are:

- Policy ABC-123, Administrative Policy for Electronic Records; Electronic Signatures
- Corporate, Plan for Compliance with 21 CFR Part 11, Electronic Records; Electronic Signatures
- SOP-456, Electronic Records; Electronic Signatures
- FDA, 21 CFR Part 11, Electronic Records; Electronic Signatures
- FDA, General Principals of Software Validation; Final Guidance for Industry and FDA Staff, January 11, 2002

Assumptions / Exclusions / Limitations

Any and all assumptions, exclusions or limitations within the context of the SOP should be discussed. Specific issues should be directly addressed in this section of the Gap Analysis and Retrospective Validation SOP. Examples are:

"Legacy systems are not grandfathered under 21 CFR Part 11, there is no limitation with respect to legacy systems."

"Networks that have been previously shown to be sustainedly compliant with 21 CFR Part 11 may be excluded from re-assessment when a new or existing system is evaluated."

Glossary of Terms

A Glossary of Terms should be included that defines all of the terms and acronyms that are used in this document or that are specific to the operation being described and may not be generally known to the reader. The reader may also be referred to the FDA document, "Glossary of Computerized System and Software Development Terminology."

Roles and Responsibilities (SOP)

The assessment of the state of validation and compliance with current regulation is, at a minimum, the joint effort of a Lead Auditor, the System Owner, a Reviewer, and a member of QA(U) being the final signatory of both the protocol and the Summary Report.

This section should clearly specify the participants by name, title, department or company affiliation and role in a table. The information in this section should conform to the corresponding information in the Project Plan, if applicable.

The System Owner is responsible for the development and understanding of the audit and remediation. He or she should acquire and coordinate support from in-house consultants, QA(U), auditors and/or contracted external support services.

The System Owner is responsible for the writing of a Summary Report and for ensuring that it is reviewed and approved by the System Owner, CVG and QA(U).

See Chapter 25 – "Summary Reports."

The QA(U) is responsible to verify that the details in the audit protocol and remediation plan are compliant with the applicable departmental, interdepartmental and corporate practices and policies. They also assure that the approach meets appropriate cGxP regulations, good business practices, and compliance with 21 CFR Part 11.

The qualifications of the individuals involved in the project are kept on file and made available as needed in accordance with cGxP. The qualifications are kept with the project documentation or are available from the Human Resources Department. The location of the qualification documentation is stated in this section.

Procedure

The protocol contains instructions and procedures that enable the assessment of infrastructure, a software application, computer or computerized system and its remediation, when indicated by the outcome of the audit. These include, but are not limited to the following items.

Title

Specify the name of the system under assessment and that it is an assessment in accordance with 21 CFR Part 11 (e.g. – 21 CFR Part 11 Assessment for the ACX-043 Mass Spectrometry Computerized System). It should be on every page of the document.

Introduction

Provide an introduction to the protocol that is an overview of the tasks at hand and explains the background of what is to be performed. This introduction concisely describes the document's purpose in sufficient detail so that a non-technical reader can understand it. For example:

> "21 CFR Part 11 went into effect on August 20, 1997. It provides the criteria for FDA acceptance of electronic records to be equivalent to paper records and electronic signatures to be equivalent to handwritten signatures. The rule must be interpreted and applied to both new and legacy systems. Systems in use that predate August 20, 1997 are not exempt from the rule. The FDA recognizes that it will take time for existing systems to attain full compliance. The FDA expects steps to be taken toward achieving full compliance of these systems to 21 CFR Part 11. This document is an assessment method of the state of validation and compliance with current regulation that will identify areas in the system and system operation that require change to achieve full compliance.

The outcome of this audit combined with the outcome of the 21 CFR Part 11 assessment provide the basis for remediation of the system to bring it into sustainable compliance with current regulation."

Purpose (protocol)

The Purpose section should clearly state that the intention of the document is to provide a means to be used to assess the state of validation and the level of compliance of a software application, computer or computerized system or infrastructure with current company requirements and regulations. This introduction to the protocol should be concise and describe the document's purpose in sufficient detail so that a non-technical reader can understand its purpose. For example:

> "The purpose of this document is to evaluate the ACX-043 Mass Spectrometry Computerized System to determine the state of validation and compliance with current company standards and requirements that are necessary to allow sustainable compliance with current regulation."

Scope (protocol)

The Scope section identifies the limitations of a specific Gap Analysis and Retrospective Validation protocol with respect to the procedure to be performed in the assessment and retrospective validation. For example, if only a part of the computer system will be validated, then this section is used to explain the reason for this decision.

Limit the scope to a specific system or software application and a specific Site, Location and/or Department within the company. It includes electronic records, electronic signatures and handwritten signatures executed to electronic records. It applies to all records that are created, modified, maintained, archived, retrieved or transmitted in electronic form under any requirements set forth in FDA regulations. For example:

> "The scope of this document covers regulations pertaining to 21 CFR Part 11, Electronic Records; Electronic Signatures and cGxP. This protocol is designed to assist with the assessment of computerized systems that create, modify, maintain, archive, retrieve, or transmit electronic records(s) that are required to demonstrate compliance with FDA regulations, and that are used in lieu of paper records. It is applicable to the ACX-043 Mass Spectrometry Computerized System located in Department 600, Bioanlytical Laboratory located in building 35, Anywhere, US."

Applicable Requirements

The protocol identifies all applicable requirements necessary to perform the assessment. These are generally the same as those specified in the governing SOP and any additional items deemed necessary.

It specifies the qualifications of the people who are to perform the assessment.

Overview

The Overview provides a general description of the system, associated software applications, and system interfaces. The execution of the protocol assesses the state of validation and compliance with current regulation.

Roles and Responsibilities (protocol)

The assessment of the state of validation and compliance with current regulation is, at a minimum, the joint effort of a Lead Auditor, the System Owner, a Reviewer, and QA(U), which is the final signatory of both the protocol and the Summary Report.

This section should clearly specify the participants by name, title, department or company affiliation and role in a table. The information in this section should conform to the corresponding information in the Project Plan, if applicable.

The System Owner is responsible for the development and understanding of the audit and remediation. He or she acquires and coordinates support from in-house consultants, QA(U), auditors and/or contracted external support services.

The System Owner is responsible for the writing of a Summary Report and for ensuring that it is reviewed and approved by the System Owner, CVG and QA(U).

QA(U) is responsible to verify that the details in the audit protocol and remediation plan are compliant with the applicable departmental, interdepartmental and corporate practices and policies. They also assure that the approach meets appropriate cGxP regulations, good business practices and compliance with 21 CFR Part 11.

The qualifications of the individuals involved in the project are kept on file and made available as needed in accordance with cGxP. The qualifications are kept with the project documentation or are available from the Human Resources Department. The location of the qualification documentation is stated in this section.

Instructions on Auditing Procedures

This section describes the procedures that the auditor is expected to follow in performing the first part of a retrospective validation – gap analysis. It includes how the audit will be documented, how deviations will be handled and other administrative items relevant to the execution of the protocol. The audit procedures are in accordance with the SOPs governing validation and compliance with cGxP.

Identification

The protocol provides a means to identify the participants in the audit. Each person participating in the audit enters the following information in a table for purposes of identification:

- Full Name – printed
- Signature
- Initials
- Affiliation

Inventory

Identify what is being audited. In a table or other suitable format, inventory items, such as, the System Name, System ID, Start Date, System Location, System Owner and System Type.

Checklist

This is the main section of the protocol and lists a series of audits that are to be performed as part of the assessment. Each audit section consists of the following:

- Category Assessed
- Assessment Question
- Response - Yes / No / Not Applicable
- Observation
- Auditor Section with Signature and Date
- Reviewer Section with Signature and Date
- Approval Section with Signature and Date

Audit the following categories of the infrastructure, computer or computerized system designated in order to assess the state of validation and compliance with current regulation and the need for remedial action based on the SOP governing the specified category:

Validation

- Project Plan
- Validation Plan
- User Requirements
- Functional Requirements
- Vendor Audit
- Software Design Specification
- Hardware Design Specification
- Design Qualification
- Unit Testing

- Factory Acceptance Testing
- End-User Testing
- Hardware Installation Qualification
- Hardware Operational Qualification
- Software Installation Qualification
- Software Operational Qualification
- Performance Qualification
- 21 CFR Part 11 Assessment
- Summary Reports
- Traceability Matrix
- Change Control
- Document Management and Storage
- Configuration Management
- Periodic Review
- Manuals and SOPs
- Logs
- Training
- Project Validation Final Report

Infrastructure

- Infrastructure Installation and Qualification
- Security
- Production Environment Management
- Asset Management
- Back-up, Archival and Restoration
- Disaster Recovery
- Maintenance
- Maintenance of Electronic Records
- Periodic Review
- Change Control
- Configuration Management
- Problem Reporting

Organization

- Overview of the IS Unit Formal Organizational Structure
- IS Steering Committee
- cGxP Procedures, Systems and Training
- Standard Operating Procedures
- Guidelines, Policies, SOPs and Training in Support of SDLC
- GxP Training for IT/IS Staff and Contractors
- Training and Training Documentation
- Computer Validation Group
- Vendors and Contractors

Summary Report

The Gap Analysis Summary Report indicates the deficiencies or gaps identified. It does not contain recommendations for remediation. This is for a separate team to determine and propose in a project plan. In addition to the information specified in Chapter 25 – Summary Reports, the report should contain the following information:

- A brief history of the computer system
- System use and critical functions
- Summation of gap analysis findings
- Gap Analysis Worksheet Inventory
- Listing of documentation deficiencies
- Conclusions

Approval Signatures with a statement that approval of the Gap Analysis Report indicates that the computer system is in a validated and compliant state, or will be upon successful completion of the listed action plans

Remediation / Retrospective Validation

The results of the Gap Analysis are analyzed along with the results of the 21 CFR Part 11 Assessment. Based on this information and analysis, make a determination of the ability of the system to meet current best industry practice for validation in accordance with cGxP and 21 CFR Part 11 requirements.

If deficiencies are indicated, prepare a Project Plan in both narrative and chart forms detailing the requirements needed to remedy the deficiencies in the system, its documentation, qualifications, operation, and use. The plan should contain a list of documents to be prepared, testing to be performed and activities that will not to be performed along with a risk assessment.

Perform a retrospective validation to close the identified validation gaps. Retrospective validation involves following the method of prospective validation as a guide to determine and perform the qualification tasks that are missing or incomplete in the current validation package. It is intended to establish documented evidence that a system does what is purports and is intended to do based on an analysis of historical records and current method of operation.

Retrospective validation involves comparing documentation to System Development Life Cycle requirements for accuracy and completeness. Using the information acquired in the Gap Analysis phase, prepare missing documents, validate the unvalidated parts of the system, and prepare the Final Validation Report. After retrospective validation, the system is considered a validated system and subject to all the rules and requirements as such in order that it be maintained in a sustainable compliant state.

In cases where the deficiencies are great in number or severe in nature, it may be more suitable to perform a complete prospective validation.

Document Management and Storage (protocol)

A Document Management and Storage section is included in each Gap Analysis and Retrospective Validation. It defines the Document Management and Storage methodology that is followed for issue, revision, replacement, control and management of each Gap Analysis and Retrospective Validation. This methodology is in accordance with an established Change Control SOP and a Document Management and Storage SOP.

Revision History (protocol)

A Revision History section is included in each Gap Analysis and Retrospective Validation protocol. The Gap Analysis and Retrospective Validation protocol Version Number and all Revisions made to the document are recorded in this section. These changes are in accordance with a Change Control Procedure SOP.

The Revision History section consists of a summary of the changes, a brief statement explaining the reasons the document was revised, and the name and affiliation of the person responsible for the entries. It includes the numbers of all items the Gap Analysis and Retrospective Validation protocol replaces and is written so that the original Gap Analysis and Retrospective Validation protocol is not needed to understand the revisions or the revised document being issued.

When the Gap Analysis and Retrospective Validation protocol is a new document, a "000" as the Revision Number is entered with a statement such as "Initial Issue" in the comments section.

Regulatory or audit commitments require preservation from being inadvertently lost or modified in subsequent revisions. All entries or modifications made as a result of regulatory or audit commitments require indication as such in the Revision History and should indicate the organization, date and applicable references. A complete and accurate audit trail must be developed and maintained throughout the project lifecycle.

Attachments (protocol)

If needed, an Attachments section is included in each Gap Analysis and Retrospective Validation protocol in accordance with the SOP on SOPs. An Attachment Index and Attachment section follow the text of a Gap Analysis and Retrospective Validation protocol. The top of each attachment page indicates "Attachment # x," where 'x' is the sequential number of each attachment. The second line of each attachment page contains the attachment title. The number of pages and date of issue corresponding to each attachment are listed in the Attachment Index.

Attachments are placed directly after the Attachment Index and are sequentially numbered (e.g., Attachment # 1, Attachment # 2) with each page of an attachment containing the corresponding document number, page numbering, and total number of pages in the attachment (e.g., Page x of y). Each attachment is labeled consistent with the title of the attachment itself.

Attachments are treated as part of the Gap Analysis and Retrospective Validation protocol. If an attachment needs to be changed, the entire Gap Analysis and Retrospective Validation protocol requires a new revision number and is re issued. New attachments or revisions of existing attachments are in accordance with an established Change Control SOP and a Document Management and Storage SOP.

Document Management and Storage *(SOP)*

A Document Management and Storage section is included in the Gap Analysis and Retrospective Validation SOP. It defines the Document Management and Storage methodology that is followed for issue, revision, replacement, control and management of this SOP. This methodology is in accordance with an established Change Control SOP and a Document Management and Storage SOP.

Revision History *(SOP)*

A Revision History section is included in the Gap Analysis and Retrospective Validation SOP. The Gap Analysis and Retrospective Validation SOP Version Number and all Revisions made to the document are recorded in this section. These changes are in accordance with a Change Control Procedure SOP.

The Revision History section consists of a summary of the changes, a brief statement explaining the reasons the document was revised, and the name and affiliation of the person responsible for the entries. It includes the numbers of all items the Gap Analysis and Retrospective Validation SOP replaces and is written so that the original Gap Analysis and Retrospective Validation SOP is not needed to understand the revisions or the revised document being issued.

When the Gap Analysis and Retrospective Validation SOP is a new document, a "000" as the Revision Number is entered with a statement such as "Initial Issue" in the comments section.

Regulatory or audit commitments require preservation from being inadvertently lost or modified in subsequent revisions. All entries or modifications made as a result of regulatory or audit commitments require indication as such in the Revision History and should indicate the organization, date and applicable references. A complete and accurate audit trail must be developed and maintained throughout the project lifecycle.

Attachments *(SOP)*

If needed, an Attachments section is included in the Gap Analysis and Retrospective Validation SOP in accordance with the SOP on SOPs. An Attachment Index and Attachment section follow the text of the Gap Analysis and Retrospective Validation SOP. The top of each attachment page indicates "Attachment # x," where 'x' is the sequential number of each attachment. The second line of each attachment page contains the attachment title. The number of pages and date of issue corresponding to each attachment are listed in the Attachment Index.

Attachments are placed directly after the Attachment Index and are sequentially numbered (e.g., Attachment # 1, Attachment # 2) with each page of an attachment containing the corresponding document number, page numbering, and total number of pages in the attachment (e.g., Page x of y). Each attachment is labeled consistent with the title of the attachment itself.

Attachments are treated as part of the Gap Analysis and Retrospective Validation SOP. If an attachment needs to be changed, the entire Gap Analysis and Retrospective Validation SOP requires a new revision number and is reissued. New attachments or revisions of existing attachments are in accordance with an established Change Control SOP and a Document Management and Storage SOP.

Further Reading

FDA, 21 CFR Part 11, *Electronic Records; Electronic Signatures*

FDA, 21 CFR Part 54, *Financial Disclosure By Clinical Investigators*

FDA, 21 CFR Part 58, *Good Laboratory Practice for Nonclinical Laboratory Studies*

FDA, 21 CFR Part 207, *Registration of Producers of Drugs And Listing of Drugs in Commercial Distribution*

FDA, 21 CFR Part 210, *Current Good Manufacturing Practice In Manufacturing, Processing, Packing, or Holding of Drugs; General*

FDA, 21 CFR Part 211, *Current Good Manufacturing Practice for Finished Pharmaceuticals*

FDA, 21 CFR Part 493, *Laboratory Requirements*

FDA, 21 CFR Part 820, *Quality System Regulation*

FDA, *General Principals of Software Validation; Final Guidance for Industry and FDA Staff*, January 11, 2002

FDA, Glossary of Computerized System and Software Development Terminology

FDA, *Guide to Inspections of Computerized Systems in the Food Processing Industry*

FDA, *Guide To Inspection of Computerized Systems in Drug Processing*, February 1983

National Archives and Records Administration, *Records Management Guidance for Agencies Implementing Electronic Signature Technologies*, October 18, 2000

ISO/IEC 12207, *Information technology - Software life cycle processes*

ISPE - GAMP 4, *Good Automated Manufacturing Practices*

ISPE - GAMP, *Supplier Guide for Validation of Automated Systems in Pharmaceutical Manufacture*

Automation and Validation of Information in Pharmaceutical Processing, Marcel Dekker, ISBN 0-8247-0119-4

Computer Systems Validation for the Pharmaceutical and Medical Device Industries, Richard Chamberlain, Alaren Press, ISBN, 0-9631489-0-8

New South Wales Security, Office of Information Technology, *Security of Electronic Information*

Project Management, *A Systems Approach to Planning, Scheduling, and Controlling*, Sixth Edition by Harold Kerzner, Ph.D. published by John Wiley & Sons, Inc.

How to Write a Job Aid, Tony Moore, Moore Performance Improvement, Inc.: www.expertojt.com

Configuration Management

ANSI/EIA 649, ANSI/EIA *Standard, Configuration Management*

ANSI/IEEE 1042, *Guide to Software Configuration Management*

EIA 632, *Processes for Engineering a System*

MIL-STD-973, *Military Standard, Configuration Management*

MIL-STD-2549, *Department of Defense Interface Standard, Configuration Management Data Interface*

Appendix

Appendix 1 - 21 CFR Part 11

21 CFR PART 11—ELECTRONIC RECORDS; ELECTRONIC SIGNATURES

Subpart A — General Provisions

21 CFR

Sec.

11.1 Scope.

11.2 Implementation.

11.3 Definitions.

Subpart B — Electronic Records

11.10 Controls for closed systems.

11.30 Controls for open systems.

11.50 Signature manifestations.

11.70 Signature/record linking.

Subpart C — Electronic Signatures

11.100 General requirements.

11.200 Electronic signature components and controls.

11.300 Controls for identification codes/passwords.

Authority: Secs. 201-903 of the Federal Food, Drug, and Cosmetic Act (21 U.S.C. 321-393); sec. 351 of the Public Health Service Act (42 U.S.C. 262).

Subpart A — General Provisions

Sec. 11.1 Scope.

(a) The regulations in this part set forth the criteria under which the agency considers electronic records, electronic signatures, and handwritten signatures executed to electronic records to be trustworthy, reliable, and generally equivalent to paper records and handwritten signatures executed on paper.

(b) This part applies to records in electronic form that are created, modified, maintained, archived, retrieved, or transmitted, under any records requirements set forth in agency regulations. This part also applies to electronic records submitted to the agency under requirements of the Federal Food, Drug, and Cosmetic Act and the Public Health Service Act, even if such records are not specifically identified in agency regulations. However, this part does not apply to paper records that are, or have been, transmitted by electronic means.

(c) Where electronic signatures and their associated electronic records meet the requirements of this part, the agency will consider the electronic signatures to be equivalent to full handwritten signatures, initials, and other general signings as required by agency regulations, unless specifically excepted by regulation(s) effective on or after August 20, 1997.

(d) Electronic records that meet the requirements of this part may be used in lieu of paper records, in accordance with Sec. 11.2, unless paper records are specifically required.

(e) Computer systems (including hardware and software), controls, and attendant documentation maintained under this part shall be readily available for, and subject to, FDA inspection.

Sec. 11.2 Implementation.

(a) For records required to be maintained but not submitted to the agency, persons may use electronic records in lieu of paper records or electronic signatures in lieu of traditional signatures, in whole or in part, provided that the requirements of this part are met.

(b) For records submitted to the agency, persons may use electronic records in lieu of paper records or electronic signatures in lieu of traditional signatures, in whole or in part, provided that:

(1) The requirements of this part are met; and

(2) The document or parts of a document to be submitted have been identified in public docket No. 92S-0251 as being the type of submission the agency accepts in electronic form. This docket will identify specifically what types of documents or parts of documents are acceptable for submission in electronic form without paper records and the agency receiving unit(s) (e.g., specific center, office, division, branch) to which such submissions may be made. Documents to agency receiving unit(s) not specified in the public docket will not be considered as official if they are submitted in electronic form; paper forms of such documents will be considered as official and must accompany any electronic records. Persons are expected to consult with the intended agency receiving unit for details on how (e.g., method of transmission, media, file formats, and technical protocols) and whether to proceed with the electronic submission.

Sec. 11.3 Definitions.

(a) The definitions and interpretations of terms contained in section 201 of the act apply to those terms when used in this part.

(b) The following definitions of terms also apply to this part:

(1) Act means the Federal Food, Drug, and Cosmetic Act (secs. 201-903 (21 U.S.C. 321-393)).

(2) Agency means the Food and Drug Administration.

(3) Biometrics means a method of verifying an individual's identity based on measurement of the individual's physical feature(s) or repeatable action(s) where those features and/or actions are both unique to that individual and measurable.

(4) Closed system means an environment in which system access is controlled by persons who are responsible for the content of electronic records that are on the system.

(5) Digital signature means an electronic signature based upon cryptographic methods of originator authentication, computed by using a set of rules and a set of parameters such that the identity of the signer and the integrity of the data can be verified.

(6) Electronic record means any combination of text, graphics, data, audio, pictorial, or other information representation in digital form that is created, modified, maintained, archived, retrieved, or distributed by a computer system.

(7) Electronic signature means a computer data compilation of any symbol or series of symbols executed, adopted, or authorized by an individual to be the legally binding equivalent of the individual's handwritten signature.

(8) Handwritten signature means the scripted name or legal mark of an individual handwritten by that individual and executed or adopted with the present intention to authenticate a writing in a permanent form. The act of signing with a writing or marking instrument such as a pen or stylus is preserved. The scripted name or legal mark, while conventionally applied to paper, may also be applied to other devices that capture the name or mark.

(9) Open system means an environment in which system access is not controlled by persons who are responsible for the content of electronic records that are on the system.

Subpart B — Electronic Records

Sec. 11.10 Controls for closed systems.

Persons who use closed systems to create, modify, maintain, or transmit electronic records shall employ procedures and controls designed to ensure the authenticity, integrity, and, when appropriate, the confidentiality of electronic records, and to ensure that the signer cannot readily repudiate the signed record as not genuine. Such procedures and controls shall include the following:

(a) Validation of systems to ensure accuracy, reliability, consistent intended performance, and the ability to discern invalid or altered records.

(b) The ability to generate accurate and complete copies of records in both human readable and electronic form suitable for inspection, review, and copying by the agency. Persons should contact the agency if there are any questions regarding the ability of the agency to perform such review and copying of the electronic records.

(c) Protection of records to enable their accurate and ready retrieval throughout the records retention period.

(d) Limiting system access to authorized individuals.

(e) Use of secure, computer-generated, time-stamped audit trails to independently record the date and time of operator entries and actions that create, modify, or delete electronic records. Record changes shall not obscure previously recorded information. Such audit trail documentation shall be retained for a period at least as long as that required for the subject electronic records and shall be available for agency review and copying.

(f) Use of operational system checks to enforce permitted sequencing of steps and events, as appropriate.

(g) Use of authority checks to ensure that only authorized individuals can use the system, electronically sign a record, access the operation or computer system input or output device, alter a record, or perform the operation at hand.

(h) Use of device (e.g., terminal) checks to determine, as appropriate, the validity of the source of data input or operational instruction.

(i) Determination that persons who develop, maintain, or use electronic record/electronic signature systems have the education, training, and experience to perform their assigned tasks.

(j) The establishment of, and adherence to, written policies that hold individuals accountable and responsible for actions initiated under their electronic signatures, in order to deter record and signature falsification.

(k) Use of appropriate controls over systems documentation including:

(1) Adequate controls over the distribution of, access to, and use of documentation for system operation and maintenance.

(2) Revision and change control procedures to maintain an audit trail that documents time-sequenced development and modification of systems documentation.

Sec. 11.30 Controls for open systems.

Persons who use open systems to create, modify, maintain, or transmit electronic records shall employ procedures and controls designed to ensure the authenticity, integrity, and, as appropriate, the confidentiality of electronic records from the point of their creation to the point of their receipt. Such procedures and controls shall include those identified in Sec. 11.10, as appropriate, and additional measures such as document encryption and use of appropriate digital signature standards to ensure, as necessary under the circumstances, record authenticity, integrity, and confidentiality.

Sec. 11.50 Signature manifestations.

(a) Signed electronic records shall contain information associated with the signing that clearly indicates all of the following:

(1) The printed name of the signer;

(2) The date and time when the signature was executed; and

(3) The meaning (such as review, approval, responsibility, or authorship) associated with the signature.

(b) The items identified in paragraphs (a)(1), (a)(2), and (a)(3) of this section shall be subject to the same controls as for electronic records and shall be included as part of any human readable form of the electronic record (such as electronic display or printout).

Sec. 11.70 Signature/record linking.

Electronic signatures and handwritten signatures executed to electronic records shall be linked to their respective electronic records to ensure that the signatures cannot be excised, copied, or otherwise transferred to falsify an electronic record by ordinary means.

Subpart C — Electronic Signatures

Sec. 11.100 General requirements.

(a) Each electronic signature shall be unique to one individual and shall not be reused by, or reassigned to, anyone else.

(b) Before an organization establishes, assigns, certifies, or otherwise sanctions an individual's electronic signature, or any element of such electronic signature, the organization shall verify the identity of the individual.

(c) Persons using electronic signatures shall, prior to or at the time of such use, certify to the agency that the electronic signatures in their system, used on or after August 20, 1997, are intended to be the legally binding equivalent of traditional handwritten signatures.

(1) The certification shall be submitted in paper form and signed with a traditional handwritten signature, to the Office of Regional Operations (HFC-100), 5600 Fishers Lane, Rockville, MD 20857.

(2) Persons using electronic signatures shall, upon agency request, provide additional certification or testimony that a specific electronic signature is the legally binding equivalent of the signer's handwritten signature.

Sec. 11.200 Electronic signature components and controls.

(a) Electronic signatures that are not based upon biometrics shall:

(1) Employ at least two distinct identification components such as an identification code and password.

(i) When an individual executes a series of signings during a single, continuous period of controlled system access, the first signing shall be executed using all electronic signature components; subsequent signings shall be executed using at least one electronic signature component that is only executable by, and designed to be used only by, the individual.

(ii) When an individual executes one or more signings not performed during a single, continuous period of controlled system access, each signing shall be executed using all of the electronic signature components.

(2) Be used only by their genuine owners; and

(3) Be administered and executed to ensure that attempted use of an individual's electronic signature by anyone other than its genuine owner requires collaboration of two or more individuals.

(b) Electronic signatures based upon biometrics shall be designed to ensure that they cannot be used by anyone other than their genuine owners.

Sec. 11.300 Controls for identification codes/passwords.

Persons who use electronic signatures based upon use of identification codes in combination with passwords shall employ controls to ensure their security and integrity. Such controls shall include:

(a) Maintaining the uniqueness of each combined identification code and password, such that no two individuals have the same combination of identification code and password.

(b) Ensuring that identification code and password issuances are periodically checked, recalled, or revised (e.g., to cover such events as password aging).

(c) Following loss management procedures to electronically deauthorize lost, stolen, missing, or otherwise potentially compromised tokens, cards, and other devices that bear or generate identification code or password information, and to issue temporary or permanent replacements using suitable, rigorous controls.

(d) Use of transaction safeguards to prevent unauthorized use of passwords and/or identification codes, and to detect and report in an immediate and urgent manner any attempts at their unauthorized use to the system security unit, and, as appropriate, to organizational management.

(e) Initial and periodic testing of devices, such as tokens or cards, that bear or generate identification code or password information to ensure that they function properly and have not been altered in an unauthorized manner.

Appendix 2 – Good Laboratory Practices - Attachment A

FOOD AND DRUG ADMINISTRATION

COMPLIANCE PROGRAM GUIDANCE MANUAL **PROGRAM** 7348.808

Date of Issuance: 02/21/01

ATTACHMENT A

Computerized Systems

The intent of this attachment is to collect, in one place, references to computer systems found throughout Part III. Computer systems and operations should be thoroughly covered during inspection of any facility. No additional reporting is required under this Attachment.

In August 1997, the Agency's regulation on electronic signatures and electronic recordkeeping became effective. The Regulation, at 21 CFR Part 11, describes the technical and procedural requirements that must be met if a firm chooses to maintain records electronically and/or use electronic signatures. Part 11 works in conjunction with other FDA regulations and laws that require recordkeeping. Those regulations and laws ("predicate rules') establish requirements for record content, signing, and retention.

Certain older electronic systems may not have been in full compliance with Part 11 by August 1997 and modification to these so called "legacy systems" may take more time. Part 11 does not grandfather legacy systems and FDA expects that firms using legacy systems are taking steps to achieve full compliance with Part 11.

If a firm is keeping electronic records or using electronic signatures, determine if they are in compliance with 21 CFR Part 11. Determine the depth of part 11 coverage on a case by case basis, in light of initial findings and program resources. At a minimum ensure that: (1) the firm has prepared a corrective action plan for achieving full compliance with part 11 requirements, and is making progress toward completing that plan in a timely manner; (2) accurate and complete electronic and human readable copies of electronic records, suitable for review, are made available; and (3) employees are held accountable and responsible for actions taken under their electronic signatures. If initial findings indicate the firm's electronic records and/or electronic signatures may not be trustworthy and reliable, or when electronic recordkeeping systems inhibit meaningful FDA inspection, a more detailed evaluation may be warranted. Districts should consult with center compliance officers and the Office of Enforcement (HFC-240) in assessing the need for, and potential depth of, more detailed part 11 coverage. When substantial and significant part 11 deviations exist, FDA will not accept use of electronic records and electronic signatures to meet the requirements of the applicable predicate rule. See Compliance Policy Guide (CGP), Sec. 160.850.

See IOM sections 594.1 and 527.3 for procedures for collecting and identifying electronic data.

Personnel - Part III, C.1.c. (21 CFR 58.29)

Determine the following:

- Who was involved in the design, development, and validation of the computer system?

- Who is responsible for the operation of the computer system, including inputs, processing, and output of data?
- If computer system personnel have training commensurate with their responsibilities, including professional training and training in GLPs
- Whether some computer system personnel are contractors who are present on-site full-time, or nearly full-time. The investigation should include these contractors as though they were employees of the firm. Specific inquiry may be needed to identify these contractors, as they may not appear on organization charts.
- QAU Operations - Part III, C.2 (21 CFR 58.35(b-d))
- Verify SOPs exist and are being followed for QAU inspections of computer operations.

Facilities - Part III, C.3 (21 CFR 58.41 - 51)

- Determine that computerized operations and archived computer data are housed under appropriate environmental conditions.

Equipment - Part III, C.4 (21 CFR 58.61 - 63)

For computer systems, check that the following procedures exist and are documented:

- Validation study, including validation plan and documentation of the plan's completion.
- Maintenance of equipment, including storage capacity and back-up procedures.
- Control measures over changes made to the computer system, which include the evaluation of the change, necessary test design, test data, and final acceptance of the change.
- Evaluation of test data to assure that data is accurately transmitted and handled properly when analytical equipment is directly interfaced to the computer. and
- Procedures for emergency back-up of the computer system, (e.g., back-up battery system and data forms for recording data in case of a computer failure or power outage).

Testing Facility Operations - Part III, C.5 (21 CFR 58.81)

- Verify that a historical file of outdated or modified computer programs is maintained.

Records and Reports (21 CFR 58.185 - 195) (PART III C.10.b.)

- **Verify** that the final report contains the required elements in 58.185(a)(1-14), including a description of any computer program changes.

Storage and Retrieval of Records and Data - Part III, C.10.c. (21 CFR 58.190)

- Assess archive facilities for degree of controlled access and adequacy of environmental controls with respect to computer media storage conditions.
- **Determine** how and where computer data and backup copies are stored, that records are indexed in a way to allow access to data stored on electronic media, and that environmental conditions minimize deterioration.
- **Determine** how and where original computer data and backup copies are stored.

Appendix 3 - Roles and Responsibilities – Validation

Document	SOP	Template	Example	Training & Assessment	Prepare	Review	Approve - Pre-Execution	Execute	Approve - Post-Execution
Letter of Notification / System Proposal	X	X	X	N	U	O, Q, G	O, Q, G	N	N
Change Control Documents	X	X	X	X	U	O, Q, G	O, Q, G	N	N
Project Team Charter	X	X	X	N	U	O, Q, G	O, Q, G	N	N
Project Plan	X	X	X	X	U	O, Q, G	O, Q, G	N	N
Risk Assessment	X	X	X	X	U	O, Q, G	O, Q, G	N	N
Validation Plan	X	X	X	X	U	O, Q, G	O, Q, G	N	N
User Requirements	X	X	X	X	U	O, Q, G	O, Q, G	N	N
Functional Requirements	X	X	X	X	U	O, Q, G	O, Q, G	N	N
Vendor Audit	X	X	X	X	Q	Q	Q	N	N
Vendor Audit Report	X	X	X	X	Q	Q	Q	N	N
Hardware Design Specification	X	X	X	X	D	U, O Q	O, Q	N	N
Software Design Specification	X	X	X	X	D	U, O Q	O, Q	N	N
Unit Testing Documentation	X	X	X	X	D	U, O Q	O, Q	D	O, Q
Factory Acceptance Test Documents	X	X	X	X	D	U, O Q	O, Q	U, D	O, Q
IQ Protocol	X	X	X	X	U, D or C	O, Q, G	O, Q, G	U, D or C	O, Q
IQ Summary Report	X	X	X	X	U, D or C	O, Q	O, Q	N	N
OQ Protocol	X	X	X	X	U, D or C	O, Q, G	O, Q, G	U, D or C	O, Q
OQ Summary Report	X	X	X	X	U, D or C	O, Q	O, Q	N	N
PQ Protocol	X	X	X	X	U, D or C	O, Q, G	O, Q, G	U	O, Q
PQ Summary Report	X	X	X	X	U, D or C	O, Q	O, Q	N	N
21 CFR Part 11 Assessment	X	X	X	X	U, D or C	O, Q, G	O, Q, G	U, D or C	O, Q
21 CFR Part 11 Assessment Report	X	X	X	X	U, D or C	O, Q	O, Q	N	N
Traceability Matrix Document	X	X	X	X	U, D or C	O, Q, G	O, Q, G	N	N
Manuals	N	N	N	N	V, D or C	O, Q, G	O, Q, G	N	N
Miscellaneous Documents	N	N	N	N	U, V, D or C	O, Q, G	N	N	N
Training Documents	N	N	N	N	U, D or C	O, Q, G	O, Q, G	N	N
SOPs	N	N	N	N	U, D or C	O, Q, G	O, Q, G	N	N
Project Validation Final Report	X	X	X	X	U, D or C	O, Q, G	O, Q, G	N	N
Decommissioning Plan & Report	X	X	X	X	O	O, Q, G	O, Q	U, M	O, Q

Key

C = Consultant, D = System Developer, G = Computer Validation Group, M = IS/IT Department Management, N = Not Applicable, O= System Owner, P = Prospective Validation, Q = Quality Assurance (Unit),

R = Retrospective Validation, U = End-User, V = Vendor/Supplier, X = Applies

Computer	Peripheral	Equipment	WAN/LAN	Operating System	Firmware	COTS	Config-uration	Bespoke	Consultant
P, R	P, R	P, R	P, R	P, R	P, R	P, R	P, R	P, R	N
P, R	P, R	P, R	P, R	P, R	P, R	P, R	P, R	P, R	N
P, R	P, R	P, R	P, R	P, R	P, R	P, R	P, R	P, R	N
P, R	P, R	P, R	P, R	P, R	P, R	P, R	P, R	P, R	N
P, R	P, R	P, R	P, R	P, R	P, R	P, R	P, R	P, R	P, R
P, R	P, R	P, R	P, R	P, R	P, R	P, R	P, R	P, R	N
N	N	N	N	P, R	P, R	P, R	P, R	P, R	N
N	N	N	N	P, R	P, R	P, R	P, R	P, R	N
P	P	P	P	P	P	P	P	P	P, R
P	P	P	P	P	P	P	P	P	P, R
P, R	P, R	P, R	P, R	N	N	N	N	N	N
N	N	N	N	N	N	N	P, R	P, R	N
N	N	N	N	N	N	N	N	P	N
P, R	P, R	P, R	P, R	P, R	P, R	P, R	P, R	P, R	N
P, R	P, R	P, R	P, R	P, R	P, R	P, R	P, R	P, R	N
P, R	P, R	P, R	P, R	P, R	P, R	P, R	P, R	P, R	N
P, R	P, R	P, R	P, R	P, R	P, R	P, R	P, R	P, R	N
P, R	P, R	P, R	P, R	P, R	P, R	P, R	P, R	P, R	N
N	N	N	N	N	N	P, R	P, R	P, R	N
N	N	N	N	N	N	P, R	P, R	P, R	N
P, R	P, R	P, R	P, R	P, R	P, R	P, R	P, R	P, R	N
P, R	P, R	P, R	P, R	P, R	P, R	P, R	P, R	P, R	N
P, R	P, R	P, R	P, R	P, R	P, R	P, R	P, R	P, R	N
P, R	P, R	P, R	P, R	P, R	P, R	P, R	P, R	P, R	N
P, R	P, R	P, R	P, R	P, R	P, R	P, R	P, R	P, R	N
P, R	P, R	P, R	P, R	P, R	P, R	P, R	P, R	P, R	N
P, R	P, R	P, R	P, R	P, R	P, R	P, R	P, R	P, R	N
P, R	P, R	P, R	P, R	P, R	P, R	P, R	P, R	P, R	N
P, R	P, R	P, R	P, R	P, R	P, R	P, R	P, R	P, R	N

Index

Lightning Source UK Ltd.
Milton Keynes UK
UKOW010913130212

187194UK00003B/3/A

9 781410 737441